Materials Issues in Applications of Amorphous Silicon Technology

MATERIALS RESEARCH SOCIETY SYMPOSIA PROCEEDINGS

ISSN 0272 - 9172

MATERIALS RESEARCH SOCIETY SYMPOSIA PROCEEDINGS

MATERIALS RESEARCH SOCIETY SYMPOSIA PROCEEDINGS

MATERIALS RESEARCH SOCIETY SYMPOSIA PROCEEDINGS   VOLUME 49

Materials Issues in Applications of Amorphous Silicon Technology

Symposium held April 15-17, 1985, San Francisco, California, U.S.A

EDITORS:

D. Adler
Massachusetts Institute of Technology, Cambridge, Massachusetts, U.S.A

A. Madan
Glasstech Solar, Inc., Wheat Ridge, Colorado, U.S.A.

M. J. Thompson
Xerox Palo Alto Research Center, Palo Alto, California, U.S.A.

MRS MATERIALS RESEARCH SOCIETY
Pittsburgh, Pennsylvania

CAMBRIDGE UNIVERSITY PRESS
Cambridge, New York, Melbourne, Madrid, Cape Town,
Singapore, São Paulo, Delhi, Mexico City

Cambridge University Press
32 Avenue of the Americas, New York NY 10013-2473, USA

Published in the United States of America by Cambridge University Press, New York

www.cambridge.org
Information on this title: www.cambridge.org/9781107405585

Materials Research Society
506 Keystone Drive, Warrendale, PA 15086
http://www.mrs.org

First published 1985
First paperback edition 2012

Single article reprints from this publication are available through
University Microfilms Inc., 300 North Zeeb Road, Ann Arbor, MI 48106

CODEN: MRSPDH

ISBN 978-0-931-83714-2 Hardback
ISBN 978-1-107-40558-5 Paperback

Contents

PART III: INTERFACES

PART IV: MULTILAYERED STRUCTURES

PART V: Si-Ge ALLOYS

Preface

Over the past few years, the present and potential applications of amorphous–silicon–alloy thin films have mushroomed, and interest in the materials as well as in the device has grown accordingly. The field has evolved rapidly since the accidental discovery of hydrogenated amorphous silicon (a-Si:H), an alloy whose semiconducting properties proved much superior to those of pure amorphous silicon because the overall reduction of network strains due to the univalency of hydrogen taken together with the strength and polarity of the Si–H bond leads to sharp reductions in the concentrations of both traps and recombination centers in a-Si:H. The initial application of a-Si:H and related alloys such as a-Si:F:H was in the area of photovoltaic energy conversion, since here the cost of devices based on crystalline silicon (c-Si) precluded their widespread use on economic grounds. The development of large area photovoltaic devices further stimulated the field, and even a continuous–web process has been demonstrated. Since it is also possible to tune the bandgap using, for example, Si-Ge and Si-C alloy films, the possibility of ultra-high-efficiency multiple-gap stacked solar cells is particularly exciting. Unfortunately, high-quality hydrogenated amorphous Si-Ge and Si-C alloys initially proved to be more difficult to develop than a-Si:H, but recent progress especially with alloys also incorporating fluorine has been very encouraging. Another potential cloud has been the metastable instabilities that characterize a-Si:H films exposed to sunlight for extended periods, but again recent advances in both understanding the effect and judicious processing of the materials and the overall device geometry has suggested that any stability problem can be overcome.

The demonstration of large-area processing of high-quality films has also revived interest in the fabrication of thin-film transistors, especially for use in flat-screen displays. Another potentially enormous application of amorphous silicon alloys, viz. as the photoreceptor for electrostatic copying, requires rapid deposition rates, and here again recent progress has been extremely exciting. The development of amorphous-silicon-alloy photodiodes for use in image sensors and compact scanners has also evolved rapidly, as has the materials use in switching and memory applications, in laser printers, and in an array of other possibilities.

Concomitant with this exponential growth in the device potential, there has been further progress in materials processing. Novel approaches to chemical vapor deposition have been developed and periodic multilayer structures have been shown to exhibit unique properties such as persistent photoconductivity (for several hours at room temperature). Clearly, the entire area is in a period of rapid growth.

It was in this milieu that the MRS Symposium on Materials Issues in Applications of Amorphous Silicon Technology took place in April 1985. Over 200 scientists from around the world participated. The present volume contains 60 papers, including all the invited presentations as well as the vast majority of both the oral and poster contributions. The volume is divided into 14 parts, and is concerned with both materials processing and characterization as well as device applications. Part I contains nine papers devoted to the deposition, growth, and crystallization of amorphous-silicon-alloy films. In particular, there has been a great deal of recent interest in alloys prepared by chemical vapor deposition, especially from disilane (to keep the temperature sufficiently low to insure that significant concentrations of residual hydrogen are incorporated in the films. Since the quality of the material is directly related to the defect concentration, the overall issue of defects is of vital importance. The two papers of Part II are concerned with both the nature and the physical consequences of defects in amorphous—silicon—alloy films.

All devices involve interfaces between the amorphous silicon alloy and other semiconductors, metals, or insulators. The five papers in Part III deal with the properties of a variety of such interfaces, all of potential commercial importance. As an interesting extension of some of these considerations, Part IV consists of two papers on novel periodic multilayer devices employing alternating amorphous—silicon—alloy films of individual thicknesses as low as 10 Å.

The 14 papers comprising Parts V through VIII are all concerned with alloys, especially those useful for bandgap modulation. In particular, Si–Ge alloys are the subject of Part V, Si–C alloys of Part VI, Si–N alloys are discussed in Part VII, and the properties of some miscellaneous alloys are reviewed in Part VIII. The large fraction of the present volume devoted to alloys reflects the excitement attendant with the possibility of stacked high–efficiency solar cells for use in central–power generation.

The remainder of the volume is directly concerned with specific applications. The six papers of Part IX make clear the striking advances in solar-cell technology that have occurred over the last year, while the six papers of Part X demonstrate how our understanding of photo–induced effects in these films has progressed concomitantly. Part XI consists of eight papers, more than one–quarter of all the papers deal with devices, all concerned with the physics and performance of thin–film transistors and diodes involving amor-

phous-silicon-alloy films. The papers in Part XII are devoted to photoreceptor applications, while image sensors and related devices are discussed in
Part XIII. Finally, the two papers of Part XIV analyze amorphous-crystalline
heterojunction transistors and thin-film devices involving polycrystalline
silicon. Taken together, the 29 papers concerned with devices eloquently
demonstrate the current state of the art.

We the organizers were pleasantly surprised with the strong positive
response to the symposium. It was unanimously decided to hold a similar
symposium at the 1986 MRS Spring Meeting. From our present perspective, it
would appear that even a more enthusiastic response may well result next
year.

David Adler

Arun Madan

Malcolm J. Thompson

PART I

Growth and Crystallization

Thermodynamic Study of Silane-hydrogen
Chemical System

Rajendra I. Patel
Electronic and Information Sector Laboratories
3M Company
St. Paul, Minnesota 55144

ABSTRACT

The thermodynamic calculations of chemical equilibrium
compositions of various species were performed for silane-hydrogen
system at different temperatures [500-6000K], pressures [0.01-10
Torr], and mixing ratios [various silane fractions]. The effect of
dopant impurities such as diborane and phosphine on equilibrium
concentration profiles of various gaseous, ionic, and condensed
species was also investigated. These data will be very useful in
understanding the deposition chemistry of silane chemical systems and
will be a valuable aid in optimization studies.

INTRODUCTION

Recently, hydrogenated amorphous silicon [a-Si:H] has become an
increasingly attractive material for wide variety of applications such
as solar cells, field effect transistors, photoreceptors, and imaging
devices. [1] Plasma assisted chemical vapor depositon and etching
are potentially important techniques for fabricating such novel
devices and synthesizing various kinds of electronic materials. In
order to understand the deposition chemistry, it is necessary to
obtain a detailed knowledge of thermodynamic properties of a chemical
system which in turn, permits one to calculate chemical equilibrium
compositions for the system. A comprehensive thermodynamic study of
the various ways of preparing silicon for photovoltaic applications,
suggested that SiH_4 is probably the best raw material for obtaining
high-purity silicon.[2]

The knowledge of thermodynamic properties of a chemical system
permits one to calculate chemical equilibrium compositions for the
system. A computer program capable of calculating equilibrium
properites of plasmas [mixtures containing ionized species] was
obtained from NASA Lewis Research Center.[3] This Fortran IV program
was written for calculations such as: 1] Chemical equilibrium
compositions for assigned thermodynamic states [T,P], [H,P], [S,P],
[T,V], [U,V], or [S,V], 2] Theoretical rocket performance for both
equilibrium and frozen composition during expansion, 3] Incident and
reflected shock properties, and 4] Chapman-Jouguet detonation
properties. A free energy minimization technique is used for the
calculations of thermodynamic equilibrium compositions.

The program considers gaseous species as well as condensed
species. Thus, it is ideal for calculating equilibrium compositions
for complex mixtures in plasma at high temperatures for a given
pressure. However, in performing these calculations, the plasma is
considered ideal, i.e., no coulombic interactions are considered.
Therefore, the results of these plasma calculations will be valid for

systems with small ionic density so that coulombic effects are unimportant. To a great extent, this assumption is true of glow discharge plasmas. In equilibrium calculations, it is also assumed that gases obey ideal gas law, homogeneous mixing, and that the composition attains equilibrium instantaneously during expansion.

The program code uses all the species with known thermodynamic functions as available in its thermodynamic library. For species whose thermodynamic functions [such as heat capacity, enthalpy, entropy, and free energy] are not available, another Fortran IV Program [4] [PAC code] can be used to calculate the needed functions from spectroscopic constants.

These programs are used to obtain chemical equilibrium compositions of various gaseous, condensed, and ionic species at temperatures ranging from 500 to 6000K, and at various pressures from 0.01 Torr to 10 Torr, and for different mixing ratios of silanes and hydrogen. The effect of dopant impurities phosphine [PH_3] and diborane [B_2H_6] on the equilibrium compositions was also studied at various temperatures.

RESULTS AND DISCUSSIONS

Undoped SiH_4-H_2 system:

Typical equilibrium concentration profiles for various silicon and hydrogen containing species are plotted in Figures 1-3 as a function

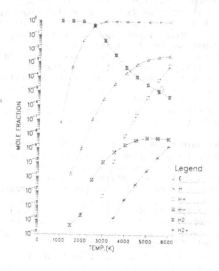

Figure 1

Temperature dependance of calculated equilibrium concentrations of electrons, H atom, H^+, H^-, H_2, and H_2^+ for $SiH_4/H_2=0.1$ and total pressure 1 Torr.

of temperature for silane to hydrogen ratio of 0.1 at total pressure of 1 Torr. The hydrogen atom concentration rises monotonically as a function of temperature and attains maximum above 3000K. Similarly, silicon atom concentration in gas phase attains maximum above 2000K, and slowly decreases above 5000K, probably forming polysilicons or polysilanes. Another very interesting feature to note is that all

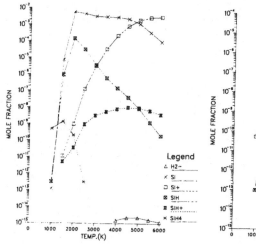

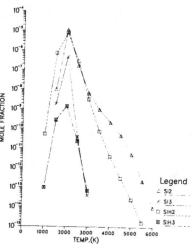

Figure 2.

Temperature dependance of
calculated equilibrium
compositions of H_2^-, Si atom,
Si^+, SiH, SiH^+, and SiH_4 for
$SiH_4/H_2=0.1$ and total pressure
of 1 Torr.

Figure 3.

Temperature dependance of
calculated equilibrium
compositions of Si_2, Si_3,
SiH_2, and SiH_3 for
$SiH_4/H_2=0.1$ and total
pressure of 1 Torr.

are at maximum at 2000K. However, ionic species [Si^+, SiH^+, H^+, H_2^+]
increase monotonically above 2000K, thus contributing significantly
only at high temperatures.

Similar concentration profiles were also obtained at different
pressures and silane-hydrogen mixing ratios. For a silane-hydrogen
system, pressure dependence on concentration profiles of various
species is shown in Figure 4 at SiH_4/H_2 ratio of 0.1 and temperature
of 2000K. Again all silicon containing species rise monotonically
with pressure and reaches maximum at 5 Torr.

In order to optimize SiH_4/H_2 ratio, concentration profiles of
silicon containing species were obtained for different silane
fractions in hydrogen at a temperature of 2000K and total pressure of
1 Torr as depicted in Figure 5. Again, all silicon species show
monotonically increasing function, reaching a maximum above 30% silane
mixtures. Thus, if silicon containing species were to be maximized in
a deposition system for higher deposition rate, then thermodynamic
studies predict a system with > 30% silane fraction in H_2 mixture at
total pressure of 5 Torr and 2000K.

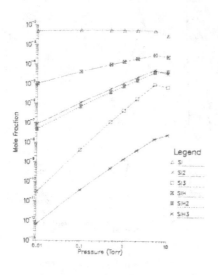

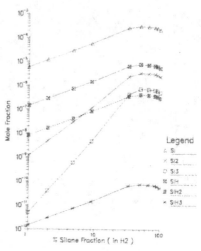

Figure 4.

Pressure dependance of calculated
equilibrium compositions of Si,
Si_2, Si_3, SiH, SiH_2, and SiH_3 for
$SiH_4/H_2=0.1$ and T=2000K

Figure 5.

Silane fraction dependance on
equilibrium concentrations of
Si, Si_2, Si_3, SiH, SiH_2, and
SiH_3 at T=2000K and total
pressure of 1 Torr.

Doped SiH_4-H_2 System:

Phosphorous doped impurities:

For n^+ materials, phosphorous is used as a dopant. Typical
experimental conditions involved in making n^+ microcrystalline
materials are $SiH_4/H_2=0.01$, $PH_3/SiH_4=2\%$ and total pressure of 1 Torr.
Figure 6 shows concentration profiles of phosphorous containing species as a function of temperature in above mentioned conditions.
Silicon and hydrogen containing species [including all ionic species]
show similar trends as mentioned in undoped materials. The
Phosphorous concentration reaches a maximum at 2000K and remains at
this constant value, however, PH goes through a maximum at 2000K.
Thus, if PH has a deleterious effect on material properties then by
operating above 3000K, only phosphorous atoms can be selectively
doped.

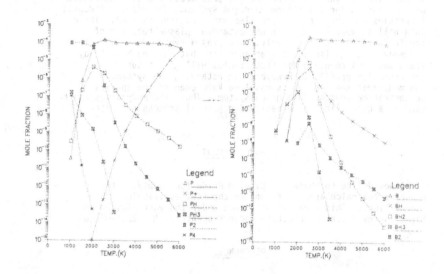

Figure 6.

Temperature dependance of
equilibrium compositions of
P, P$^+$, PH, PH$_3$, P$_2$, and P$_4$,
for SiH$_4$/H$_2$=0.01, P=1 Torr
and PH$_3$/SiH$_4$=2%.

Figure 7

Temperature dependance of
equilibrium compositions of
B, BH, BH$_2$, BH$_3$, and B$_2$ for
SiH$_4$/H$_2$=0.1, P=1 Torr, and
B$_2$H$_6$/SiH$_4$=2%.

Boron doped impurities:

For p$^+$ materials, diborane is used as a source for boron dopant.
The equilibrium compositions for boron containing species are plotted
in Figure 7 as a function of temperature. Concentation profiles were
obtained under usual conditions of SiH$_4$/H$_2$=0.1, B$_2$H$_6$/SiH$_4$=2%, and
total pressure of 1 Torr. Again, atomic boron concentration rises
rapidly and reaches a maximum at 2500K, while BH and BH$_3$ goes through
a maximum at 2500K and BH$_2$ at 2000K. However, BH, BH$_2$ and BH$_3$ drop
off quickly above 2500K. Thus, knowledge of concentration - temperature profiles will be a valuable aid in optimization studies for
selective doping.

SUMMARY

The thermodynamic calculations of chemical equilibrium compositions at various temperatures, pressures, and silane fraction were performed for the silane-hydrogen system. The effect of dopant impurities on concentration profiles was also investigated. These data will be very valuable in understanding plasma behavior of chemical systems, qualitatively. The plasma system does not attain thermodynamic equilibrium instantaneously, but may reach a steady state and thus coupling kinetics studies will give actual concentration profiles. However, understanding the system in thermodynamic sense gives us baseline knowledge of the species concentration present in a given system. Thus, thermodynamics can be used as a guide line for establishing general process parameters.

ACKNOWLEDGEMENT

I would like to thank Dr. B. J. McBride for supplying the computer programs for calculations of complex chemical equilibrium compositions. This work was performed under the auspicies of the Solar Energy Research Institute under subcontract No. ZB-4-03056-2.

REFERENCES

1. J. I. Pankove, Editor, Semiconductors and semimetals, Vol.21, Part D, Academic Press, Inc., 1984

2. J. M. Mexmain, D. Morvan, E. Bourdin, J. Amourous, and P. Fauchais, Plasma Chem. and Plasma. Processing 3, 393 (1983)

3. S. Gordon, and B. J. McBride, "Computer Program of Calculations of Complex Chemical Equilibrium Compositions, Rocket Performance, Incident and Reflected Shocks, and Chapman-Jouquet Detonations," NASA SP-273, 1973

4. B. J. McBride, and S. Gordon, "Fortran IV Program for Calculation of Thermodynamic Data, "NASA TN-D-4097, 1967

ELECTRICAL AND OPTICAL PROPERTIES OF PECVD AMORPHOUS SILICON
GROWN AT LOW FREQUENCIES.

W.W.PIPER and G.E.POSSIN
G. E. Corporate Research and Development, PO Box 8, Schenectady,NY 12301.

ABSTRACT

The semiconducting properties of a-Si are ideally suited for making a
switch to control the matrix addressing of an array of liquid crystal
pixels. This investigation explores the differences between a-silicon
deposited by plasma enhanced chemical vapor deposition at 13.56MHz and at
60KHz. The bonded hydrogen concentration, the index of refraction and the
optical band gap of a-Si have been measured. The device used to explore the
field effect mobility of a-Si is an inverted staggered FET fabricated on a
glass substrate with conventional photolithography.

INTRODUCTION

Since plasma-enhanced chemical vapor deposition (PECVD) entails a
synergistic combination of free radical chemistry in the plasma and the
physical deposition of kinetic energy at the growth surface, it is quite
reasonable to expect that significant differences in the properties of
deposited materials might occur below a transition frequency which permits
constant acceleration between collisions as ions penetrate the plasma
sheath. Other system parameters such as power and concentration of bonded
hydrogen are of course also significant.

The semiconducting properties of a-Si are ideally suited to the system
requirements of a FET switch used to control the matrix addressing of an
array of liquid crystal pixels [1]. Response time, saturation on-current,
contact voltage drop, threshold voltage and off-current are all important
device parameters that are dependent upon bulk a-Si properties and on
interface characteristics. Previous reports have described a-Si FETs
fabricated with PECVD done at high frequency (13.56MHz). We have earlier
reported FETs fabricated from amorphous silicon and silicon nitride grown in
a low frequency plasma discharge [2]. This investigation explores whether
there exist any differences between material deposited at 13.56MHz and that
obtained in a plasma generated at 60KHz.

EXPERIMENTAL PROCEDURE

The plasma depositions were done in two commercially available reactors
both of which are capacitatively coupled, Reinberg-configured chambers. Each
is configured with circular, parallel electrodes. The bottom electrode is
heated and grounded and the top electrode is powered and temperature
controlled with circulating coolant. Gas flow is radially inward. On both
machines the electrode spacing was set at 25mm. The Technics PlasmaEtch II
reactor has an electrode diameter of 20cm and the PlasmaTherm PD3011 reactor
has electrodes 70cm in diameter. Each machine could be powered with an ENI
PL-1 at 60KHz or with properly matched and shielded power supplies at
13.56MHz. All of the depositions reported here were made at a platen
temperature of 300°C and a counterelectrode temperature of 60°C. The chamber
gas for the a-Si was 10% silane diluted with argon. All gases were of
electronic grade purity. Depositions were made at a power level of 30mw/cm²
in the smaller chamber and at 12 mw/cm² in the bigger chamber. Gas residence
time was about three seconds in each reactor.

Measurements to determine the bonded hydrogen content of the films were made on a Nicolet Fourier transform IR spectrometer. The hydrogen concentration [H] was calculated by integrating under the bond—wagging mode at 620 cm^{-1} first identified by Brodsky et al.[4].

$$[H] = A_\omega \int \alpha(\omega)/\omega \; d\omega \qquad (1)$$

The calibration constant, $A_\omega = 1.6 \times 10^{19}$cm^{-2}, determined by Fang et al.[5], is used in this paper.

Measurements to determine the optical band gap and index of refraction were made on a Cary14 spectrophotometer. Films of a—Si were deposited on 22mil thick wafers of Corning 7059 glass. Film thicknesses ranged from 0.2 to 1.2 microns. Data was analyzed using the standard equations for transmission through a thin film on a transparent substrate [3].

The FET device structure is shown in Fig. 1. After a thin layer of gate metal was patterned on Corning 7059 glass a triple layer of amorphous materials was plasma deposited in single pumpdown. The gate dielectric was 150nm of silicon nitride, the intrinsic a—Si layer was 200nm thick and the n$^+$ layer 50nm thick. N$^+$ material was obtained by mixing silane and phosphene to yield a source—drain contact layer with a bulk conductivity of about 10^{-2}/ohm-cm. A layer of Mo was deposited to contact the n$^+$ layer after which the Mo and n$^+$ were patterned back to form the FET structure. After plasma etching a thickness of 100 nm of a—Si remained in the channel. Channel lengths of 5 and 25 microns were fabricated which varied in width from 20 to 1000 microns. Gate to source and drain overlap in these research devices was large enough to insure the absence of overlap limited behavior [2].

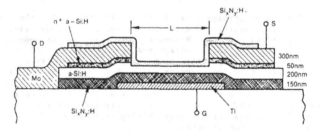

Fig.1. Cross—section of a inverted, staggered gate FET.

The field effect effective mobility was measured in two ways. At low drain bias voltage, we have assumed the validity of

$$I_d = \mu_e \; C \; (W/L) \; (V_g - V_t) \qquad (2)$$

and in the saturation region the usual square law model

$$I_d = \mu_e \; C \; (W/2L) \; (V_g - V_t)^2 \qquad (3)$$

is employed where C is the gate insulator capacity per unit area, W and L the channel dimensions, and V_g and V_t are the gate and threshold voltages.

RESULTS

Films of a-Si 0.6 and 1.2 microns thick were deposited on polished single crystal wafers and were examined optically for absorption bands between 4000 and 400cm⁻¹. According to the Brodsky et al. [4] analysis, a single hydrogen atom attached to a Si atom gives rise to a band peaking at 2000cm⁻¹, whereas multiple H-bonds on one Si shifts this band to 2100cm⁻¹. Our samples grown at both frequencies indicated a large prepondrance of isolated H-bonds although the low frequency samples did have an asymmetry suggesting that a much as 10% of the bonding was due to multiple bonds at one Si atom. Integration of the wagging band at 620cm⁻¹ yielded an estimate of a bonded hydrogen concentration of 14±2 % for the high frequency samples and 17±2 % for the films grown at low frequency.

A summary of the shape of the band edge of a-Si films is displayed in Fig. 2. There is no significant difference between the low and high frequency samples. The optical band gap is noted to be 1.78ev and the slope of the square root of (αE) function is 750 per eV. The value of the index of refraction was significantly altered at different frequencies. At one micron n=3.55 for the low frequency samples and n=3.70 for the high frequency samples.

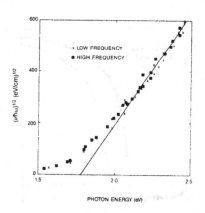

Fig. 2. Energy dependence of the square root absorption function.

Figure 3 compares the typical transfer characteristics obtained using the two different deposition frequencies. We observe very little differences between the device characteristics except for a very consistent difference in the field effect mobility. This is typically between 0.45 and 0.6 for the high frequency depositions and about 0.3cm²/volt-sec for the low freqency material. Threshold voltages are typically between 0.5 and 2.5 volts and somewhat larger for the low frequency depositions. The currents in the OFF state(-5 volts gate) are typically less than 5pA per 100 microns of channel width for drain voltages of less than 15 volts. Above 15 volts source drain breakdown causes a nonlinear increase in the OFF current. Long-length (25 microns) channel devices were for the most part relied on to minimize any contact voltage drop effects. The gate capacitance is 3.9x10⁻⁸ f/cm². Since the nitride thickness is known to be 150nm, the nitride dielectric constant is 6.4 at 1 KHz for both deposition frequencies. For the material deposited at low frequency, FETs with short channel lengths (5 microns) exhibit some contact drop indicating that the midgap bulk density of states is somewhat larger.

The measurements of field effect mobility as a function of temperature are summarized in TABLE I. Fig. 4. displays typical data used to calculate mobilities in the linear region.

TABLE I. Summary of effective mobility data
at low and high frequencies.

Measurement	Frequency KHz.	Mobility($25°C$) cm^2/volt-sec	Activation Energy (ev)	Prefactor cm^2/v-sec
Linear	60	0.22 ± 0.05	0.14	48 ± 20
Saturation	60	0.35 ± 0.05	0.09	11 ± 3
Linear	13560	0.54 ± 0.06	0.09	18 ± 6
Saturation	13560	0.54 ± 0.06	0.09	18 ± 6

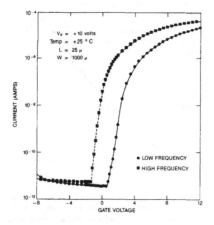

Fig. 3. Transfer characteristic
of a typical FET device.

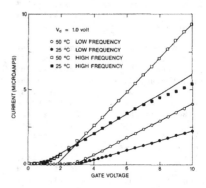

Fig. 4. Gate voltage dependence of
drain current in the linear region.

DISCUSSION AND SUMMARY

The lower index of refraction at low frequency may be due to larger hydrogen content however McLeod et al.[6] report a change in index with bonded H concentration which would account for only one-third of the difference measured in this study. Incorporation of additional amounts of molecular H_2, which was not measured, could account for the remaining difference without altering the electrical properties. The value of the optical band gap is in good agreement with other reported values [7,8,9]. Within experimental error the lack of any noticeable change of E_0 with [H] does not contradict the rate of change reported by Ross and Jaklik [7]. The absence of any significant difference in the shape and position of the band edge would indicate that the bulk density of states function just below the conduction band is independent of the deposition frequency at a constant concentration of bonded hydrogen.

The field effect mobilities are comparable to or larger than those obtained by others (see ref.[10] for a review). In fact the value of 0.54 for the high frequency material is higher than most other reported values. For the low frequency material the difference in the calculated mobility using data in the linear region and in the saturation region is due to voltage drop at the contacts [2]. We have more confidence in the mobility values calculated in the saturation region since simulations indicate this data is less affected by contact effects. The activation energy for the mobility is an indication of the position of the quasi-Fermi level in strong accumulation relative to the free carrier edge. Because the density of states is changing rapidly in this region it is not necessarily equal to the energy difference. Values of about 0.1 eV have been reported [11]. Both the high(HF) and low (LF) frequency materials have the same activation energy and almost the same prefactor. This indicates that the density of states near the band edges is similiar for both materials. This is consistent with the optical data which also indicates the same tail state distribution for both materials.

The increased sensitivity to contact effects for those devices fabricated using the LF material indicates the mid gap density of states is larger for the LF material. Previously reported measurements [2] on similiar LF material analyzing space charge limited currents indicated a mid gap density of states of $5x10^{15}$ #/cm3/eV. The reduced sensitivity of the HF devices to contact effects indicates a lower bulk mid-gap figure for the HF material. The exponential slope of the drain current versus gate voltage in the subthreshold region is 3 decades/volt for both HF and LF devices. This is consistent with either $3x10^{17}$ #/cm3/eV of bulk mid-gap states or $1.3x10^{11}$ #/cm2/eV of interface states. Hence in the subthreshold region both devices are not limited by bulk mid-gap states but by a similiar density of interface states.

ACKNOWLEDGMENTS

The authors wish to thank J. M. Anostario for his assistance in depositing the amorphous films and T. J. Soltys for processing the FET structures. IR spectroscopy was carried out with the help of P. J. Codella. Encouragement and support by J. E. Bigelow and R. C. Hodge is appreciated.

REFERENCES

[1] A.J.Snell,K.D.MacKenzie,W.E.Spear,P.G.LeComber, Appl.Phys.24,357(1981).

[2] G.E.Possin,D.E.Castleberry,W.W.Piper,H.G.Parks,pp.151-54 4th Research Display Conf.-EURODISPLAY'84 Sept. 18-20,Paris (to be published in Proc. Society for Information Display).

[3] O.S.Heavens,''Optical Properties of Thin Solid Films'',p.156ff. London, Butterworths (1955).

[4] M.H.Brodsky,Manuel Cardona,and J.J.Cuomo,Phys.Rev.B16,3556 (1977).

[5] C.J.Fang,K.J.Gruntz,L.Ley,M.Cardona,F.J.Demond,G.Muller,and S.Kalbitzer. J.Non-Cryst,Solids35β36,255 (1980).

[6] R.D.McLeod,W.Pries,H.C.Card,and K.C.Kao, Appl.Phys.Lett.45,628 (1984).

[7] R.C.Ross and J.Jaklik,Jr., J.Appl.Phys.55,3785 (1984).

[8] J.B.Webb and S.R.Das, J.Appl.Phys.54,3282 (1983).

[9] F.B.Ellis,R.G.Gordon,W.Paul, and B.G.Yacobi,J.Appl.Phys.55,4309 (1984).

[10] M.Powell, MRS Symposium on Comparison of Thin Film Transistor and SOI Technologies, p.247. North Holland (1984).

[11] K.D.MacKenzie,A.J.Snell,I.French,P.G.LeComber, and W.E.Spear,Appl.Phys.A31,87 (1983).

PROPERTIES OF INTRINSIC a-Si FILMS DEPOSITED FROM
HIGHER ORDER SILANES BY CHEMICAL VAPOR DEPOSITION

RICHARD E. ROCHELEAU, STEVEN S. HEGEDUS AND BILL N. BARON
Institute of Energy Conversion, University of Delaware, Newark, DE 19716

ABSTRACT

Amorphous silicon films have been deposited by chemical vapor deposition using disilane at temperatures between 360 and 525°C at growth rates up to 50 A/s. Intrinsic films have the following properties: σ_p less than 5 x 10^{-6} $(\Omega\text{-cm})^{-1}$; σ_d less than 5 x 10^{-11} $(\Omega\text{-cm})^{-1}$ with E_a = 0.7 to 0.8 eV; diffusion length around 0.1 µm; Urbach energy 48 to 55 meV; and mid-gap density of states greater than 5 x 10^{16} cm^{-3} eV^{-1}. Boron compensation improved collection efficiency by lowering the mid-gap density of states, not by improving the µτ product. Pin cells with efficiencies of 4% and J_{sc} = 10.9 mA/cm^2 (87.5 mW ELH) were fabricated.

INTRODUCTION

Amorphous silicon hydrogen films (a-Si:H) are of interest for solar cells, photoreceptors, and thin-film transistors. Most films of commercial value today are deposited by glow discharge (GD) from silane. Alternate techniques are of interest for materials optimization and to study mechanisms of film growth. One new process being evaluated is chemical vapor deposition from higher order silanes which achieves higher growth rates than LPCVD from silane; and which generates reactive species thermally rather than by electron and ion impact processes as in glow discharge. As expected for these process differences, some properties of films deposited by CVD from higher order silanes are quantitatively different than GD deposited films.

The object of this paper is to describe the film growth process and materials properties of intrinsic a-Si:H films deposited by CVD using disilane, with emphasis on properties of interest for solar cells.

REACTOR AND FILM GROWTH

Intrinsic, boron doped, and phosphorus doped a-Si:H layers were grown in a low pressure CVD system utilizing a horizontal tubular reactor (1 cm ID) heated in a tube furnace. Key features of the CVD system include: laminar flow through the deposition zone; radial temperature uniformity at all axial positions; and a well-defined axial temperature gradient.

The intrinsic film growth rate depends on deposition temperature (T_d), pressure (P), gas flow rate (Q), and axial position (Z). Growth rates up to 50 A/s have been achieved without formation of powder in the gas phase by selection of the deposition conditions within the following range: T_d = 360 to 525°C, P = 2 to 50 torr, and Q = 0.3 to 3 sccm disilane. The dependence of film growth rate on these parameters was reported earlier [1]. A chemical reaction engineering model based on component mass balances which describes intrinsic film growth has been developed and verified by experiment [2]. A key result of the model is quantitative verification that film growth is principally due to the longer chain silanes for growth rates exceeding 0.5 A/s. Figure 1 shows reactor temperature, film growth rates, and mole fraction of key gaseous species as a function of position in the reactor for depositions at two temperatures each with a gas holding time in the 30 cm reactor of approximately 40 seconds at 24 torr. In each case, the amount of film growth attributable to SiH_2 is less than 10% of the total. As expected for an activated surface reaction, the growth rate is higher at 430° than at 400° despite the somewhat lower silanes concentrations. Similar plots of gas composition for different gas flow rates at constant temperature and pressure yield concentration, temperature and growth rate data which show that the film growth rate is not well correlated to any single gaseous species and is not at all correlated with disilane concentration. The best agreement between the measured and calculated effluent composition is obtained when all species Si_2H_6 and higher are used as film precursors.

The behavior represented in Figure 1 is typical of results over a much wider range of conditions. In particular the concentrations of the silanes vary widely with reactor condition and with position in the reactor. At conditions yielding high growth rate, the effluent concentration of hydrogen is higher due to the surface reaction. The mono and disilane mole fractions monotonically increase and decrease, respectively, with axial position while the high order species exhibit a maximum about midway through the reactor.

ABSORPTION AND CONDUCTIVITY

Optical measurements were made on intrinsic films deposited on 7059 glass. The absorption coefficient (α) was obtained assuming only first surface reflection. The linear portion of the data plotted as $(\alpha/E)^{1/2}$ vs E was extrapolated to obtain an estimate of the gap. For CVD samples the optical gap from this method was 0.05 to 0.1 eV lower than was obtained from $(\alpha E)^{1/2}$ vs E. The α/E representation was used since it yielded the best linear fit over the widest energy range for the CVD samples. As summarized in Figure 2, the optical gap increased with decreased deposition temperature and increased film growth rate but did not change significantly with source of disilane or boron compensation up to 30 vppm. The changes with growth rate and temperature are consistent with qualitative infrared absorption measurements which showed increased mono and dihydride absorption in films deposited at lower temperature or higher rates. Total hydrogen in material grown under conditions yielding gaps above 1.5 eV was 8 to 10 mole % based on SIMS analysis.

Dark conductivity as a function of temperature (25-200°C) and photoconductivity at approximately 90 mW/cm^2 were measured on intrinsic films deposited on 7059 glass with 2000A thick molybdenum contacts 1 mm apart in a gap cell configuration. Arrhenius plots of $\sigma_d(T)$ were linear with slopes ranging from 0.7 to 0.8 eV. Dark conductivities at 25°C were 0.5 to 5 x 10^{-11} (Ω-cm)$^{-1}$ and photoconductivities were 0.5 to 5 x 10^{-6} (Ω-cm)$^{-1}$. The conductivity was independent of deposition temperature, film growth rate, disilane source and boron compensation. A more detailed study of conductivity was not undertaken because of the insensitivity to deposition conditions. Rather, solar cells and devices were fabricated to allow analysis of device related materials properties and optimization of cell performance.

DEVICE FABRICATION AND PERFORMANCE

Solar cells (ITO/pin/metal) and analytic devices (ITO/nin/metal) were fabricated to optimize conversion efficiency and for analysis of intrinsic layer properties. The range of i-layer deposition conditions included: four sources of disilane; deposition temperatures from 380 to 460°C; growth rates from 0.1 to 50 Å/s; and boron compensation up to 30 vppm.

Typical photovoltaic parameters (at 87.5 mW/cm^2, ELH) were: V_{OC} = .65 to .72 V, J_{sc} = 8 to 10.9 ma/cm^2, FF = 40 to 50, and η = 2.5 to 4.0%. The diode quality factor was 1.3 to 1.5 independent of boron compensation. Efficiency in the CVD cells has been limited by a high series resistance (> 20 Ω-cm^2) and high density of mid-gap states (> 5 x 10^{16} cm^{-3} eV^{-1}).

The series resistance consists of a light independent contact resistance, an i-layer thickness independent but light dependent term; and for thicker i-layers, a contribution due to the low i-layer photoconductivity. Except for the photoconductive term it has been difficult to separate limitations due to materials properties and those due to poor interfaces attributable to the fabrication process.

The high density of mid-gap states produces a poor field profile which inhibits minority carrier collection. In the remainder of this paper, we will discuss the effect of i-layer deposition conditions on the collection properties and density of states of pin solar cells.

COLLECTION EFFICIENCY ANALYSIS

While boron compensation did not affect the bulk electro-optical properties of intrinsic films, it significantly affected cell performance.

The collection efficiencies (η_C) of two compensated and two uncompensated pin cells deposited at 440°C from the same disilane are shown in Figure 3. The compensated devices have twice the response at 700 nm as the uncompensated ones, and 10 to 20% larger J_{sc} for the same i-layer thickness. Furthermore, the red response and J_{sc} of the compensated cells increased with i-layer thickness up to 5500 Å indicating an increase in collection width. Analysis of the long wavelength collection efficiency [3] (Figure 4) shows that boron compensation increases the collection width from ~ 0.2 μm (at 0 vppm) to ~ 0.3 μm (at 9 vppm). Measurement of η_c as a function of voltage bias [4] indicates that holes are the limiting carrier in CVD i-layers, even with boron levels up to 30 vppm. Measurement of the field-free diffusion length and modeling of the collection efficiency were used to determine if boron compensation improved the hole $\mu\tau$ product or the field profile.

The field-free diffusion length, L_D, was determined using surface photovoltage (SPV). As shown in Table I, L_D was approximately .1 μm independent of boron compensation and deposition temperature (400 and 440°C). Field-aided collection in the depletion region, characterized by L_C = $\mu\tau E(x)$ is linked to the field-free diffusion in the neutral region, characterized by L_D = $(kT\mu\tau/q)^{1/2}$, by their dependence on the hole $\mu\tau$ product. Good agreement to the collection efficiency of uncompensated cells was reported using $\mu\tau$ = 2-4 x 10^{-9} cm^2/V-s [4]. The 0.1 μm diffusion length found from SPV implies a hole $\mu\tau$ product of 3.5 x 10^{-9} cm^2/V-s, which is in excellent agreement with the above value. The best fit to the measured η_c for boron compensated pin cells was obtained with the same $\mu\tau$ values as uncompensated cells but a lower space charge density. Based on these results, boron lowers the space charge density, increasing field-aided collection throughout the compensated i-layer, but does not effect hole $\mu\tau$. This was verified by density of states measurements.

Table I. Diffusion Length from SPV

CVD Sample	i-Layer Conditions Temp. (°C)	Boron (vppm)	Gr. Rate (Å/s)	L_D (µm)
545.72	400	12	2.0	.12
546.72	440	12	4.1	.11
551.61	440	0	11.6	.10
553.51	440	0	27.0	.07

DENSITY OF STATES

Tail-state absorption was determined from the sub-bandgap primary photoconductivity spectra [5]. The Urbach energy, E_0, was calculated from ln (n_c) vs hν for chopped light hν between 1.0 and 1.7 eV on pin devices with 5 V reverse bias and light bias. Figure 4 shows E_0 as a function of i-layer deposition temperature between 380 and 460°C. A minimum E_0 = 48 meV was found with an i-layer deposited at 440°C and 7 Å/s. The values of E_0 ~ 50 meV are quite low for a material deposited at such high temperatures indicating steeply falling band tail-state edges. E_0 was independent of growth rate and boron compensation. The latter is surprising since boron is expected to increase the shallow acceptor trap levels near the valence band.

Mid-gap density of states (g_0) was determined from the space charge limited current (SCLC) on nin devices [6] and by steady-state capacitance temperature spectroscopy (CT) on pin devices [7]. In this latter method, the slope of $C^2/(dC/dT)$ vs T yields g_0. All results reported below are for i-layers greater than 5000 Å thick. Boron compensated i-layers could only be measured with the CT method since nin and pip devices with boron compensated i-layers displayed rectifying behavior. Figure 5 summarizes g_0 from CT measurements and from SCLC at $E_c - E_f$ = 0.7 eV as a function of i-layer deposition temperature. The two methods are in good agreement. Values of g_0 between 1 and 3 x 10^{17} cm^{-3} eV^{-1} were measured for depositions between 380°C and 440°C, and are independent of growth rate. However, boron compensation reduced g_0 by a factor of 2 to 3. This supports the earlier conclusion that boron compensation improves the collection width by reducing the space charge density, which increases the field strength; not by improving µτ.

ACKNOWLEDGEMENT

We are grateful for the collaboration of W. Buchanan, J. Cebulka, A. Moore and J. Dick. This work was supported under SERI Subcontract Number XB-4-04061-1.

REFERENCES

1. Final Report, SERI Subcontract No. XB-3-03089-1.
2. R. J. Bogaert, PhD Dissertation, University of Delaware, 1985.
3. Wronski et al., Solar Cells 2 (1980) 245.
4. S. Hegedus, J. Non-Crystalline Solids, 66 (1984) 369.
5. W. Jackson et al., Phys. Rev. B, 27 (1983) 4861.
6. W. den Boer, J. Physique C4 (1981) 451.
7. J. D. Cohen et al., Phys. Rev. B, 25 (1982) 5321.

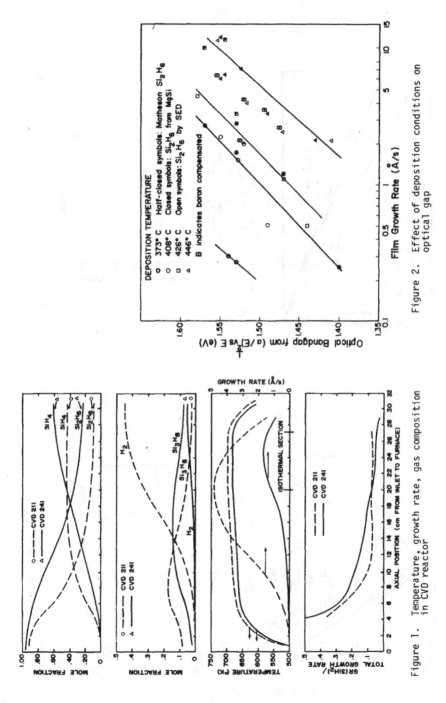

Figure 2. Effect of deposition conditions on optical gap

Figure 1. Temperature, growth rate, gas composition in CVD reactor

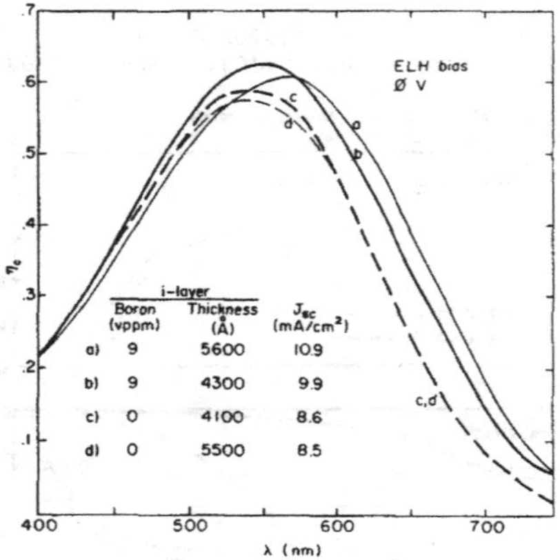

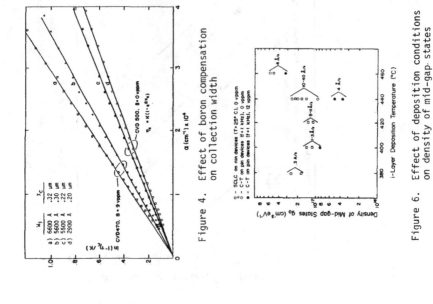

Figure 4. Effect of boron compensation on collection width

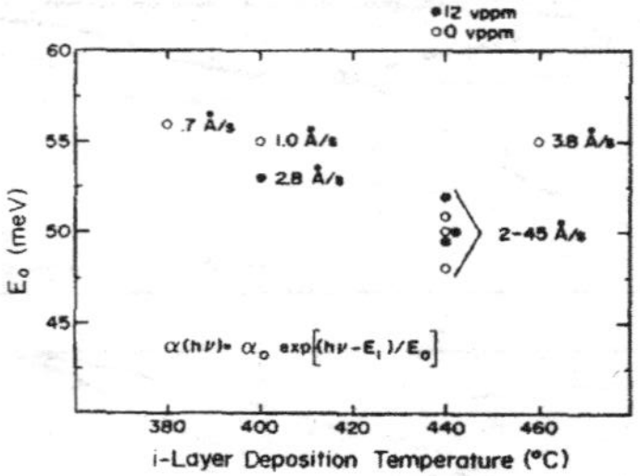

Figure 6. Effect of deposition conditions on density of mid-gap states

Figure 3. Effect of boron compensation on collection efficiency

Figure 5. Effect of deposition conditions on Urbach energy

HYDROGENATED AMORPHOUS SILICON FILMS BY THE PYROLYSIS OF DISILANE[1]

T. L. CHU,*SHIRLEY S. CHU,*S. T. ANG,* D. H. LO,** A. DUONG,**
C. G. HWAUNG,** AND L. BOOK*
*Southern Methodist University, Dallas, TX 75275
**Poly Solar Incorporated, Garland, TX 75041

ABSTRACT

The thermal decomposition of disilane in a hydrogen or helium flow has been used for the deposition of hydrogenated amorphous silicon (a-Si:H) films on the surface of several substrates at 450°-500°C. The concentration of disilane in the reaction mixture has been found to affect strongly the deposition rate and the photoconductivity of a-Si:H films. The AM1 conductivity of a-Si:H films increases with increasing disilane concentration and approaches 10^{-5}(ohm-cm)$^{-1}$ at disilane concentrations higher than about 4%, and the conductivity ratio is about 10^5. The density of gap states in CVD a-Si:H films have been determined by the photothermal deflection spectroscopy, capacitance-temperature, capacitance-frequency, and space-charged-limited current measurements with similar results.

INTRODUCTION

Chemical vapor deposition (CVD) appears to be promising for the preparation of hydrogenated amorphous silicon (a-Si:H) films. The CVD of a-Si:H films has been under investigation during the past few years. The thermal decomposition of monosilane (SiH$_4$) at substrate temperatures below about 650°C yields amorphous silicon films which contains less than 1% hydrogen and exhibit essentially no photoconductivity due to the high density of gap states, about 10^{20}/cm^3 eV. The hydrogen content in these films may be increased appreciably by plasma hydrogenation at 400°C, and the reduction of dangling bond density has been shown by electron spin resonance measurements. The thermal decomposition of disilane (Si$_2$H$_6$) requires considerably lower temperatures than that of SiH$_4$, and under proper conditions, the resulting films show high photoconductivity. The deposition and properties of a-Si:H films by the Si$_2$H$_6$ process has been studied in a static system, under reduced pressure, and under atmospheric pressure in a gas flow system. The atmospheric pressure flow process is preferred because of its simplicity in operation and the high deposition rates obtainable (up to about 100 Å/sec).

In this work, the thermal decomposition of Si$_2$H$_6$ in a hydrogen or helium flow has been used for the deposition of a-Si:H films on Corning 7059 glass and other substrates, and their properties, with emphasis on electrical conductivity and density of gap states, investigated. The experimental procedures and results are summarized in this paper.

DEPOSITION OF a-Si:H FILMS

Hydrogenated amorphous silicon films were deposited on Corning 7059 glass, single crystalline silicon, and Mo/glass substrates by the thermal decomposition of disilane in a hydrogen or helium atmosphere. Disilane was

---

[1]Supported by the Electric Power Research Institute under contract RP 1193-2

prepared by the glow-discharge technique followed by fractionation or purchased commercially; the commercial disilane contains ethylsilane as the major impurity. The substrates were supported on a silicon susceptor in a horizontal fused silica reaction tube, and the susceptor heated externally with an rf generator. During the deposition process, the substrate was maintained at 450°-500°C, and hydrogen (or helium) containing 1-10% disilane was introduced into the reaction tube at 4ℓ/min. At a given composition, the deposition rate increases exponentially with temperature, indicating that the surface reaction is the rate-determining step. The activation energy deduced from the Arrhenius plot is about 3 eV. Deposition rates of up to 120 Å/sec have been obtained in a hydrogen atmosphere, and the deposition rate in helium is about twice of that in hydrogen.

To determine the composition of the decomposition-products of Si_2H_6 at 480°-490°C, the effluent from the fused silica reaction tube was collected in a stainless steel sampling bottle and was analyzed using a gas chromatograph. The concentrations of SiH_4, unreacted Si_2H_6, Si_3H_8, and unreacted $C_2H_5SiH_3$ are estimated to be 19.3%, 65.6%, 10.6%, and 4.5%, respectively. In comparison with about 5% C_2H_5-SiH_3 in commercial disilane, an appreciable fraction of $C_2H_5SiH_3$ has decomposed. The reaction by-products, SiH_4 and Si_3H_8, are presumably formed through the following reactions:

$$Si_2H_6 = SiH_2 + SiH_4$$

$$SiH_2 + Si_2H_6 = Si_3H_8$$

Thus, the SiH_2 radical plays an important role in the deposition of a-Si:H films from Si_2H_6.

The composition profile in a-Si:H films deposited from commercial Si_2H_6 on Corning 7059 glass substrates at 490°C was determined by the SIMS technique. A typical SIMS depth profile is shown in Fig. 1. The semi-quantitative composition of the a-Si:H film is: Si : 89.7%, H : 9.4%, C : 0.7%, O : 0.8%,

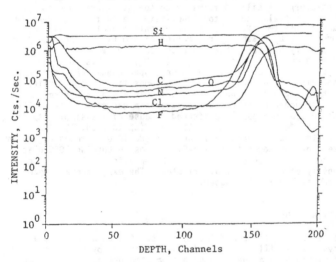

Fig. 1. SIMS profile of CVD a-Si:H films on Corning 7059 glass substrate.

$C\ell$: 0.004%, F < 0.002%, and N : 0.09%. The high carbon content is due to the $C_2H_5SiH_3$ contamination in commercial Si_2H_6.

The optical absorption of a-Si:H films deposited on Corning 7059 glass substrates by the thermal decomposition of H_2-Si_2H_6 mixtures has been measured at room temperature using a Cary 17D spectrophotometer. The optical bandgap of the a-Si:H films deduced from the $(\alpha E)^{\frac{1}{2}}$ (where α is the absorption coefficient and E is the photon energy) versus E plot is 1.65-1.68 eV.

ELECTRICAL CONDUCTIVITY OF a-Si:H FILMS

The dark conductivity and AM1 (simulated by using quartz-halogen lamps) conductivity of a-Si:H films deposited on Corning 7059 substrates at 460°-500°C using a H_2-Si_2H_6 or He-Si_2H_6 mixture containing various concentrations of Si_2H_6 have been measured at room temperature using a coplanar configuration. Contacts to the undoped a-Si:H films were made by evaporating successively 200 Å of antimony and 1 μm of aluminum onto the film surface, followed by heat treatment in a hydrogen atmosphere at 200°C. While the room temperature dark conductivity of undoped films, $(2-4) \times 10^{-10}$ $(ohm-cm)^{-1}$, is essentially independent of the substrate temperature and the composition of the reaction mixture, the AM1 conductivity depends strongly on the reactant composition at Si_2H_6 concentrations lower than about 4%. As an illustration, Fig. 2 shows the dark

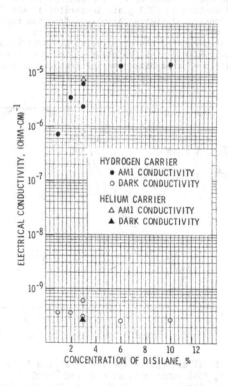

Fig. 2 The dark and AM1 conductivities of a-Si:H films deposited at 500°C as a function of the concentration of disilane in the reaction mixture.

and AM1 conductivities of a-Si:H films deposited on Corning 7059 glass substrates at 500°C using H_2-Si_2H_6 and He-Si_2H_6 mixtures containing various concentrations of Si_2H_6. It is noted that at low Si_2H_6 concentrations, the AM1
conductivity of a-Si:H films increases rapidly with increasing concentration
of Si_2H_6 and levels off at Si_2H_6 concentrations higher than about 4%. The
poor photoconductivities of a-Si:H films deposited at low disilane concentration may be due to the low hydrogen content in the films. Similar relations
were obtained at 475°C. The use of hydrogen or helium as the diluting gas
has no appreciable effects on the conductivities of a-Si:H films. The conductivity ratio, $\sigma_{AM1}/\sigma_{dark}$, in better quality films is about 10^5.

The dark conductivity of a-Si:H films deposited on Corning 7059 glass
substrates at 480°-490°C from a H_2- 6% Si_2H_6 mixture has also been measured as
a function of temperature. The log σ versus 1/T plot of an a-Si:H film yields
an activation energy of 0.79 eV.

DENSITY OF GAP STATE IN a-Si:H FILMS

The density of gap states in CVD a-Si:H films was determined by four
techniques: (1) photothermal deflection spectroscopy (PDS), (2) capacitance-
temperature relation of Schottky barriers, (3) capacitance-frequency relation
of Schottky barriers, and (4) space-charge-limited current (SCLC) measurements.

The PDS technique measures directly the subgap absorption and is the
most accurate technique used in this work. The absorption spectra of two CVD
a-Si:H films deposited on Corning 7059 glass substrates at 480°C and 500°C are
shown in Fig. 3. The density of states calculated from the subgap absorption
coefficients is approximately 3×10^{16}/cm^3 in both samples.

The other techniques require the preparation of Schottky barrier structures such as Pd/(i) a-Si:H/(n$^+$) μC-Si/Mo/glass. Figure 4 shows the capacitance-temperature relation of a Schottky barrier prepared from a CVD a-Si:H

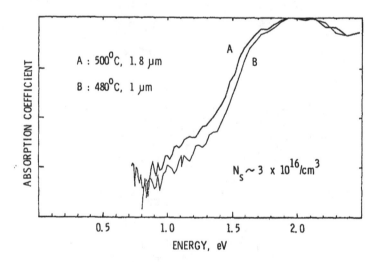

Fig. 3   Subgap optical absorption spectra of a-Si:H films deposited
         on Corning 7059 substrates at 480° and 500°C (Courtesy of
         Dr. John L. Crowley of Electric Power Research Institute).

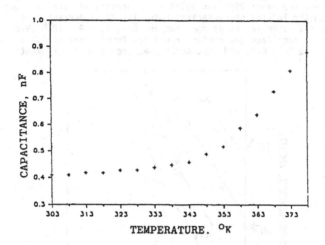

Fig. 4 Capacitance of an a-Si:H Schottky barrier at 1 kHz as a
function of temperature.

film deposited at 490°, where the ac modulated signal was 1 kHz at 5 mV, and
the Schottky barrier was reverse biased at 0.5 V. The density of localized
gap states around the Fermi level, deduced from the slope of $C^2(T)/(dC/dT)$, is
$5.2 \times 10^{16}/eV\text{-}cm^3$. Figure 5 shows the capacitance-frequency relation of the
same Schottky barrier, where the capacitance saturates at a modulated frequency

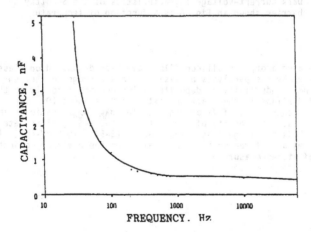

Fig. 5 Capacitance of the Schottky barrier shown in Fig. 4 as
a function of frequency.

of 700 Hz. The density of localized gap states around the Fermi level, deduced from the built-in potential and Debye length, is 5.3×10^{16}/eV-cm^3. Figure 6 shows the dark forward current-voltage characteristics of a Schottky barrier in the temperature range 299° and 353°K. The density of localized gap states, deduced from the lnJ-lnV characteristics and its first derivative m(V) in the space-charged-limited regime at room temperature, is 4.4×10^{16}/eV-cm^3. Thus, the density of localized gap states around the Fermi level in CVD a-Si:H films measured by the C-T, C-f, and SCLC techniques are in good agreement.

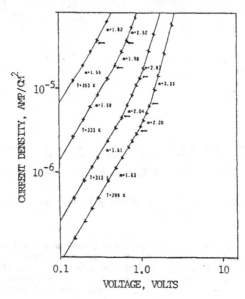

Fig. 6   Dark current-voltage characteristics of the Schottky
          barrier shown in Fig. 4 as a function of temperature.

SUMMARY

Hydrogenated amorphous silicon films have been deposited on glass and other substrates by the pyrolysis of disilane in a hydrogen or helium atmosphere. The AM1 conductivity of deposited films depends strongly on the concentration of disilane in the reaction mixture and is about 10^{-5} (ohm-cm)$^{-1}$ at disilane concentrations of 4% or higher. The $\sigma_{AM1}/\sigma_{dark}$ ratio is about 10^5 in most films. The density of gap states in CVD a-Si:H films determined by several methods are in good agreement, about $(3-5) \times 10^{16}$/eV-cm^3. Thus, CVD a-Si:H films are of reasonably good quality on the basis of conductivity and density of states measurements.

DUAL-CHAMBER PLASMA DEPOSITION OF A-Si:H
SOLAR CELLS AT HIGH RATES USING DISILANE

G. RAJESWARAN, P.E. VANIER, R.R. CORDERMAN, AND F.J. KAMPAS
Metallurgy and Materials Science Division, Brookhaven National Laboratory,
Upton, NY 11973

ABSTRACT

The use of a separated chamber deposition system for the fabrication of a-Si:H solar cells from disilane at high deposition rates results in a substantial improvement in short circuit current compared to that obtained from a single-chamber system. The spectral responses of cells fabricated in the dual-chamber mode are compared to those made in the single-chamber mode. The results are interpreted by assuming that the rate of removal of boron contaminants from the chamber is independent of deposition rate.

INTRODUCTION

The p/i interface of a glass/tin oxide/p/i/n/Al amorphous thin film solar cell has to be abrupt for satisfactory cell performance. It is necessary to reduce the level of boron atoms from $10^{20}/cm^3$ in the p-type layer to less than $5 \times 10^{16}/cm^3$ (1 ppm) at the beginning of the intrinsic layer. It has also been shown by many groups that boron profiling of the i-layer at the ppm level can help to eliminate low-field regions in the device and aid in efficient carrier collection [1,2]. In single-chamber solar cell deposition systems there are two possible sources of boron contaminants. First, there can be residual diborane left over from the p-layer deposition whose removal rate is characteristic of the deposition system design. This removal rate can be substantially accelerated by injecting the diborane close to the deposition chamber, baking the walls of the chamber to minimize adsorption and by reducing the volume of the deposition chamber relative to the discharge volume. In addition, purging procedures can be introduced after the p-layer deposition to enhance the removal of residual diborane. The second source of boron impurities may be the increase in boron concentration due to the sputtering of the previously deposited boron-doped film on the walls of the chamber and on the opposite electrode. This sputtering may be caused by energetic ion bombardment of exposed surfaces at the beginning of the i-layer discharge. In some designs of single-chamber systems, when low deposition rates (1-2 Å/s) are utilized for the i-layer, it is possible to arrive at process variables and purge procedures that will actually use the background boron concentration to advantage in solar cells. The 10.1% solar cell reported by the RCA group [3] was deposited in a single-chamber system. In this paper, the problems of solar cell fabrication at high deposition rates (>15 Å/s) in a single-chamber system are explored. It has been found that at >15 Å/s deposition rates, an unique dual-chamber deposition system results in substantial improvements in carrier collection.

EXPERIMENTAL

All the depositions reported in this study were performed in a dual-chamber system shown in Fig. 1. Ultrahigh vacuum components are used in both chambers and heating tapes are used for baking the walls of the chambers above $100^{\circ}C$. The chamber on the right in Fig. 1 is called the dopant (D) chamber and is used for making B- or P-doped amorphous silicon alloy films. The intrinsic (I) chamber on the left in Fig. 1 is used only

for intrinsic amorphous silicon films. The I chamber is not exposed to high
levels of dopant gases such as diborane or phosphine. Copper gaskets seal
all the flanges except for the top electrode assembly of the D-chamber where
a Viton o-ring is used. Each chamber is independently pumped by a diffusion
pump and liquid nitrogen cold trap assembly. Typical base pressures are
5×10^{-7} Torr for the I-chamber and 8×10^{-7} Torr for the D-chamber. The
samples are mounted on a 10 cm aluminum slider with stainless steel sliding
surfaces. The slider dovetails into position in the top electrode assembly
of either chamber. The heater and thermocouple are external to the vacuum
system and heating is regulated by a Eurotherm feedback controller. The
temperature at the substrate surface was calibrated with reference to the
controller setting using thin film platinum resistors. A magnetically
coupled rotary-linear motion manipulator is used to engage onto the slider
and transfer it from the D-chamber to the I-chamber or vice-versa. The
transfer is performed under helium flow from I- to D-chamber so that the
contamination of the I-chamber is minimized. Viewports in each chamber make
the sample transfer easier and serve as windows for plasma diagnostics. The
slider hangs from a grounded top electrode assembly and is the anode. The
driven electrode (the cathode) is surrounded by a plasma shield ring to
confine the discharge.

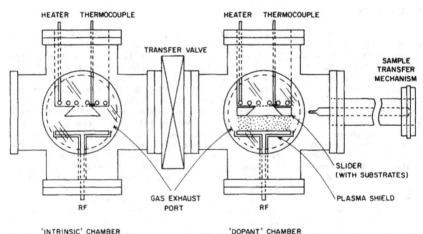

Fig. 1 Dual-Chamber Deposition System

Gas flow is perpendicular to the plane of Fig. 1, that is, from the
front of the system towards the back. Gas connections were made so that
intrinsic layers from monosilane or disilane could be deposited in the
D-chamber in addition to doped layers. Only intrinsic layers can be
deposited in the I-chamber. These connections ensured that p/i/n solar cell
structures could be deposited either in the single-chamber mode (Mode 1) or
in the separated chamber mode (Mode 2). The critical deposition parameters
used for >15 Å/s disilane intrinsic layers are listed in Table 1. This
table also lists the conditions used for the p-type silicon carbide and the
n-type amorphous silicon layers.

A series of devices was fabricated using tin-oxide coated glass from
Cherry Display Products. The coating of this glass is milky in appearance,
which suggests that it should help in trapping light in the cells. The
sheet resistance of this coating was 10-15 ohms/square. Aluminum back
contacts were vacuum evaporated using contact masks to define cell areas.
In order to eliminate edge collection effects when measuring cell
performance, the remaining exposed silicon regions were etched away in a

$CF_4 + O_2$ plasma. The tin-oxide coating and the aluminum serve as etch stops. The cells were tested under a Xenon arc lamp source with AM1 filtering and the intensity of illumination was calibrated with a crystalline silicon solar cell secondary standard. This secondary standard is based on an amorphous silicon cell previously measured at SERI under AM1 solar simulation. Spectral response measurements were carefully made using a Tektronix photometer whose readings are accurate to $\pm 5\%$ without correction. The photometer spectral response (provided by Tektronix) which is traceable to NBS was used to correct the calibration to higher accuracy. Errors due to non-uniformity of the light beam were avoided by masking of the photometer sensor area to match that of the cell under test and by precise alignment. All spectral responses were measured under uniform white light bias using a chopped monochromatic beam and a lock-in analyzer.

Table I: Deposition Conditions

Frequency $=13.56$ MHz, Area of electrode $= 80$ cm^2

Layer	p	i	n
Substrate Temperature	$250^{\circ}C$	$250^{\circ}C$	$250^{\circ}C$
rf power	10 W	15 W	5 W
Gas pressure	250 mTorr	1000 mTorr	250 mTorr
Flow rate	SiH_4 20 sccm	Si_2H_6 50 sccm	SiH_4 2.5 sccm
	CH_4 30 sccm	He 450 sccm	Ar 37.5 sccm
Dopant concentration	0.5%		0.4%

RESULTS AND DISCUSSION

The properties of intrinsic a-Si:H layers made by plasma decomposition of Si_2H_6-He mixtures in the I-chamber of the dual-chamber system described above were reported elsewhere [4]. In summary, the ratio of AM1 photoconductivity to dark conductivity was found to be 10^6 and in one case, as high as 10^7. These films were deposited at 20 Å/s and may have minimal incorporated background impurities (oxygen, hydrocarbons, etc.) per unit volume. The optical gaps could be varied over a wide range (1.8 - 2.2 eV) by changing the substrate temperature T_s. This variation resulted from increasing hydrogen incorporation at lower temperatures. For a given T_s, an i-layer deposited from disilane absorbs fewer photons and can generate less current under AM1.5 global solar illumination than a similar film produced from monosilane. For instance, at $T_s = 250^{\circ}C$, the extrapolation of a Tauc plot gives 1.9 eV for an i-layer from disilane compared to 1.7 eV for a film from monosilane. Spin density measurements indicate that a minimum spin count of $5 \times 10^{15}/cm^3$ occurs for $T_s \approx 260^{\circ}C$ [5]. These films exhibit some photo-induced degradation but this behavior is similar to that of films made from monosilane-hydrogen mixtures at low deposition rates. At optimum T_s, platinum-Schottky barrier diodes were fabricated on films deposited on highly doped crystalline silicon substrates and were found by capacitance measurements to have depletion widths as great as 2 μm even at zero bias [6].

Preliminary solar cells of a glass/SnO$_2$/p/i/n/Al structure were fabricated utilizing the intrinsic layers described above. Devices were fabricated with i-layers deposited either in Mode 1 or Mode 2, while all other conditions were kept constant. The devices deposited using Mode 2 showed remarkably different characteristics from those made in Mode 1. In particular, the Mode 2 cells show much higher short circuit currents (J_{sc}) and lower open circuit voltages (V_{oc}). Figure 2 shows a comparison of two cells prepared in the above manner. The single-chamber or Mode 1 cell exhibits good values for V_{oc} (0.85 V), moderate values of fill factor

(0.56), but very low values of J_{sc} (6 mA/cm^2). The separated chamber or Mode 2 cell exhibits better values of J_{sc} (12.2 mA/cm^2), moderate values of fill factor (0.53), and a reduction of V_{oc} to 0.74 V.

   Spectral response measurements were performed on both types of cells to compare their detailed behavior. The results are shown in Fig. 3. It is immediately apparent that the Mode 1 cell has a lower response at all wavelengths, but is particularly weak on the blue side of the spectrum. On the other hand, the Mode 2 cell has excellent blue response, and also reasonable red response considering that the disilane-deposited i-layers have wider gaps than the usual a-Si:H deposited at low rates from monosilane. By digitally convoluting the measured spectral responses with the SERI-tabulated AM1.5 global solar spectrum, values were obtained for J_{sc} which were consistently higher than the values measured with the Xenon lamp solar simulator. This suggests that the efficiencies reported here are conservative estimates.

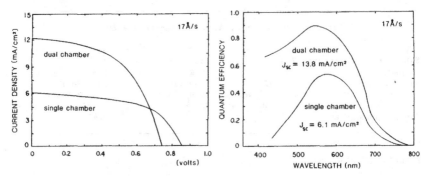

Fig. 2   Effect of Dual-Chamber   Fig. 3   Effect of Dual-Chamber
         Mode on J-V Curves   Mode on Spectral Response

   A tentative explanation of these results rests on the reasonable assumption that boron tailing in the intrinsic layer is particularly severe at high deposition rates. A schematic representation of this explanation is shown in Fig. 4. As mentioned in the introduction, there can be two major sources of boron contamination in a single-chamber system. First, the gaseous remnants from the p-layer deposition (indicated by the solid curve in Fig. 4 and labelled as residual diborane) and second, the increase in boron containing species evolving from the surfaces exposed to the i-layer plasma (indicated by the dashed line in Fig. 4 and labelled as plasma-induced boron). The levels of contaminants and the removal rates shown in Fig. 4 were chosen arbitrarily. But, every deposition chamber has its characteristic boron background concentration and elimination rate. Consider case B shown in Fig. 4 (5000 Å of i-layer at 1.7 Å/s). This is typical of high efficiency cells reported in the literature [3]. Assume that the residual diborane removal rate has been accelerated by optimizing chamber design. The resulting solar cell structure is affected only by the elimination rate of plasma-induced boron. These effects are felt only in the vicinity of the p/i interface, since the deposition rate is low. The p/i interface is therefore abrupt and may appear as shown in Fig. 5B. The carrier collection width W is almost as large as the thickness of the i-layer. In addition, by optimization of deposition conditions, it is possible to tailor the i-layer with beneficial boron-grading. Now consider case A (5000 Å at 17 Å/s) shown in Fig. 4. A major fraction of the i-layer is deposited before the boron concentration in the chamber falls low enough to allow the deposition of truly intrinsic material. Therefore, much of the light (especially the blue component) is absorbed in a thick layer of lightly boron-doped material deposited from disilane on top of the intentionally doped carbide p-layer. The carriers generated in this region of the cell recombine rapidly and do not contribute to the external current. The carrier collection width W for this case shrinks towards the back of the cell (away from the p/i interface) as shown in Fig. 5A. Cells prepared in Mode 1 correspond to the structure of Fig. 5A and hence have low J_{sc} and poor blue response (Figs. 2 & 3). The extra thick p-type layer consists of a fairly wide-gap material containing a large number of ionized acceptors, which can support a field across the thin i-layer that remains and this results in high V_{oc}. Cells prepared in Mode 2 have much less boron contamination and produce structures similar to Fig. 5B. With large collection widths, the J_{sc} in Mode 2 cells are high and so is the blue response (Figs. 2 & 3).

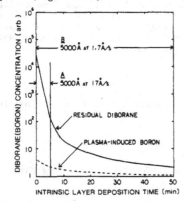

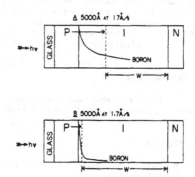

Fig. 4 Deposition Rate Affects Boron Tailing in Cell

Fig. 5 Deposition Rate Affects p/i Interface Position

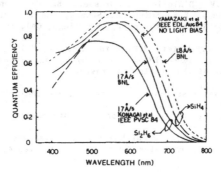

Fig. 6 Comparison of BNL Spectral Responses with Published Data

Spectral response measurements were compared with published data from other groups. In particular, recent data on an 8% cell deposited from disilane at 17 Å/s [7], is compared with a similar BNL cell in Fig. 6 (solid lines). The shape of the curves are very similar, but the BNL cell actually generates more current at all wavelengths. A similar comparison is made in Fig. 6 between the spectral responses of two cells made at low deposition rates from monosilane in separated chamber systems. In this case, the reported cell efficiency from the other group [8] was 10.5%. It seems that the use of the dual-chamber system for both monosilane (1.8 Å/s) and disilane (17 Å/s) depositions results in cells whose quantum efficiencies are comparable with some of the best cells reported in the literature.

CONCLUSIONS

Comparing the disilane-deposited cells with the monosilane-deposited cells (Fig. 6) highlights the problem of the lower red response obtained with disilane. This is a direct result of the higher concentration of hydrogen incorporated in the i-layers when disilane is used. A higher concentration of H leads to a wider gap and reduced absorption of light. Evidently, the efforts to reduce the gap of the i-layers must be continued, either by raising the substrate temperature or by mixing monosilane with disilane. Currently, the best BNL cell deposited at 17 Å/s from disilane has the following parameters: $V_{oc} = 0.79$ V, $J_{sc} = 15.2$ mA/cm^2, FF = 0.52 and $\eta = 6.2\%$. In conclusion, it seems that the use of a dual-chamber deposition system overcomes the problem of boron contamination which becomes particularly severe at high deposition rates.

ACKNOWLEDGEMENTS

This research was supported by the Solar Energy Research Institute under Contract No. DE-AC02-76CH00016.

REFERENCES

1. H. Haruki, H. Sakai, M. Kamiyama, and Y. Uchida, Solar Energy Materials 8, 441 (1983).

2. P. Sichanugrist, M. Konagai, and K. Takahashi, Solar Energy Materials 11, 35 (1984).

3. A. Catalano, R.V. D'Aiello, J. Dresner, B. Faughnan, A. Firester, J. Kane, H. Schade, Z.E. Smith, G. Swartz, and A. Triano, Proc. 16th IEEE Photovoltaic Specialists Conf., (San Diego, CA), 1421 (1982).

4. G. Rajeswaran, P.E. Vanier, R.R. Corderman, and F.J. Kampas, Mat. Res. Soc. Fall Meeting, Nov. 1984; Proc. Symp. on Plasma Synthesis and Etching of Electronic Materials (in press).

5. W. E. Carlos, Naval Research Laboratory, private communication.

6. H. Okushi, Electrotechnical Laboratory, private communication.

7. M. Konagai, T. Matsushita, P. Sichanugrist, K. Takahashi, and K. Komori, Proc. 17th IEEE Photovoltaics Specialists Conference, Kissimmee, May 1984, 347-52.

8. S. Yamazaki, A. Mase, K. Urata, K. Shibata, H. Shinohara, S. Nagayama, M. Abe, T. Hamatani, and K. Suzuki, IEEE Elect. Dev. Lett., Vol. EDL-5, No. 8, Aug 1984.

AMORPHOUS SILICON FILMS AND SOLAR CELLS PREPARED BY MERCURY-SENSITIZED PHOTO-CVD OF SILANE AND DISILANE

ALAN E. DELAHOY, B. DOELE, F.B. ELLIS, JR., K.R. RAMAPRASAD, T. TONON, AND
J. VAN DINE
Chronar Corp., Princeton, NJ 08542, USA

ABSTRACT

The low temperature deposition of a-Si:H films and devices by mercury-sensitized photo-decomposition of silanes is described. The problem of window deposition is ameliorated by the use of a perfluoropolyether coating to give deposition rates of 0.3 Å s^{-1}. Highly photosensitive undoped layers can be produced. These films exhibit many properties in common with glow discharge films including thermal quenching of photoconductivity between 125-200°K and the Staebler-Wronski effect. A wide gap silicon carbide p layer is described, and is usefully employed in p-i-n structures to give 6% efficient solar cells. SIMS analysis of these devices shows that it is possible to achieve sharp p-i interfaces in a single chamber reactor. Deposition rates greater than 2 Å s^{-1} have recently been achieved.

INTRODUCTION

Photovoltaic panels consisting of monolithically interconnected cells of hydrogenated amorphous silicon are now available commercially from several manufacturers both in the US and in Japan. Solar conversion efficiencies of the order of 8% have been achieved for 100 cm^2 submodules (see for example [1]). In all cases the amorphous silicon is prepared by glow discharge decomposition of silane. Alternative a-Si:H deposition techniques that might ultimately prove superior (either in deposition uniformity or rate, cell efficiency, material stability, reproducibility, or lower cost) have been under intensive investigation at Chronar Corporation and elsewhere. Disilane is often employed as a means of increasing deposition rate.

Amorphous silicon prepared by thermal CVD of disilane apparently possesses an unavoidable minimum density of defects ($\sim 10^{17}$ Si dangling bonds cm^{-3}) as a result of deposition temperatures in excess of 400 C.. This leads to low photoconductivity and poor device fill factors [2]. Amorphous silicon prepared by mercury sensitized decomposition of disilane, however, is grown at considerably lower temperatures (200-250 C), is highly photoconductive, and can yield device fill factors greater than 0.66. Advantages not yet apparent may also accrue from the absence of both film and chamber bombardment by ions, and a deposition chemistry different from other techniques.

The deposition is effected as follows. Ultra-violet light from a low pressure mercury lamp is resonantly absorbed by mercury atoms carried into the deposition chamber by one or more silanes. The excited Hg atom (Hg*) abstracts hydrogen from the silane [3] resulting in a silyl type radical that is the precursor to a-Si:H deposition.

$$
\begin{aligned}
&\text{Hg} + h\nu \longrightarrow \text{Hg}^* \\
&\text{Hg}^* + \text{Si}_2\text{H}_6 \longrightarrow \text{HgH} + \text{Si}_2\text{H}_5 \\
&\text{HgH} \longrightarrow \text{Hg} + \text{H} \\
&\text{H} + \text{Si}_2\text{H}_6 \longrightarrow \text{H}_2 + \text{Si}_2\text{H}_5 \\
&\text{or} \longrightarrow \text{SiH}_4 + \text{SiH}_3
\end{aligned}
\tag{1}
$$

Note that atomic hydrogen is produced by dissociation of the short-lived intermediate HgH. Furthermore, higher silanes (Si_3H_8, Si_4H_{10}.) will be formed by recombination of silyl radicals and subsequent reactions; the larger hydrogen deficient "silanes" as well as powder particles may condense on the cooler window. The atomic hydrogen participates in many reactions both in the gas phase and on the growing film surface, and is expected to strongly influence both the film growth mechanism and the resulting a-Si:H properties.

The low pressure mercury lamp is typically operated at 40 C at which temperature the vapor pressure of Hg is 7 mTorr. A few Torr of a gas such as Ar is added to aid initiation of the discharge, and to lower recombination losses at the wall. (The lowest excited state of Ar is metastable and exceeds the ionization energy of Hg, thus forming a Penning mixture). Almost all of the radiation emitted by the low pressure Hg lamp is resonance radiation at 253.7 nm and 185.0 nm resulting from the transitions $6\,^3P_1 \rightarrow 6\,^1S_0$ ("forbidden") and $6\,^1P_1 \rightarrow 6\,^1S_0$. The energy emitted at 253.7nm is greater than that at 185nm.

A serious hinderance to early progress in this field was the deposition of a highly absorbing a-Si:H film on the window of the deposition system that prevented the UV light from penetrating the chamber. If no attempt is made to address this problem films with a limiting thickness of only a few hundred Angstroms can be grown. As mentioned by Inoue et al. [4] it was found that coating the inside surface of the window with certain low vapor pressure and UV transparent oils leads to much thicker films. In the configuration shown in Fig. 1 films up to about 0.4 μm can be grown from disilane at an average deposition rate of about 0.3 Å s^{-1}.

EXPERIMENTAL

The chamber used for these depositions is shown in Fig. 1. It has an inside diameter of 6.1 cm, a height of 4.8 cm, and is externally heated. To allow the 185 μm radiation to enter the chamber, a Suprasil quartz window is used and the space between the window and lamp is flushed with nitrogen to exclude oxygen and prevent ozone formation. By trial and error, perfluoro-polyether fluids have so far proved most effective in inhibiting window deposition. Mercury is carried into the chamber by passing the silanes over a mercury reservoir immersed in a water bath at 30-70 C. For the most part electronic grade disilane produced by silent electric discharge of monosilane has been used. A flushing gas (He) is introduced just under the window in an attempt to reduce deposition on the window. During deposition, the gases are exhausted by a mechanical pump, and the system is throttled to achieve the desired pressure. Deposition conditions are typically pressure 5 Torr, substrate temperature 240 C, mercury reservoir temperature 70 C, He flow rate 180 sccm, disilane flow rate 16 sccm.

RESULTS AND DISCUSSION

Window Deposition

Clearly, some of the crucial variables determining the film deposition rate are light intensity, window transparency, absorption coefficient of the gas, and window-substrate distance. As part of our program in understanding these factors we have monitored film thickness as a function of deposition time by recording the intensity of a HeNe laser beam reflected from the

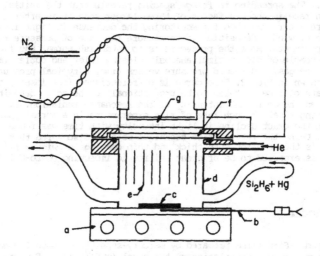

Fig. 1. Schematic diagram of mercury-sensitized photo-CVD a-Si:H deposition system.

a. Aluminum block heater
b. Thermocouple
c. Substrate
d. Stainless steel chamber
e. Baffles (now discarded)
f. Suprasil quartz window
g. Low pressure mercury lamp

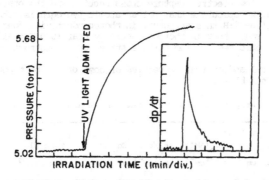

Fig. 2. Static rate of pressure rise upon irradiation.

substrate. The spreading in fringe spacing reveals that the initial deposition rate is about a factor of three larger than it is after 2 hours.

A more accurate method for monitoring the decrease in light intensity as a function of window deposition is to measure the rate of pressure rise in static experiments. Here the procedure is to establish normal gas flows, capture a sample of gas by simultaneously closing inlet and outlet valves, irradiate the gas, and record pressure versus time. A typical pressure rise curve is shown in Fig. 2. This curve is differentiated (see insert) and the maximum rate of rise is taken to be proportional to the total deposition chemistry and hence to the intensity of the resonance radiation. By interspersing a set of such rate of rise experiments in a normal photo-CVD deposition the exact decline in light intensity with time was found, as shown in Fig. 3. Using the rate of rise technique we also discovered that the light intensity is the same for both thick and thin coatings of window fluid. This technique is expected to be of great utility in improving photo-CVD deposition rates.

Film Properties

Undoped a-Si:H films prepared by mercury-sensitized photo-CVD have band gaps, determined by extrapolation of the usual $(\alpha E)^{1/2}$ versus E plot, in the range 1.75 to over 1.85 eV depending largely on substrate temperature. Such films have an AM1 photoconductivity between 1×10^{-5} and 2×10^{-4} ohm^{-1} cm^{-1}. The films discussed here exhibit the well known Staebler-Wronski effect, their AM1 photoconductivity dropping by a factor of 19 after a 14 hour light soak at AM1, and recovering after a 3 hour anneal at 142 C. In contrast, it has been reported that undoped a-Si:H films prepared at atmospheric pressure by direct photo-decomposition of disilane at 185nm do not exhibit the Staebler-Wronski effect [5]. The temperature dependence of the photoconductivity as measured in a coplanar configuration is shown in Fig. 4. Note that the photoconductivity exhibits a low temperature peak and thermal quenching between 125-200K. These features are usually observed in undoped glow discharge films (and never in thermal CVD films) suggesting that photo-CVD a-Si:H possesses a gap state distribution similar to that of glow discharge a-Si:H. This implies the existence of two types of state in the gap [6], one having a low electron capture cross section and the other being the Si dangling bond. We speculate that unless atomic hydrogen is present during the deposition of a-Si:H this structure in photoconductivity does not occur. The cause of the small hump in photoconductivity near $1000/T = 4K^{-1}$ is unknown, but it could result from surface states or reconstruction of a specific defect.

The electrical and optical properties of some of our best photo-CVD films are summarized in Table I.

TABLE I   Properties of a-Si:H films prepared by photo-CVD

Type	σ_d $\Omega^{-1}cm^{-1}$	σ_p(AM1) $\Omega^{-1}cm^{-1}$	σ_p/σ_d $\Omega^{-1}cm^{-1}$	Eg eV	Tauc slope (cm eV)$^{-\frac{1}{2}}$	L_p μm
p*	7.5 E-6	–	–	1.97	952	–
i	1.3 E-10	2.1 E-4	1.6 E6	1.77–1.80	810–845	0.27
n	5.2 E-3	1.8 E-4	3.5 E-2	1.76	805	–

*a-Si,C:H

The films may be substitutionally doped n type using phosphine and p type using diborane. The wide band gap silicon carbide p layer reported in Table I was prepared using equal flow rates of disilane and dimethlysilane, and a diborane to total silane ratio of 0.3%. Its properties are comparable if not better than those of most glow discharge window layers. Omission of the dimethylsilane reduces the band gap to 1.66eV. SIMS analyses of a-Si:H films prepared by three different techniques are compared in Table II. It is worth noting that the hydrogen concentration in the photo-CVD film is lower than in the other two types of film, and that no mercury can be detected in the film ($<10^{18}$ cm^{-3}).

TABLE II SIMS analysis of a-Si:H films prepared at Chronar by rf glow discharge, LPCVD, and Hg-sensitized photo-CVD.

| Species | Concentration (atoms cm^{-3} or %) | | |
	GD	LPCVD	photo-CVD
H	16%	16%	10%
C	7×10^{18}	5×10^{18}	3×10^{19}
O	2×10^{19}	3×10^{18}	2×10^{19}
N	5×10^{17}	1×10^{17}	1×10^{18}
Cl	1×10^{16}	3×10^{16}	1×10^{17}
Hg	NA	NA	ND
transition elements	ND	ND	ND

NA = not analyzed, ND = not detected

Device Fabrication and Analysis

Solar cells having the structure glass/SnO$_2$/p-i-n/Al have been prepared by photo-CVD using both disilane and monosilane source gases. At the time of writing our highest conversion efficiencies are 6.0% and 3.8%, respectively. Efficiencies as high as 8.6% have been reported in the literature [7]. An example of a current-voltage curve for a p-i-n device prepared from disilane is shown in Fig. 5. The device employs the wide band gap silicon carbide p layer described in the previous section, and has a conversion efficiency of 5.5%. When a similar device was fabricated with the mercury lamp turned off during the p layer deposition the Voc was only 0.44V, showing that the p layer is photochemically, and not thermally, derived. A SIMS depth profile for a thin photo-CVD p-i-n cell is shown in Fig. 6. A sharp p-i interface with a 0.05μm wide boron shoulder is revealed. The ratio of the boron background in the i layer to the boron peak in the p layer is 10^{-4}. Dissociation of the tin oxide is not apparent. The spectral response of a typical device is shown in Fig. 7 under both zero bias (curve A) and -1V bias (curve B) conditions. A very high blue response is observed, which is further evidence of minimal boron tailing. The ratio of curve A to curve B is also plotted, its flatness implying that carrier collection is not strongly limited by either electrons or holes and its magnitude (close to unity) implying a high fill factor.

Amorphous silicon solar cells produced by photo-CVD exhibit light-induced changes qualitatively similar to changes observed in glow discharge cells. For example, two p-i-n cells were first annealed at 85 C for 1 hour, measured, light soaked under open circuit conditions at 30 C under AM1 illumination, and re-measured. One cell degraded in efficiency by 4.5% and the other by 2.1%. Most of the lost cell efficiency was recovered after annealing at 85 C for 2 hours. Because of the sparseness of the data we cannot conclude that the photo-CVD devices exhibit less light induced degradation than their glow discharge counterparts.

38

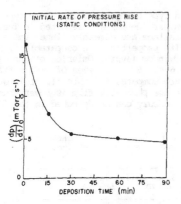

Fig. 3. Results of five static rate of rise experiments interspersed in a 90 minute photo-CVD deposition of an a-Si:H layer.

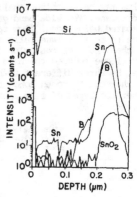

Fig. 6. SIMS depth profile of a photo-CVD p-i-n solar cell before metallization.

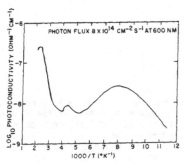

Fig. 4. Temperature dependence of the photoconductivity for a photo-CVD a-Si:H i layer.

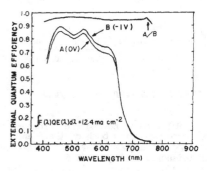

Fig. 7. External quantum efficiency versus wavelength for a photo-CVD p-i-n a-Si:H solar cell.

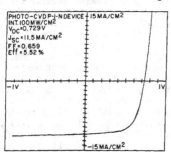

Fig. 5. Current-voltage curve for a photo-CVD p-i-n solar cell under AM1 (100 mW cm^{-2}) illumination.

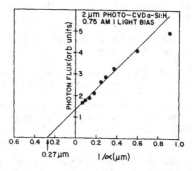

Fig. 8. Surface photovoltage measurement of a 2µm photo-CVD a-Si:H i layer showing a hole diffusion length of 0.27µm.

PROSPECTS

We have recently discovered that appropriate modification of the deposition system virtually eliminates the fall-off in growth rate. Growth rates greater than 2 Å s^{-1} are now observed, and an upper limit to film thickness has not been determined. This development has enabled us to deposit a 2µm thick a-Si:H film for measurement of the hole diffusion length using the constant surface photovoltage technique. The data is shown in Fig. 8 and indicates an effective diffusion length of 0.27µm. The respectable values of deposition rate, diffusion length and conversion efficiency obtained in this work suggest that photo-CVD (or a hybrid process) may supplant the established glow discharge technique as the preferred manufacturing method for high quality a-Si:H. The critical deposition parameters that permit these higher deposition rates are currently under study.

ACKNOWLEDGEMENTS

We are indebted to Mr. J. Dick (SERI) for performing SIMS analyses, to Dr. A. Moore (University of Delaware) for helpful discussions concerning surface photovoltage measurements, and to Dr. A. Lichter for disilane preparation. This work was supported in part by SERI under subcontract no. XB-5-04092.

REFERENCES

1. A.E. Delahoy, F.B. Ellis, Jr., E. Eser, S. Gau, H. Volltrauer, and Z. Kiss, Proceedings of the Sixth European Photovoltaic Solar Energy Conference, London, April 1985.

2. A.E. Delahoy, "Properties of Hydrogenated Amorphous Silicon Prepared by Chemical Vapor Deposition from Higher Silanes", Photovoltaics for Solar Energy Applications II, D. Adler, Editor, Proc. SPIE 407, 47 (1983).

3. T.L. Pollock, H.S. Sandhu, A. Jodhan, and O.P. Strausz, J. Am. Chem. Soc. 95, 1017 (1973).

4. T. Inoue, M. Konagai, and K. Takahashi, Appl. Phys. Lett., 43, 774 (1983).

5. Y. Mishima, Y. Ashida, and M. Hirose, J. Non-Crystalline Solids 59 & 60, 707 (1983).

6. P.E. Vanier, A.E. Delahoy, and R.W. Griffith, J. Appl. Phys., 52, 5235 (1981).

7. T. Tanaka, W.Y. Kim, M. Konagai, and K. Takahashi, Technical Digest of International PVSEC-1, Kobe, Japan (1984), p. 563.

FABRICATION OF AMORPHOUS SILICON DEVICES
ON PLASTIC SUBSTRATES

F. R. JEFFREY, G. D. VERNSTROM, F. E. ASPEN, R. L. JACOBSON
3M Electrical and Communications Technologies Laboratory
3M Center, St. Paul, Minnesota 55144

Abstract

As Amorphous silicon devices spread into more products, the
advantages inherant in continuous processing utilizing a plastic roll
substrate will assume greater importance. Advantages include cost, ease
of handling, shipping weight, steady state processing conditions and
feedback control available in continuous processing. A new set of
problems and constraints also comes with plastic substrates. These
problems include mechanical strength with flexing, substrate surface
defects, differential expansion coefficients, shrinkage, and contaminant
control. These problems along with efforts to evaluate and overcome them
are discussed.

Discussion and Results

As amorphous silicon devices move out of the laboratory and into the
marketplace, the substrate on which the devices are fabricated will assume
greater importance. Plastic substrates hold promises of lower cost and
greater versatility in applications and handling. They also present a new
set of problems and constraints which must be considered. Some of the
advantages of plastic substrates will first be discussed, followed by
results of our recent efforts to evaluate and overcome the constraints.

One of the major advantages of plastic substrates is that they can be
processed, handled, and shipped in the form of a continuous flexible web
rolled onto spools. Thin metal substrates can also be handled this way,
but are less suitable for monolithically integrated devices. Complexity
of loading is reduced because of the area of substrate loaded at one time.
Also, whenever outside the processing machine, the substrate is rolled up
tightly, thus the surface is always protected from dust and dirt
contamination. Transport of the devices is also easier because plastics
weigh less and are less fragile than glass or ceramic substrates.

Plastic web substrates give a processing advantage in that they are
well suited to continuous processing. In the continuous processing system

discussed later in this paper, all depositions occur in a steady state. Thus there are no transient power conditions or gas turbulances to stir up dust. On line monitoring is possible in this system with a feedback loop controlling the deposition parameters. A more uniformly high quality product can be expected from such an arrangement.

The flexibility of plastic substrates also gives a possible application advantage. The final device can be formed and trimmed to meet the demands of the specific use. Solar cells can be shaped to fit non-flat surfaces, electrophotographic devices can be wrapped around any size of drum or used in belt form, and thin film transistor arrays for driving displays can be part of a flexible circuit board.

One of the first difficulties that comes to mind is mechanical strength and the limits of flexibility. Results from solar cells and electrophotographic devices fabricated on polyimide substrates show tolerance of a high level of flexing without noticable damage. Solar cells built on polyimide were tested over a curvature of 1 cm radius without failure. The flexing limit for electrophotographic devices appears limited mainly by the elastic limits of the constituant layers. Tests which included using curvatures of less than 5 mm radius and cycling over a 1 cm radius for 10,000 cycles showed no damage.

Fig. 1a shows a micrograph of a polyimide substrate used to make solar cells. Clearly the surface is much rougher and has a greater defect density than glass. If these defects were to propagate into the silicon device causing shunts, the polyimide would definitely be unacceptable as a substrate. The devices, however, show a much lower shunt defect density than the density of surface defects on the polyimide. EBIC[1] studies (electron beam induced current) have been undertaken to look for a correlation between the existing shunts and any particular type of substrate defect. Fig. 1b is an EBIC image of a PV device showing a moderate resistivity shunt. Portions of the device where electron beam generated carriers are collected and brought to the contact show up as bright areas. The shunt shows up as a fuzzy dark area. In the dark center area, close to the physical shunt path, nearly all collected carriers are routed through the shunt while near the edge of the dark spot some of the carriers go to the external contacts and some through the shunt, giving the characteristic fuzzy look. The equivalent SEM image of this same area shows no physical surface defect corresponding to the shunt. Figs. 2a and 2b are EBIC and SEM images of another device. The EBIC image clearly shows open circuit or masking type defects correlating

with surface defects on the SEM image. These type defects are well
defined areas where no carrier collection takes place, characterized by
sharp EBIC edges rather than the fuzz of shunt defects. There are also
some surface defects which show no effect on the EBIC image. The initial
conclusion is that the majority of surface defects on these substrates do
not cause serious device defects. Shunt defects which do occur in the
devices may be caused by a specific type of surface defect, but are more
likely caused by small dust particles or loose metalization which can
move, causing a pinhole. There is no evidence yet of surface roughness or
straitions causing loss of collection efficiency as seen on stainless
steel substrates.[2]

 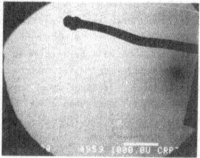

Fig. 1. a) a 75x micrograph of the polymide substrate surface showing
line and point defects. b) An EBIC image of a PV device showing a
shunt defect. These defects appear uncorrelated with substrate defects.

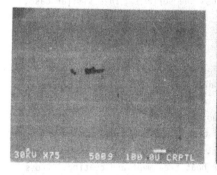 

Fig. 2. a) EBIC image of a PV device showing an open or masking type
defect. b) SEM image of same device showing surface damage causing the
defect.

Differential expansion between silicon and substrate material has a large effect on the handling properties of the finished device and also some effect on the device performance itself. With plastic substrates there are usually two separate problems: the thermal expansion coefficient and shrinkage due to extended process heating. The shrinkage, which can be a result of further crosslinking of the polymer or loss of absorbed water or other solvent, may be minimized by proper heat treatment of the polymer to completely preshrink it before processing begins.

Differences in the thermal expansion coefficients is quite a different matter and forms a limitation on the choice of polymers. Fig 3 shows a picture of two PV devices deposited on 2 mil polyimide. The coefficient of expansion for the polyimide on the left was $21 \times 10^{-6} C^{-1}$ while that on the right was $6 \times 10^{-6} C^{-1}$. The curled device on the left is quite functional and may be flattened by mounting on a rigid backing plate, but still has a number of difficulties associated with it. Two problems which have been observed result from the creation of a fine fracture pattern in the substrate metalization. The first is a separation of individual islands of metal, causing a high resistivity back contact layer. This occurs when thick silicon layers such as those used for elctrophotographic applications were deposited on brittle metalization layers like chromium. The second problem exists when the stress is not severe enough to separate the metal islands, but causes fractures which serve as conduits for reactive species, i.e. water, to move along the metal-substrate interface. The stress at this interface causes debonding of the metal layer after the reactive species loosens the bonds. As one section becomes debonded, more of the reactive species is admitted, thus allowing the process to propagate across the surface. In a similar case, the stresses can be focused when a crack begins in the silicon layer causing the crack to propagate and branch in the same manner as tempered glass.

One final consideration which has been found to be highly important is impurity control[3,4]. Two main contamination sources are inherant in the use of rolled plastic substrates, outgassing of the plastic material itself and gas trapped in the rolled up substrate. If a device such as a solar cell is to be made, there is the additional problem of cross contamination between deposition zones. Outgassing of the material has been adequately reduced by proper prior heat treatment and coating. Cross contamination and contaminants from web rolls have been attacked by use of a system shown schematically in fig. 4. Conductance limiting slits allow the web to pass through each deposition stage, but minimize diffusion of contaminants into the deposition zone. Between each deposition zone there is a pumped region which keeps the partial pressure of contaminants at the slit opening at a low level. At the typical operating pressure of 1 torr, the gas outflow through the slits adds a further impediment to contaminants diffusing into the deposition area. This system has been shown capable of providing isolation of up to 5 orders of magnitude between zones.

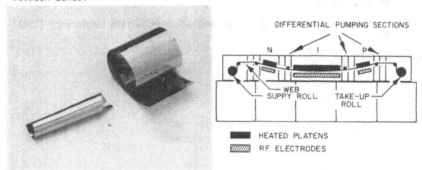

Fig. 3. PV devices fabricated on polyimides with different expansion coefficients. Left: 21 x 10 exp -6 right: 6 x 10 exp -6.

Fig. 4. Schematic diagram of multichamber, continuous coating system using conductance limiting slits and differentially pumped zones.

In summary, advantages and problems associated with the use of rolled plastic substrates and continuous deposition processes using these substrates have been considered. From the present data, it appears that the problems are solvable and the advantages are strong enough to warrant the effort necessary to find the solutions.

ACKNOWLEDGEMENT

This work was performed under the auspices of SERI and the U.S.D.O.E. under subcontract No. ZB-4-03056-2.

REFERENCES

1. H. J. Leamy, J. Appl. Phys. 53 R 51 (1982).

2. B. G. Yacobi, T. J. McMahon, A. Madon, Solar Cells.

3. D. E. Carlson, A. Catalano, R.V. D'Aiello, C. R. Dickson, R. S. Oswald. Optical Effects in Amorphous Semiconductors, AIP Conference Proceedings, Snowbird, Utah, 1984, ed. by P. C. Taylor and S. G. Bishop. P. 234.

4. Y. Kuwano et al, 16th IEEE Photovoltaic Specialists Conference (1981) 698.

HIGHLY CONDUCTIVE AND WIDE BAND GAP MICROCRYSTALLINE SILICON FILMS PREPARED BY PHOTOCHEMICAL VAPOR DEPOSITION AND APPLICATIONS TO DEVICES

S. NISHIDA, H. TASAKI, M. KONAGAI, AND K. TAKAHASHI
Department of Physical Electronics, Tokyo Institute of Technology,
2-12-1, Ohokayama, Meguro-ku, Tokyo 152, Japan

ABSTRACT

Doped hydrogenated microcrystalline silicon (μc-Si:H) and fluorinated hydrogenated microcrystalline (μc-Si:F:H) films were prepared by the mercury photosensitized decomposition of a disilane-hydrogen or a difluorosilane-hydrogen gas mixture, respectively. The maximum dark conductivity and optical band gap of μc-Si:H films were respectively 20 $S\cdot cm^{-1}$ and $\sim$2.0 eV for n-type and 1 $S\cdot cm^{-1}$ and 2.3 eV for p-type. A higher dark conductivity as much as 50 $S\cdot cm^{-1}$ and a wide gap of 2.0 eV were obtained for n-type μc-Si:F:H. It is most significant that the gaseous ratio of hydrogen to disilane should be enhanced to obtain such a highly conductive and wide gap film. The crystallinity of the photo-deposited μc-Si:H films appeared to be improved in comparison with that of films by the conventional plasma glow discharge technique.

Introduction

Amorphous-microcrystalline mixed-phase silicon (μc-Si) films have been studied intensively by several groups[1,2] as a promising material for window layers of amorphous silicon (a-Si) solar cells. Other unique applications of μc-Si for power and pressure sensors have also been proposed[3]. Usually the glow discharge (GD) process is used for depositing μc-Si as well as a-Si.

In a conventional capacitively coupled GD system, a self bias is generated between the substrate electrode and plasma[4], so that ionized particles impinge on the growing surface of the films and affect the electrical quality of a-Si[5] and the crystallinity of μc-Si[6]. This ion damage raises significant problems. For example, limitation of the efficiency of a-Si solar cells due to the poor film quality and interface states between p and i or i and n layers, and the degradation of gauge factors in piezoresistive effects in μc-Si films.

Recently, photochemical vapor deposition (photo-CVD) processes have been applied to produce a-Si films[7,8] and even p-i-n a-Si solar cells[9]. In the photolysis process, ultraviolet photons excite and dissociate the reactant molecules in a gas mixture and the photon energy is not sufficient to ionize the molecules. So using this technique, ion-damage free a-Si films of high quality and far improved interfaces, together with μc-Si films of fine crystalline quality, are expected.

In this study, we report on the performance of doped highly conductive, wide gap μc-Si:H and μc-Si:F:H films produced by photo-CVD, and the piezoresistive effects of the films for application to a pressure sensor are presented.

Experiment

The schematic diagram of the photo-deposition apparatus used in this study is shown in Fig.1. The mercury photosensitization method was used to enhance the dissociation of reactant gases[10]. A low pressure mercury lamp was employed as a light source, radiating intense 1849 and 2537Å resonance lines ($\sim$ 30 mW/cm^2 at 3-cm distance). The reactant gas of

Table I. Preparation conditions for µc-Si:H and µc-Si:F:H by photo-CVD.

	type	T_{sub} (°C)	pressure (torr)	Si_2H_6 (sccm)	H_2/Si_2H_6	PH_3/Si_2H_6 (ppm)	B_2H_6/Si_2H_6 (ppm)
µc-Si:H	n	100~300	2	1~2.5	40~80	1500~20000	
	p	150~300	2	0.4	150~350		3000~50000

	type	T_{sub} (°C)	pressure (torr)	SiH_2F_6 (sccm)	H_2/SiH_2F_6	PH_3/Si_2H_6 (ppm)
µc-Si:F:H	n	150~300	5	5	2~7	100~2000

disilane (Si_2H_6) or difluorosilane (SiH_2F_2), with addition of phosphine (PH_3) for n-type or diborane (B_2H_6) for p-type, was diluted with hydrogen and premixed with a very small amount of mercury vapor in a thermally controlled mercury vaporizer, before being introduced into the reactor.

A synthesis quartz-glass (Suprasil) window (12.5 cm in diameter and 6mm in thickness), which transmits 1849Å resonance line, was coated with low vapor pressure oil[8], and was placed between the light source and substrate. Films were deposited on Corning 7059 glass substrates for electrical and optical measurements. Preparation conditions for µc-Si:H and µc-Si:F:H are listed in Table 1. The temperature of the mercury vaporizer was kept at 50°C.

Results and discussion
µc-Si:H

Figure 2 shows the variation of dark conductivity and optical gap, determined from a $(\alpha h\nu)^{1/2}-h\nu$ plot, of n-type and p-type µc-Si:H films versus the dilution ratio of H_2 to Si_2H_6. In Fig.2(a), the conductivity and optical gap for n-type µc-Si:H were about 10^{-3} S•cm^{-1} and 1.8 eV, respectively, at a dilution ratio of 40, increasing up to 13 S•cm^{-1} and 1.98 eV as the ratio was enhanced to 70. Notably, the change of conductivity is particulary rapid and remarkable. This variation coincides with the phase transition of the deposited films from amorphous to amorphous-microcrystalline mixed-phase state, which was confirmed from reflective high energy electron diffraction (RHEED). As for p-type films, a similar dependence of conductivity and optical gap on the amount of hydrogen was observed (Fig.2(b)). The maximum conductivity of about 1 S•cm^{-1} and wide optical gap of 2.16 eV, were obtained at a dilution ratio of 300.

In our experimental conditions, there exist large amounts of hydrogen radicals generated by UV photons (because of relatively large quenching cross section of H_2 with respect to Hg^* [11]), and they are significant in realizing the microcrystallization of a-Si. It is reported that in the

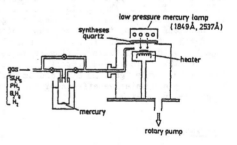

Fig.1 Schematic diagram of the photo-CVD system.

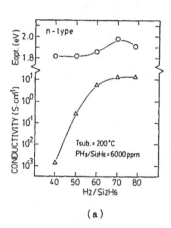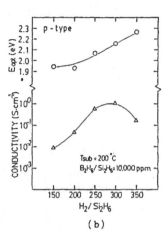

(a)   (b)

Fig.2 Conductivity and optical gap versus the dilution ratio
       of H_2/Si_2H_6 for (a) n-type and (b) p-type μc-Si:H
       prepared by photo-CVD.

mechanism for microcrystallization of undoped a-Si prepared by GD
process, the enhancement of the hydrogen content in reactant gases is
important because hydrogen radicals cover the growing surface of the film
and play a role in decreasing the barrier height against the surface
migration of the precursors[6] to form a crystalline structure. The same
growth process may be available for the photo-CVD and the result of Fig.2
supports this. Thus, the enhancement of the dilution ratio is most
significant to obtain such a wide-gap and high-conductivity
microcrystalline film.

Figure 3 represents the conductivity and optical gap as a function
of gaseous impurity ratio (a) PH_3/Si_2H_6 and (b) B_2H_6/Si_2H_6. As seen in
Fig.3(a), the conductivity for n-type was as high as $2 S \cdot cm^{-1}$ with a
relatively small addition of 3000 ppm PH_3 to Si_2H_6, and decreased rapidly
with an addition below 3000 ppm, whereas the optical gap decreased with
increasing impurity ratio. The activation energy for the conductivity
was about 0.1 eV at the impurity ratio of 1500 ppm, and was almost
constant in 20 - 30 meV range at a ratio above 3000 ppm, which is similar
to that for plasma-produced n-type μc-Si:H. The variations of
conductivity and optical gap for p-type μc-Si:H against the impurity
ratio of B_2H_6/Si_2H_6 were the most striking feature, as shown in Fig.3(b).
The optical gap remained at about 2.16 eV, being independent of the
doping ratio up to 10^{-2}, whereas the conductivity increased slightly,
10^{-1} to 10^0 $S \cdot cm^{-1}$, with increasing doping ratio from 10^{-3} to 10^{-2}. It
is well known that the incorporation of boron atoms into the films causes
the narrowing of the band gap in a-Si and a-SiC, prepared by the
conventional GD technique. Contrary to this, photo-deposited p-type
μc-Si:H still showed such a wide band gap of 2.05 eV with the
conductivity of 1.5 $S \cdot cm^{-1}$ even at a high doping concentration of 20000
ppm. Thus, as seen above, doped μc-Si:H films by photo-CVD are likely to
be employed as a window layer of a-Si solar cells.

The preferential orientation of the crystallites of the
photo-deposited films was found to be more pronounced, in contrast with
that by conventional GD method. Figure 4 shows the RHEED pattern for a
photo-CVD n-type μc-Si film prepared on a glass substrate at 200°C. The
ring patterns were partially cut, and the diffracted intensities of some
specific positions were enhanced. The crystallites in the film were

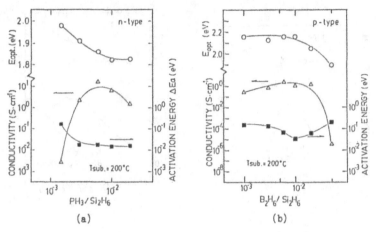

(a)   (b)

Fig.3 Conductivity and optical gap as a function of gaseous
       impurity ratio (a) PH_3/Si_2H_6 for n-type and
       (b) B_2H_6/Si_2H_6 for p-type.

mainly oriented in the <110> direction, which is contrary to the result
reported for conventional n-type μc-Si:H (polycrystal-like) films,
prepared by GD method at a low temperature below 300°C[12]. This may be
due to both the absence of ion damage and the catalytic effect of
hydrogen radicals mentioned before. The same feature was also seen in n-
and p-type films deposited at 150 and 200°C, respectively, by photo-CVD.
    The improvement of the crystallinity for photo-deposited films
should be reflected by the piezoresistive effect in μc-Si films[3]. We
measured the resistive change of n-type μc-Si:H film deposited at 150°C
against applied strains, and the gauge factor was found to be - 61
(Fig.5). This is a large value compared with that by conventional GD
technique.

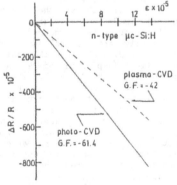

Fig.4 RHEED pattern for a photo-CVD
n-type μc-Si:H film prepared on a
glass substrate at 200°C.   The
crystallites in the film were
mainly oriented in the <110>
direction.

Fig.5 Piezoresistive effect of n-type
μc-Si:H prepared by photo-CVD at
150°C, compared with that by
plasma-CVD.

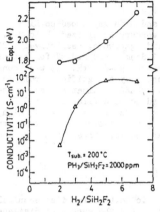

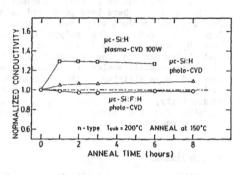

Fig.6 Conductivity and optical gap versus the dilution ratio of H₂/SiH₂F₂ for n-type μc-Si:F:H by photo-CVD.

Fig.7 Change in conductivity after succesive annealing treatments. Films prepared by photo-CVD are more stable than that by plasma-CVD.

μc-Si:F:H

It is well known that the incorporation of fluorine atoms makes the film thermally stable[13]. However, highly conductive (more than 1 $S \cdot cm^{-1}$) μc-Si:F:H can be prepared by plasma-CVD, only at a substrate temperature above 300 °C, if SiF_4 is used as the reactant gas. Such a high temperature will restrict the selection of substrates, making polymeric films unusable.

On the other hand, SiH_2F_2 is expected to have a lower threshold energy of dissociation than SiF_4[14]. So we attempted to prepare highly conductive μc-Si:F:H films from $SiH_2F_2-H_2$ gas mixture at a temperature below 300°C, using the same photo-CVD system as shown in Fig.1.

Figure 6 shows the dark conductivity and optical gap of n-type μc-Si:F:H versus the dilution ratio of H_2 to SiH_2F_2 at a substrate temperature of 200 °C. Similar results as seen in the μc-Si:H were observed for the μc-Si:F:H, but with a much smaller dilution ratio. A maximum conductivity of 50 $S \cdot cm^{-1}$ and a wide band gap of 2.0 eV were obtained at the dilution ratio of 5. The gauge factor at this dilution ratio was found to be -28, which is somewhat larger than the value obtained for n-type μc-Si:F:H by plasma-CVD[3].

In order to examine the stability of the μc-Si:F:H by photo-CVD, an annealing at 150 °C in a N_2 atmosphere was done with a n-type film prepared at 200 °C. The result of changes in conductivity of the μc-Si:F:H is represented in Fig.7, in comparison with those of n-type μc-Si:H films by poto-CVD and plasma-CVD. All the values are normalized by the individual initial value. As seen in the figure, the conductivities for both plasma- and photo-CVD produced films were changed. However, the conductivity of plasma-CVD μc-Si:H increased by about 30% after 1 hour's annealing and the change was six times larger than that of μc-Si:H by photo-CVD. The difference between them is considered to be due to the ion damage caused by the high power during the preparation in plasma-CVD process. On the other hand, the μc-Si:F:H film by photo-CVD had very little change in conductivity within 1 % aberration, after 8 hours' annealing.

So highly conductive μc-Si:F:H films prepared by photo-CVD can be used in fabrication of a high quality pressure sensor.

Conclusion

The performances of highly conductive and wide gap doped μc-Si:H and μc-Si:F:H films have been presented using mercury sensitized photo-CVD. The maximum conductivity and optical band gap were, respectively, 21 $S \cdot cm^{-1}$ and 2.0 eV for n-type μc-Si:H, 1 $S \cdot cm^{-1}$ and 2.3 eV for p-type μc-Si:H, and 50 $S \cdot cm^{-1}$ and 2.2 eV for n-type μc-Si:F:H. It was found that hydrogens in the reactant gas mixture play a significant role for the microcrystallization of doped films. RHEED examination showed the improved crystallinity of the photo-deposited μc-Si:H films compared with that of conventional GD μc-Si:H films. These films have a high conductivity, a wide gap, and a large gauge factor. So they are suitable for the material of a-Si solar cells, power sensors, and pressure sensors.

Acknowledgment

The authors wish to thank Anritsu Electric Co.Ltd. for its technical assistance and Mitsui Toatsu Chemicals Inc. for the supply of disilane gas. This work was supported partially by the Agency of Industrial Science and Technology under the Sunshine Project, and by Grant-in-Aid for Scientific Research from the Ministry of Education, Science and Culture.

References

1) Y.Uchida, T.Ichimura, M.Ueno, and H.Haruki, Jpn. J. Appl. Phys. 21, L586 (1982)

2) H.Haruki, H.Sakai, M.Kamiyama, and Y.Uchida, Solar Energy Materials 8, 441 (1983)

3) S.Nishida, M.Konagai, and K.Takahashi, Thin Solid Films 112, 7 (1984)

4) T.Hamasaki, M.Ueda, A.Chayahara, M.Hirose, and Y.Osaka, Appl. Phys. Lett. 44, 600 (1984)

5) T.Shimada, N.Nakamura, S.Matsubara, H.Itoh, S.Muramatsu, and M.Migitaka, Technical Digest of 1st Int. Photovoltaic Science and Engineering Conf., Kobe, Japan, 445 (1984)

6) A.Matsuda, Proc. 10th International Conference on Amorphous and Liquid Semiconductors; J. Non-Cryst. Solids 59 & 60, 767 (1983)

7) T.Saitoh, S.Muramatsu, S.Matsubara, and M.Migitaka, Jpn. J. Appl. Phys. 22, Suppl. 22-1, 617 (1983)

8) T.Inoue, M.Konagai, and K.Takahashi, Appl. Phys. Lett. 43, 774 (1983)

9) T.Tatsuya, W.Y.Kim, M.Konagai, and K.Takahashi, Appl. Phys. Lett. 45, 865 (1984)

10) T.L.Pollock, H.S.Sandhu, A.Jodhan, and O.P.Strausz, J. Amer. Chem. Soc. 95, 1017 (1973)

11) K.J.Laidler, The Chemical Kinetics of Excited States, p.107, The Clarendon Press, Oxford (1955)

12) A.Matsuda, S.Yamasaki, K.Nakagawa, H.Okushi, K.Tanaka, S.Iizima, M.Matsumura, and H.Yamamoto, Jpn. J. Appl. Phys. 19, L305 (1980)

13) A.Madan, S.R.Ovshinsky and E.Benn, Phil. Mag. B 40, 259 (1979)

14) H.Koinuma, H.Natsuaki, K.Fueki, K.Sato, T.Hirano and K.Isogaya, Technical Digest of 1st Int. Photovoltaic Science and Engineering Conf., Kobe, Japan, 743 (1984)

SELECTED AREA EPITAXIAL REGROWTH OF
AMORPHOUS Si/(100) Si STRUCTURES
BY LASER ANNEALING

A. CHRISTOU*, C. VARMAZIS, T. EFTHIMIOPOULOS and C. FOTAKIS
Research Center of Crete, Institute of Electronic Structure and Laser
University of Crete, Physics Department, Iraklio, P. O. Box 470, Crete,
Greece

ABSTRACT

Excimer laser KrF (248 nm) annealing at 93 mJ/cm^2 and 175 mJ/cm^2 has
been found to recrystallize amorphous silicon on (100)Si. The major
impurities introduced by excimer laser annealing are carbon, while surface
roughness remains as a major problem. Channel mobilities measured on
MOSFETs processed on epitaxially regrown silicon were 98-115 cm^2/v.s.
Leakage currents between recrystallized silicon regions were 1-2 uA/cm^2.

INTRODUCTION

The increase in packing densities of silicon integrated circuits
requires the development of new process technologies such as SOI
structures (silicon on insulator) [1,2]. Recently Izumi et al [3] have
reported on a 1K RAM using buried oxide layers and regrown epitaxial
layers. Previously all SOI layers were formed by implanting high doses of
low energy donor or acceptor impurities. These layers were typically
regrown by annealing [4,5,6]. For instance [6] silicon on insulator
structures have been formed by implantation of 1.8×10^{18} oxygen cm^{-3} at 200
keV into (100) silicon wafers. The crystallinity of the top silicon layer
was recovered by furnace annealing at 1150°C for 2h in dry nitrogen using
a silox cap. In the present investigation we report on the epitaxial
regrowth of amorphous silicon layers deposited over selected regions of
(100) silicon. Epitaxial regrowth has been achieved by excimer laser
annealing at energy densities up to 175 mJ/cm^2. We have determined that
three discrete regions of excimer laser annealing exist: amorphized
region, epitaxial regrown region and a localized melted region.

EXPERIMENTAL DETAILS

The deposition of silicon MBE and amorphous layers used in the
present investigation have been reported previously [7]. Essentially
several immersions of the (100) silicon substrates into NH$_3$OH:H$_2$O (10:1)
were used in order to eliminate carbon and metallic ion contamination from
the substrate surface prior to insertion into the vacuum system. The MBE
vacuum system was evacuated to 1×10^{-10} Torr and the silicon substrate was
thermally desorbed at 850°C for 10 minutes. The amorphous silicon was
deposited by e-gun evaporation at a substrate temperature of 150°C. The
amorphous silicon was deposited to a thickness of 8um and onto silicon
with Si$_3$N$_4$ defined square regions. The Si$_3$N$_4$ was 3000 A° thick and stripe
widths were 100um while each square region was 500um x 500um. A
cross-section of the structure utilized is shown in Figure 1. A second
structure utilizing Mo layers approximately 3000 A° thick were also
processed in order to access the stability of buried Mo layers during
laser annealing.

*Naval Research Laboratory, Washington, D.C. 20375

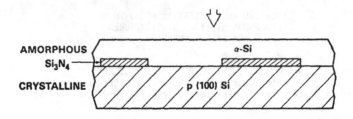

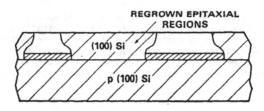

Figure 1   Cross-section of excimer laser annealing experiment utilizing Si N.

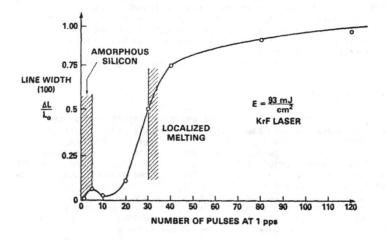

Figure 2   KrF laser annealing showing threshold regions for number of pulses.

Excimer laser Annealing

A KrF excimer laser (248 nm) was used in the present experiments. Two energy densities were utilized: 93mJ/cm^2 and 175 mJ/cm^2 at a pulse rate of one pulse per second (1 pps). In order to determine a threshold for recrystallization both the energy density and pulse rates were maintained constant while varying the number of pulses. Each sample, subjected to a varying number of pulses, was assessed by electron channeling in order to determine the transitions between amorphous and recrystallization or the on-set of localized melting. In this manner the ideal laser annealing parameters for epitaxial regrowth may be determined. Figure 2 shows the variation of the electron channeling (100) line width with number of pulses at 1 pps for the energy density of 93 mJ/cm^2. The on-set of localized melting occurs at 30 pulses. Above 30 pulses, even though the a-Si has epitaxially regrown, localized melt regions were present up to 120 pulses studied. Below 5 pulses, the films remained amorphous and hence the region of epitaxial regrowth was determined to be between 5 and 30 pulses. Similar thresholds were observed at an energy density of 175 mJ/cm^2.

Crystallinity and Channel Mobilities

Crystallinity determination was accomplished by electron channeling of a 20 keV electron beam in a scanning electron microscope. After each recrystallized region was analyzed by electron channeling, the leakage current was measured between adjacent recrystallized regions. The leakage current was determine to be 1-2 uA/cm^2 assuming a parallel conduction path. In the epitaxially regrown regions p-channel MOSFETs were processed. The n$^+$ regions were formed by ion implantation and each transistor was an enhancement mode type (normally off).

The channel mobilities were determined from the linear region of the I V curves. At pinch-off, the charge in the inversion layer becomes zero and the drain voltage and drain current is designated as $V_{d,sat}$ and $I_{d,sat}$. The saturation current is then given by [8]

$$I_{d,sat} = \frac{mW}{L} uC_i(V_g-V_t)^2 \qquad (1)$$

were m=1/2 for the doping concentration in the present experiment, W and L is the gate width and length, V_t is the threshold voltage in the linear region and C_i is the channel capacitance. By plotting the square root of $I_{d,sat}$ versus $V_g - V_t$, the channel mobility u may be determined.

EXPERIMENTAL RESULTS

Excimer laser annealing utilizing the KrF laser resulted in selected area epitaxial (100) growth of a-Si in regions where the a-Si/(100) Si interface occurred. At an energy density of 93 mJ/cm^2 the threshold was 5 pulses at 1 pps. Regions of a-Si/a-Si3N4 did not result in epitaxial regrowth but did increase the resistivity of the a-Si by decreasing the leakage currents through the a-Si from an initial value of 10-20 uA/cm^2 to 1-2 uA/cm^2. Figure 1 shows schematically the resultant structure. In the second set of samples where Si3N4 was replaced with a 3000 A molybdenum film, laser annealing at 93 mJ/cm^2 and 175 mJ/cm^2 resulted in melting of the Mo at the interface and the presence of Mo bumps on the silicon

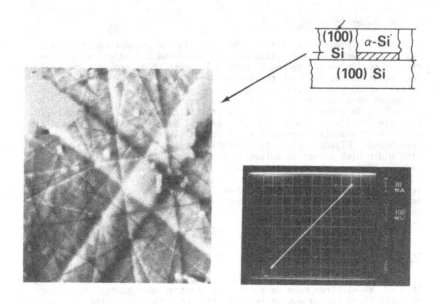

Figure 3 Electron channeling pattern for recrystallized silicon.

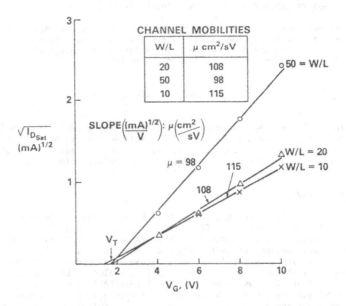

Figure 4 Channel mobility measurement from silicon MOSFETs.

surface. The silicon surrounding the Mo bumps was however (100) epitaxial Si, regrown from the a-Si. The Mo bumps were present throughout the regrown (100) Si.

The electron channeling pattern of the recrystallized silicon is shown in Figure 3, indicating a (100) orientation for the regrown silicon. The amorphous silicon indicated the absence of a channeling pattern as expected. Auger electron spectroscopy analysis of the recrystallized silicon indicated that both carbon and bulk oxygen have been introduced perhaps by a diffusion mechanism from the a-Si/(100)Si interface. However, surface oxides and contaminants were desorbed by the laser beam. Laser annealing at 93 and 175 mJ/cm^2 did leave the silicon surface with a degree of roughness which may be indicative of inhomogenities in the excimer laser beam profile.

MOSFET Results

The MOSFET characteristics showed typical p-channel behavior. The analysis of channel mobility was made by varying the gate width to length ratio (W/L) from 10 to 5 and using the saturation region of the I-V characteristics. Figure 4 shows the linear relationship of saturated drain current and gate voltage. Also shown in Figure 4 is the extrapolated value for the threshold voltage V_t (1.5 - 1.75 Volts). The threshold voltage is shown not to be V_g-dependent which indicates that the channel is lightly doped (10^{15} cm^{-3}) and thin. The threshold voltage can then be expressed as

$$V_t = 2P_b + \frac{\sqrt{2E_s q N_a \ (2P_b)}}{C_i} \qquad (2)$$

where P_b is the surface potential at inversion, E_s is the permittivity of the semiconductor, N_a is the density of the acceptors and C_i is the insulator capacitance. Using the above equation, with an oxide thickness of 1000 A$^\circ$, and 5 x 10^{15} cm^{-3} doping of the regrown silicon layer, a threshold voltage of 1.6 Volts was calculated. A threshold voltage of 1.2 V can be obtained at a doping level of 2 x 10^{15} cm^{-3}. Therefore we may conclude that the regrown silicon resulted in MOSFETs with a low threshold voltage indicative of lightly doped 5 x 10^{15} cm^{-3} channel regions. In addition fixed interfacial charges were minimum since voltage shifts corresponding to the flat band voltage were not observed.

CONCLUSIONS

The successful selected area epitaxial growth of silicon on (100) silicon has been accomplished at an energy density of 93 mJ/cm^2. MOSFETs fabricated with laser recrystallized channels resulted in mobilities of 98-115 cm/v.s. and threshold voltages of 1.5 - 1.7 V. The measured threshold voltages resulted in a calculated impurity concentration of 5 x 10^{15} cm-3.

References

1. F.M. D'Heurle, Proc. 1st International Symposium on VLSI Science and Technology, Detroit, 1982.

2. T. Hayaski, H. Okamoto and Y. Homma, Inst. of Phys. Conf. Ser. No. 57, 533 (1981).

3. K. Izumi, Y. Omura and T. Sakai, Electronic Materials Conf. Colorado, June 1982.

4. M. H. Badawi, J. Phys. D. 10 19831 (1977).

5. L. Csepregi, E.F. Kennedy, T.J. Gallagher and J.W. Mayer, J. Appl. Phys. 48, 4234 (1977).

6. H.W. Law, R. F. Pinizzotto, H.T. Yuan and D.W. Bellavance, Electron. Lett. 17, 356 (1981).

7. A. Christou, E.D. Richmond, B.R. Wilkins and A.R. Knudson, Appl. Phys. Lett. 44(8) 796 (1984).

8. S.M. Sze, in Physics of Semiconductor Devices, Wiley Inc. New York (1981) pp. 431 - 473.

PART II

Defects

DEFECTS IN TETRAHEDRALLY COORDINATED AMORPHOUS SEMICONDUCTORS

P.C. Taylor
Department of Physics, University of Utah, Salt Lake City, UT 84112

ABSTRACT

In the tetrahedrally coordinated amorphous semiconductors the dominant defects deep in the gap are attributed to dangling bonds on the group IV atoms. These defects are commonly thought to have effective electron-electron correlation energies U_{eff} which are positive, although some tight-binding estimates suggest negative U_{eff}. Defect states near the band gap edges are invoked to account for many experimental results including the usual appearance of an Urbach absorption edge. These shallow defect states are usually attributed to strained bonds but two-fold-coordinated group IV atoms have also been suggested. The application of light of near-band-gap energies alters the density of paramagnetic dangling bonds. For large spin densities ($n_s > 10^{17}$ cm^{-3}) this increase is probably due to the creation of new defects, but it is possible that at lower densities ($n_s < 10^{16}$ cm^{-3}) the rearrangement of electronic charge in existing defects is important. Impurities also contribute to the defects observed in tetrahedral amorphous semiconductors. Particular species include trapped atomic and molecular hydrogen, trapped NO_2^- molecules, singly-coordinated oxygen atoms and E' centers.

INTRODUCTION AND BACKGROUND

The successful doping of films of hydrogenated amorphous silicon (a-Si:H) [1,2] and the subsequent production of efficient, large area solar cells from this material has made the understanding of defects of fundamental importance. In addition, the potential use of a-Si$_x$Ge$_{1-x}$:H and other tetrahedrally-coordinated alloys in multilayer photovoltaic devices requires a much better understanding of the important defects in these amorphous semi-conductors. Because of this interest, these amorphous films can now be grown very reproducibly and many of their properties are very well characterized. However, several important aspects of the defects in these alloys remain controversial.

The major contribution to electronic states with energies near the center of the gap comes from dangling bonds on the group IV atoms. In a-Si:H there exists an electron spin resonance (ESR) response at g = 2.0055 which is at least 10^{15} spins cm^{-3} even in the best films. This response is attributed to an unpaired spin on a silicon dangling bond. In alloys of a-Si$_x$Ge$_{1-x}$:H there exist two ESR signals one attributed to a silicon danglig bond and one to a dangling bond on germanium. For x < 0.9 the germanium dangling bond signal is usually $> 10^{17}$ spins cm^{-3}.

The usual interpretation [3] of these ESR signals is in terms of unpaired spins which are highly localized on three-coordinated Si or Ge atoms. These spins occur at neutral sites (D^0) which have not been compensated by the addition of hydrogen to the films. The dangling bonds are commonly assumed to be neutral in the ground state as a result of the fact that the effective electron-electron correlation energy U_{eff} is a positive quantity [3]. Although it is less commonly accepted, the possibility that at least some of the Si and Ge dangling bonds form a negative U_{eff} system has also been suggested [4]. If U_{eff} is negative, then there exist both positively and negatively charged analogs of the dangling bond defect (D^+ and D^-) which form the ground state of the system. The charged defects D^+ and D^- are, of course, spinless. Either of these suggestions is consistent with the fact that the density of states deep in the gap generally exhibits only one resolved peak which occurs somewhere near the middle of the gap [3].

Near the band edges there are thought to be shallow defects which contribute to the exponential or Urbach tail which is always observed in the absorption edge of tetrahedrally-coordinated amorphous semiconductors. Once again, the most commonly accepted interpretation of these states is in terms of strained bonds in the amorphous network [6], but specific defects have also been suggested [4,7]. In particular, it has been speculated that neutral two-fold coordinated Si and Ge atoms (T_2^0) may be responsible for these shallow electronic states [7].

Thus there are two essentially opposite points of view with regard to the fundamental, "intrinsic" defects in a-Si:H and related alloys. The first, and most commonly accepted, view is that the dominant deep defects are neutral dangling bonds with positive U_{eff} and the dominant shallow defects are the result of strained bonds. The second view is that at least some of the deep defects are spinless dangling bonds with a negative U_{eff} and the shallow states are specific two-fold-coordinated defects (T_2^0).

When light of roughly band gap energies is incident on a-Si:H there exist in general three transient, optically-induced ESR signals. These three signals, which have g-values of 2.004, 2.013 and 2.0055, are attributed to electrons trapped in localized electronic states below the conduction band edge, to holes trapped in localized states below the valence band edge and to silicon dangling bonds, respectively. The dependence of these three signals on doping [6-11] has suggested that the resonances at 2.004 and 2.013 are due to electrons and holes, respectively, trapped at strained or weak bonds near the band edge [12,13]. In a second interpretation [7] these two signals are suggested to be due to electrons and holes which are trapped at T_2^0 defects yielding T_2^- and T_2^+ charge states, respectively.

The optically-induced ESR response at g = 2.0055, which is attributed to Si dangling bonds, has also been interpreted within the framework of more than one model. Whether or not this defect constitutes a negative U_{eff} system remains a controversial point.

Impurities also play important roles in determining the important defects in films of tetrahedrally-coordinated amorphous semiconductors. It is known from nuclear magnetic resonance (NMR) and ESR studies, respectively, that molecular hydrogen (H_2) and atomic hydrogen can be trapped in these films under the appropriate conditions. Molecular hydrogen, in particular, is a nearby ubiquitous impurity in a-Si:H and a-Si$_x$Ge$_{1-x}$:H films.

In addition to defects associated with hydrogen, other impurity species can create observable levels of defects when present in sufficient quantities. Such defects include trapped NO_2^- radicals, singly-coordinated oxygen atoms and triply-coordinated silicon atoms which are bonded to three oxygen atoms (E' centers). The presence of the E' centers in oxygen-doped a-Si:H indicates that these materials are highly inhomogeneous with the presence of SiO$_x$-rich regions.

METASTABLE ELECTRONIC STATES

Optically-induced metastable changes in many transport and optical properties of a-Si:H are well known. The first observation of such an effect was a decrease in the photoconductivity after optical excitation reported by Staebler and Wronski [14]. Dersch et al. [15] first observed a metastable increase in the ESR intensity of the Si dangling bond signal after irradiation with white light at 300K. An increase is also observed after x-irradiation [16], and these increases can be annealed by cycling to temperatures of ~ 500K [15,17].

Several experiments have probed the kinetics of this metastable increase in the ESR [18-21]. At high intensities of the exciting light the ESR intensity as a function of time $n_s(t)$ exhibits a region whose $n_s(t) \propto t^{1/3}$. This behavior has been interpreted [19,21] as the optically-induced creation of dangling bonds where the production of new dangling bonds is limited by the presence of the existing density of dangling bonds. This limitation occurs because the dangling bonds themselves are assumed to provide a parallel non-radiative process which competes with transitions which can generate new defects.

An example of the approximate $t^{1/3}$ dependence of the spin density with time at high exciting light intensities is given in Fig. 1. The spin densities plotted in this figure are actually only the optically-induced portion with the original "dark" spin densities subtracted. In these samples the initial spin densities are primarily due to bulk states and not surface states. The magnitudes of the dark spin densities are $\sim 3 \times 10^{15}$ spins cm^{-3}.

It is apparent from Fig. 1 that the growth rate decreases with decreasing intensity of the incident light. At times longer than those shown in Fig. 1 and at low incident intensities (< 100 mW cm^{-3}) the slope decreases continuously with time such that no unique power law can be defined in this regime. This decrease in slope for an approximate power law behavior occurs in films where the dark spin density is low ($< 5 \times 10^{15}$ spin cm^{-3}). It has been suggested [18] that there may be a second process contributing to the metastable ESR at low intensities of inducing light.

Other experiments suggest that more than one center may be involved in the optically-induced metastabilities observed in a-Si:H films [22-24]. For example, Han and Fritzsche [23] have found that at least two states are necessary to explain the inducing and annealing behavior of optically-induced charges in photoconductivity, and Guha et al. [24] have shown that the annealing behavior of the photoconductivity depends on the temperature at which the samples are originally exposed to light. Similar conclusions can be inferred from optical absorption, optically detected magnetic resonance and time resolved photoconductivity measurements.

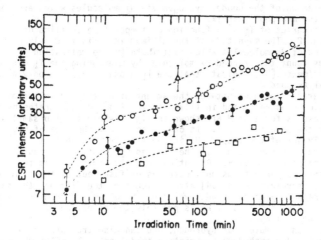

Fig. 1. Metastable, optically-induced ESR intensity in a-Si:H as a function of irradiation time at 300K (3 μm film). Open circles, closed circles and open squares represent data taken with 1.3, 0.4 and 0.1 W cm^{-2} from a tungsten source, respectively. Open triangles are data taken at 6471 Å and 700 mW cm^{-2}. The spin density at 500 min for 0.1 W cm^{-2} irradiation density (open squares) is approximately 10^{16} cm^{-3}. (After ref. 18.)

ELECTRONIC STATES AND THE URBACH EDGE

Urbach [25] first observed a tail on the band edge absorption in ionic crystals. This tail depends exponentially on the energy with a shope (on a semi-log plot) which is inversely proportional to the temperature. Although the tetrahedrally coordinated amorphous semiconductors do not, in general, exhibit this temperature dependence, they do show an exponential energy dependence.

There are several explanations of the Urbach edge in a-Si:H and related alloys. As mentioned above the most commonly accepted explanation is that this absorption results from strained bonds [6,26]. Several specific models have been proposed to calculate this effect [27,28], but the details are not yet certain. An alternative explanation suggests that the Urbach edge may involve transitions between various charge states of two-fold-coordinated silicon atoms [7].

THE ROLES OF SELECTED IMPURITIES

Impurities can sometimes play important roles in determining the important defects in tetrahedrally coordinated amorphous semiconductors. Often the most important "impurity" species is hydrogen which is added intentionally to reduce the density of states near the middle of the gap. In the process some of the hydrogen can be trapped as molecular hydrogen (H_2) in the films.

Some of the H_2 molecules trapped in films of a-Si:H or a-Si_xGe_{1-x}:H can be probed by examining the spin lattice relaxation of the bonded hydrogen atoms [29]. There exists a characteristic minimum in the temperature dependence of the spin lattice relaxation time T_1 which occurs near 40K. It is now well established [30,31] that this characteristic minimum results from a relaxation of the bonded hydrogen via H_2 molecules which are somehow trapped in the amorphous matrix.

Typical curves of T_1 as a function of temperature in films of a-Si:H are shown in Fig. 2. The magnitude of the minimum in T_1 is inversely proportional to the number of H_2 molecules which contribute to the relaxation process [30]. One can thus count the trapped H_2 molecules by monitoring the T_1 minimum. Recent experiments [32] indicate that the T_1 measurements see only a subset of the total trapped H_2.

There are differences between the amount of H_2 trapped in freshly prepared films on substrates and in films removed from the substrates by selective etching. Results for a-Si:H are shown in Fig. 3 where the two samples were prepared at the same time [33]. It can be seen from the data of Fig. 3 that approximately 40% of the trapped H_2 in the film on the substrate is lost when the substrate is dissolved away. The trapped H_2 constitutes about 0.1 at. % in typical samples of a-Si:H. Recent experiments indicate that the extra H_2 in the freshly prepared films on substrates diffuses out of the films on a time scale of several months [34] after which time the film has the same amount of trapped H_2 as the powder which has been removed from the substrate. The mechanism for this loss of H_2 is probably the releasing of internal strains as the film ages.

In addition to molecular hydrogen, one can also trap atomic hydrogen in a-Si:H albeit under much more restrictive conditions. In films which contain substantial oxygen impurities atomic hydrogen can be produced [16] by x-irradiation at low temperatures (T < 100K). The atomic hydrogen is stabilized by the oxygen at temperatures below ~ 300K, but combines to form H_2 above this temperature.

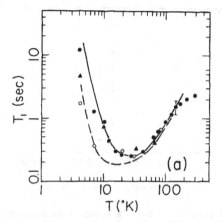

Fig. 2. T_1 as a function of temperature in two representative samples of a-Si:H of low defect density. Solid data points are taken at 42.3 MHz and open data points at 12.3 MHz. The lines are fits obtained from the model of Conradi and Norberg [30]. (After ref. 29.)

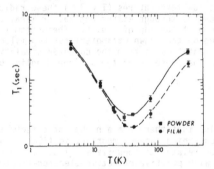

Fig. 3. T_1 as a function of temperature in flakes (solid squares) and thin-films (solid circles) of a-Si:H. The solid and dashed lines are aids to the eye. (After ref. 33.)

ESR spectra of films of a-Si:H doped with oxygen also show other defects associated with impurities. A typical ESR spectrum is shown in Fig. 4. The broad line in this figure is attributed to a hole trapped at a singly coordinated oxygen atom while the narrower line is the Si dangling bond resonance which has been distorted for technical reasons [16]. The narrowest feature at g = 2.0023 is attributed to an unpaired electron on a silicon atom which is bonded to three oxygen atoms (E' center). Even for samples purposely doped with oxygen the presence of this center requires that there is clustering of the oxygen into SiO_x-rich regions in the films. This fact is but one example of the inhomogeneous nature of a-Si:H films.

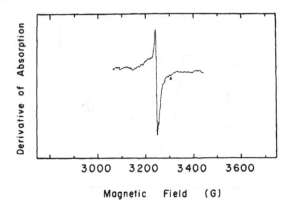

Fig. 4. Central portion of the ESR response in O-doped a-Si:H after x-irradiation at 77K. The strong central line is the Si dangling bond signal. The broad features in the spectrum correspond to an oxygen-related defect (hole center) as described in the text. The very narrow (~ 2 G) feature near g = 2.002 corresponds to silicon E' centers (see text). (After ref. 16.)

A final impurity species which one can sometimes trap in films of a-Si:H is the NO_2^- radical. At low temperatures (T < 80K) these radicals can be paramagnetic (NO_2 molecules) through either optical or x-ray excitation. A typical spectrum is shown at the top of Fig. 5. The bottom trace is a computer simulated spectrum based on the known parameters for NO_2 molecules. The departure of the experimental curve from the computer simulation in the region of g ≈ 2.0055 is due to the presence of the silicon dangling bond resonance in the experimental trace.

SUMMARY

Evidence is mounting that there are a number of defects in addition to the silicon dangling bond which can be important in films of a-Si:H. Some important defects may be related to impurity species. The dangling bonds are usually associated with positive effective electron-electron correlation energies U_{eff} although negative U_{eff} has been suggested for at least some of these states.

Shallow defect states near the band edge are usually attributed to strained bonds, but two-fold-coordinated silicon atoms have also been suggested. Both the dangling bonds and the shallow defect states may play important roles in determining the metastable changes which occur upon application of light. These metastable, optically-induced changes probably involve the creation of new defects, but some changes may also arise from the rearrangement of charge in existing defects.

ACKNOWLEDGMENTS

The author gratefully acknowledges collaboration with W.D. Ohlsen, M. Gal, C. Lee, E.D. VanderHeiden and R. Ranganathan. This research was supported in part by the National Science Foundation under grant number DMR-83-04471, by the Office of Naval Research under contract number N00014-83-K-0535, and by ARCO Solar, Inc.

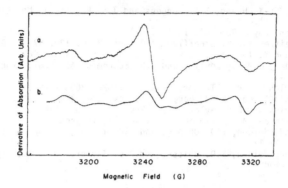

Fig. 5. Experimental (top trace) and computer-simulated (bottom trace) ESR derivative spectra of NO_2 radicals in a-Si:H after x-irradiation at 77K. The additional feature near 3240 G in the experimental trace is due to the Si dangling bond signal as described in the text. (After ref. 16.)

References

1. W.E. Spear and P.G. LeComber, Solid State Commun. 17, 1193 (1975); Philos. Mag. 33, 935 (1976).
2. D.E. Carlson, RCA Rev. 38, 211 (1977).
3. D.K. Biegelsen, Proc. Electron Resonance Soc. Symp. 3, 85 (1981).
4. D. Adler and F.R. Shapiro, Physica 117B + 118B, 932 (1983).
5. N.B. Goodman and H. Fritzsche, Phil. Mag. B42, 149 (1980).
6. R.A. Street, D.K. Biegelsen and J.C. Knights, Phys. Rev. B24, 969 (1981).
7. D. Adler, Kinam 4C, 225 (1982).
8. J.C. Knights, D.K. Biegelsen and I. Solomon, Solid State Commun. 22, 133 (1977).
9. J.R. Pawlik and W. Paul in Amorphous and Liquid Semiconductors, W.E. Spear, ed. (Univ. of Edinburgh, Edinburgh, Scotland, 1977), p. 437.
10. A. Friederich and D. Kaplan, J. Electron. Mater. 8, 79 (1979).
11. R.A. Street and D.K. Biegelsen, Solid State Commun. 33, 1159 (1980).
12. R.A. Street and D.K. Biegelsen, J. Non-Cryst. Solids 35 + 36, 651 (1980).
13. A. Friederich and D. Kaplan, J. Phys. Soc. Jpn. 49, Suppl. A., 1237 (1980).
14. D.L. Staebler and C.R. Wronski, Appl. Phys. Lett. 31, 292 (1977).
15. H. Dersch, J. Stuke and J. Beichler, Phys. Status Solidi, B105, 265 (1981); B107, 307 (1981); Appl. Phys. Lett. 38, 456 (1981).
16. W.M. Pontuschka, W.E. Carlos, P.C. Taylor, and R.W. Griffith, Phys. Rev. B25, 4362 (1982).
17. P.C. Taylor and W.D. Ohlsen, Solar Cells 9, 113 (1983).
18. C. Lee, W.D. Ohlsen, P.C. Taylor, H.S. Ullal and G.P. Ceasar, AIP Conf. Proc. 120, 205 (1984).
19. M. Stutzmann, W.B. Jackson and C.C. Tsai, AIP Conf. Proc. 120, 213 (1984).
20. C. Lee, W.D. Ohlsen and P.C. Taylor, Phys. Rev. B31, 100 (1985).
21. M. Stutzmann, W.B. Jackson and C.C. Tsai, Phys. Rev. B, in press.
22. M.H. Tanielian, N.B. Goodman and H. Fritzsche, J. de Phys. 42, C4-375 (1981).
23. D. Han and H. Fritzsche, J. Non-Cryst. Solids 59 + 60, 397 (1983).

24. S. Guha, C.-Y. Huang, S.J. Hudgens and J.S. Payson, J. Non-Cryst. Solids 66, 65 (1984).
25. R. Urbach, Phys. Rev. 42, 1324 (1953).
26. L. Schweitzer and M. Scheffler, AIP Conf. Proc. 120, 379 (1984).
27. S. Abe and Y. Toyozawa, J. Phys. Soc. Jpn. 50, 2185 (1981).
28. M.H. Cohen, C.M. Soukoulis, and E.N. Economou, AIP Conf. Proc. 120, 371 (1984).
29. W.E. Carlos and P.C. Taylor, Phys. Rev. B26, 3605 (1982).
30. M.S. Conradi and R.E. Norberg, Phys. Rev. B24, 2285 (1981).
31. W.E. Carlos and P.C Taylor, Phys. Rev. B25, 1435 (1982).
32. J.B. Boyce and M. Stutzmann, Phys. Rev. B, in press.
33. E.D. VanderHeiden, W.D. Ohlsen, and P.C. Taylor, J. Non-Cryst. Solids 66, 115 (1984).
34. E.D. VanderHeiden, W.D. Ohlsen, and P.C. Taylor, Bull. Am. Phys. Soc. 30, 354 (1985).

DETERMINATION OF DENSITY OF LOCALIZED STATES IN AMORPHOUS SILICON
ALLOYS FROM THE LOW FIELD CONDUCTANCE OF THIN N-I-N DIODES

Michael Shur and Michael Hack*
Department of Electrical Engineering, University of Minnesota,
Minneapolis, MN 55455
*ECD, Inc., 1675 West Maple Road, Troy, MI 48084

ABSTRACT

We describe a new technique to determine the bulk density of localized
states in the energy gap of amorphous silicon alloys from the temperature
dependence of the low field conductance of n-i-n diodes. This new technique
allows us to determine the bulk density of states in the centre of a device,
and is very straightforward, involving fewer assumptions than other
established techniques. Varying the intrinsic layer thickness allows us to
measure the density of states within approximately 400 meV of midgap.

We measured the temperature dependence of the low field conductance of an
amorphous silicon alloy n-i-n diode with an intrinsic layer thickness of 0.45
microns and deduced the density of localised states to be $3\times10^{16}cm^{-3}eV^{-1}$ at
approximately 0.5 eV below the bottom of the conduction band. We have also
considered the high bias region (the space charge limited current regime) and
proposed an interpolation formula which describes the current-voltage
characteristics of these structures at all biases and agrees well with our
computer simulation based on the solution of the complete system of transport
equations.

1. INTRODUCTION

The density of localized states in the mobility gap of amorphous silicon
alloys plays an important role in determining the transport and optical
properties of these materials. Several different techniques, such as field
effect studies, capacitance measurements, DLTS etc., have been used to
determine the density of states. In addition, information about the
localized states distribution has also been obtained from both
photoconductivity studies and from the characteristics of amorphous thin film
transistors.

There is, however, a considerable variation in the measured values and in
the overall shape of the density of states deduced by different techniques
and from measurements on different samples. Some of it may be related to the
doping dependence of the density of states [1,2] as well as to the dependence
of the density of states on the fabrication procedure. In particular, the
interpretation of DLTS data requires a complicated computer analysis. Field
effect studies may be quite sensitive to surface states [3] and, hence, the
measured density of states at the surface can be substantially different from
the bulk density. In capacitance versus temperature and frequency studies to
determine the density of states [4], simplifying assumptions made during the
interpretation of experimental data (such as the assumed field profile, etc.)
may lead to substantial errors.

Determination of the bulk density of states in amorphous silicon alloys
has been carried out by analyzing the space charge limited conduction in
n-i-n diodes. However, the effects of carrier diffusion near the contacts
may complicate the boundary conditions making the interpretation of the
results less straightforward.

Mat. Res. Soc. Symp. Proc. Vol. 49. ' 1985 Materials Research Society

In this paper we describe a new technique to determine the density of localized states in the energy gap of amorphous silicon alloys from the temperature dependence of the low field conductance of n-i-n diodes.

2. THEORY

At zero bias carriers diffuse from the highly doped layers into the i-layer of n-i-n diodes filling the localized states around the Fermi level. If the intrinsic layer is sufficiently thin these carriers change the conductance activation energy throughout the device (see Fig.1). This change in the activation energy in the centre of the sample depends on the thickness of the i-layer and is directly related to the density of localized states at the Fermi level in this region. Varying the film thickness and, hence, changing the position of the Fermi level in the middle of the device enables one to determine the density of states as a function of energy. This technique can only be applied to samples thin enough for carriers to diffuse from the n+-contact to the centre of the sample.

The Fermi level position throughout the device is found by solving Poisson's equation. Once the free carrier distribution has been calculated the sample resistance is related to the density of localized states, the thickness of the intrinsic layer and to temperature.

The density of localized states in the upper half of the gap may be approximated by a superposition of several exponential functions. The sample conductance is determined by the free carrier distribution and, hence, by the density of states in the middle of the sample as this is the most resistive region. The density of states, $g_A(E)$, in this energy range may be approximated by by an exponential function of energy E:

$$g_A = g_1 \exp[(E - E_{FL})/E_A] \tag{1}$$

Here g_1 is the density of states at $E = E_{FL}$ and E_A is the characteristic energy of the variation of the density of states.

Mathematically this problem is identical to that of electron spill-over in crystalline n^+-i-n^+ structures [5] where the thermal energy kT in the crystalline problem is substituted by E_A and n by n_t where n is the free electron density and n_t is the trapped electron density. In particular, for the relevant case when the concentration of carriers in the i-layer is much smaller than that in the n^+-regions we find

$$n_t(x) \approx 2\varepsilon E_A/[q^2(L/2-x)^2] \tag{2}$$

A further discussion of this equation may be found in [6]. Eq. (2) is valid for $0 < x < L/2$ where L is the length of the i-region and the carrier distribution for $-L/2 < x < 0$ is symmetrical.

Once the concentration of the trapped carriers is known the Fermi level E_F is given by:

$$E_F \approx E_{FL} + E_A \ln[n_t/(g_1 E_A)] \tag{3}$$

leading to the following expression for the free carrier density $n(x)$:

$$n(x) = N_c \exp[E_c(E_F-E_c)/kT] = n_0[qn_t(x)/(g_1 E_A)]^{1/\alpha} \tag{4}$$

Fig. 1. Charge injection into an
n-i-n device.
a. Device structure.
b. Bottom of the conduction band
vs. distance.
Dashed line - long sample
Solid line - short sample
c. Electron carrier density vs.
distance
d. Density of localized acceptor-
like states vs. energy

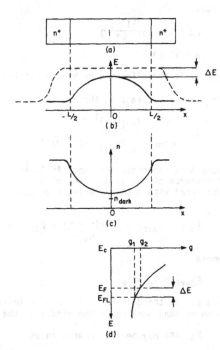

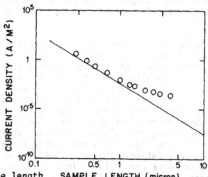

Fig. 2. Current density vs. sample length.
o - computer simulation
Solid line - analytical theory

Where

$$n_0 = N_c \cdot \exp(\frac{E_1 - E_c}{kT})$$

$$\alpha = kT/EA$$

and N_c is the effective density of states in the conduction band.

This analysis leads to the following expression for the low-field resistance

$$R = \frac{L}{q\mu A n_0} \left[\frac{\bar{q} g_1 L^2}{8\varepsilon} \right]^{1/\alpha} \frac{1}{(1 + 2/\alpha)} \tag{5}$$

Here A is the cross-section. As follows from Eq. (5), the temperature dependence of the conductance of the n^+-i-n^+ structure is approximately activated with a characteristic energy:

$$E_{act} = E_{dark} - E_A \ln(\frac{8\varepsilon}{q g_1 L^2}) = E_{dark} - \Delta E \tag{6}$$

where

$$E_{dark} = E_c - E_{FL} \tag{7}$$

and E_{FL} is the dark Fermi level in a infinitely long sample, i.e. in a sample with no band bending in the middle of the device.

Eq. (6) may be also rewritten as

$$g_1 \exp \left[\frac{E_c - E_{FL} - E_{act}}{kT} \right] = \frac{8\varepsilon}{qL^2} \tag{8}$$

Hence measuring ΔE we deduce the density of states using eq. (8).

In order to check the validity of this analytical approach we simulated the dark current-voltage characteristics of an n^+-i-n^+ structure using our comprehensive computer model of amorphous silicon devices [7]. The results of the simulation are in good agreement with the analytical solution (see Fig. 2).

To measure the density of the localized states using the new approach described above we fabricated amorphous silicon n^+-i-n^+ structures and amorphous silicon gap cell structures grown on the same substrate under identical deposition conditions. The value of the activation energy E_{dark} measured from the temperature dependence of the dark conductance of a one mm gap structure was 0.6 eV. The thickness of the i-layer was 0.45 microns. The current-voltage characteristics for the n-i-n diode were linear up to voltages of the order of 100 mV. The measured temperature dependence of the current through the n^+-i-n^+ structure at 25 mV is shown in Fig. 3. As can be seen from the figure this dependence is thermally activated (as expected) with an activation energy $\Delta E \approx 0.47$ eV, except in the low temperature region where the dark conductivity (not included into the model) dominates and the current becomes too small to measure. There is also a noticeable change in the activation temperature at larger temperatures which may be related to the change of E_A with temperature. Using eq. (8) we deduce a density of states of 3×10^{16} $cm^{-3} eV^{-1}$ at 0.47 eV below the conduction band. This number is in good agreement with the data obtained by other techniques. If we

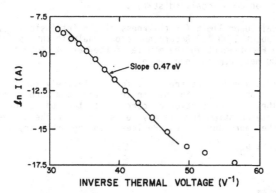

Fig. 3. Measured temperature dependence of the current.

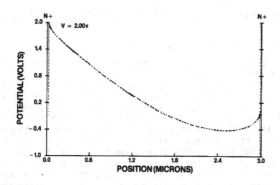

Fig. 4. Potential distribution in a-Si n-i-n structure for V = 2 volts.

estimate the minimum density of states g_1 to be approximately 5×10^{15} $cm^{-3}eV^{-1}$ then from eq. (5) we deduce $E_A/q \approx 73$ meV, also in agreement with the expected values for deep localized states.

The practical lower limit for thickness of the intrinsic layer in an n-i-n diode is about 1000 Å. According to our analysis this should lead to movement of the Fermi level in the middle of the sample close to 0.4 eV, i.e. up to 300 mV÷400 meV from the band edge.

The results obtained above are valid for a relatively short structure where the density of carriers in the center of the structure is much larger than the equilibrium concentration of carriers in the conduction band, n_0. The following interpolation formula may be used to describe the zero fields resistance of the amorphous silicon devices with the arbitrary length:

$$R_o \approx R*R_D/(R + R_D) \tag{9}$$

where

$$R_D = L/(q\mu n_0)A \tag{10}$$

is the Ohm's Law resistance and R is given by eq. (8).

B. High Field Regime

When a high voltage is applied across an a-Si n-i-n diode electrons are injected into the i-region from the cathode n-region. This leads to a space charge limited injection current. The density of localized states may then deduced from the I-V characteristics, see, for example, [8].

In this case the current-voltage characteristic may be calculated by solving Poisson's equation together with the current continuity equation. We can now propose an interpolation formula that describes the current-voltage characterstics of the n-i-n a-Si devices for all thicknesses and all biases. This formula accounts for the contributions from the equilibrium (dark) conductivity which dominates in longer samples in low electric fields, from the spill-over conductivity which is dominant in the short samples in low electric fields and the space charge limited current which is important in short examples in high electric fields:

$$j = q\mu N_c \ [(2+\alpha)^{1+\alpha}/(1+\alpha)^{2+\alpha}]V^{(1+1/\alpha)}L^{(-1-2/\alpha)}(\epsilon/qg_cE_a)^{1/\alpha} + V/R_0 \tag{11}$$

where R_0 is given by eq. (9).

In order to check the results of this analysis and to get a better insight into the device physics we also simulated the n-i-n a-Si structures at high bias voltages using our comprehensive computer model. The potential profile and the free carrier distributions for the bias voltage V = 2 volts are shown in Figs. 4-5. As can be seen from Fig. 4 the potential minimum in the i region is shifted towards the cathode contact with the increase in the bias voltage. At V = 2 volts it is still, however, some distance away from the cathode (the latter condition is implied in the simplified analytical treatment of the high bias regime given above). As can be seen from Fig. 5 in the boundary layers near the contacts the diffusion effects play an important role with the diffusion and drift components of the current being very large and nearly cancelling each other. In these regions the carrier distribution is similar to that under the equilibrium conditions and may be found using our theory for the low voltage regime. It is worthwhile to note the electron concentration is relatively uniform over most of the intrinsic region away from the contacts.

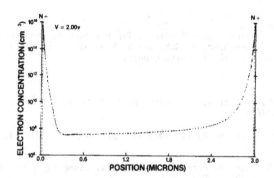

Fig. 5. Free electron distribution in a-Si n-i-n structure for V = 2 volts.

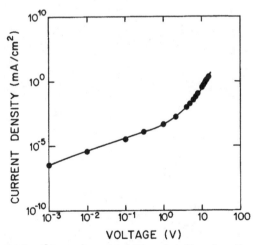

Fig. 6. Current voltage characteristic of a-Si n-i-n structure.
o - computer simulation, ... - analytical model

In Fig. 6 we compare the results of our analytical calculation (see eq. (11) with the numerical simulation. As can be seen from the figure the agreement is quite good confirming the validity of the analytical model.

Acknowledgement

We would like to thank Stan Ovshinsky for his support and encouragement and W. Czubatyj, W. Den Boer, S. Guha and S. Hudgens for useful discussions and T. Glatfelter for help in measurements.

References

1. C. R. Wronski, B. Abeles, T. Tiedje and G. D. Cody, Solid State Comm., 44, 1423, (1982).

2. R. A. Street, J. Zesch, M. J. Thompson, Appl. Phys. Lett., 43, 672, (1983).

3. C. Hyun, M. S. Shur, M. Hack, Z. Yaniv and V. Cannella, Appl. Phys. Lett., 45, 11, 1202, (1984).

4. P. Viktorovitch, J. Appl. Phys., 52, 1392, (1981).

5. Van der Ziel, M. S. Shur, K. Lee, T. H. Chen, and K. Amberadis, IEEE Trans. Electron Devices, ED-30, 128, (1983).

6. M. Shur and M. Hack, J. Appl. Phys., Aug. 1985.

7. M. Hack and M. Shur, J. Appl. Phys., 54, 5858, (1983).

8. R. L. Weisfield, J. Appl. Phys., 54 (11), 6401, (1981).

PART III

Interfaces

TRANSIENT PHOTOCONDUCTIVITY STUDIES OF a-Si:H INTERFACES

R. A. STREET
Xerox Palo Alto Research Center, Palo Alto, CA   94304

ABSTRACT

The application of transient photoconductivity to the study of contacts and interfaces with a-Si:H is reviewed. The photocurrent is shown to contain three terms – one from the drift of photogenerated carriers, and two from contact and bulk effects due to the electric field induced by the drifting carriers. For different sample configurations, each of these terms can dominate, and each gives different information about a-Si:H bulk or surface electronic properties. The effects are illustrated with data from metal contacts, dielectric interfaces, doped layers and gap cell measurements.

1.   INTRODUCTION

In recent years there has been a growing realization that many of the electronic properties of a-Si:H are governed by contact or interface effects [1]. The origin of these effects is the space charge that is inevitably present at the interface between two materials. The density of states in a-Si:H is sufficiently low that the space charge layer extends for a substantial fraction of the typical sample thickness, and hence can exert a large influence on the observed properties. Examples are the band bending due to adsorbed gas at free surfaces [2], Schottky barrier formation at metal contacts [3,4], and space charge transfer at dielectric interfaces [5,6]. The existence of the surface region is exploited and studied by a variety of experiments, including current–voltage [3], capacitance [7], and field effect [8] measurements. Recent work has found that transient photoconductivity can also be used as an effective measurement of interface electronic properties, through the information it gives about the internal electric field and width of the space charge layer [4,5,9,10,11,12,13]. This paper reviews the technique, illustrating the measurement with data on different sample structures.

Many studies of transient photoconductivity have explored the bulk properties of a-Si:H, with the measurements generally made in one of two ways. Time–of–Flight experiments use a capacitor configuration to measure the drift mobility [14], and coplanar electrode structures are used to study recombination [15]. The measurements described here are mostly with a sample in the configuration of a parallel plate capacitor with semitransparent electrodes and so are closely related to the Time–of–Flight technique. The sample thickness is typically 2–5 μm. The light excitation is a 5 nsec pulse from a nitrogen/dye laser, with a wavelength of about 5000 Å which gives an absorption depth in a–Si:H of about 1000 Å. A voltage is applied either as a d.c. bias or as a pulse, and as will be seen, the results differ for the two cases.

The light pulse excites mobile carriers of density n near one electrode. As the carriers move in the electric field of the sample, there is a charge induced on the opposite electrode of magnitude

$$Q = ne\bar{x}/d \qquad (1)$$

where $\bar{x}$ is the average distance moved and d is the sample thickness. The observed photocurrent is then given by dQ/dt. In practice the photocurrent is the measured quantity and Q the charge collection is obtained by numerical integration.

However, eq. 1 is not a complete description of the photocurrent. In general two other contributions are possible so that the total can be expressed as

$$I_{pc}(t) = dQ/dt + \Delta I(t)_{contact} + \sigma_{bulk} \Delta E(t)_{bulk} \qquad (2)$$

The second two terms in eq. 2 arise indirectly from the motion of the photoinduced charge, as follows. The drift of carriers represented by eq. 1 results in a space charge in the sample. Consequently there will be an additional field both at the contacts and in the bulk. If there is a significant injection current at one of the contacts, then this distortion of the field will change the magnitude of the current. Similarly, the additional field can cause a modification of the current in the bulk of the sample. Although these two terms are indirect, they can be very important for the correct interpretation of the measurements, as discussed below. Furthermore these effects cannot be eliminated by reducing the density of excited carriers n, because, to a first approximation, all three terms in eq. 2 are linear in n. The remainder of this paper outlines the conditions under which each term in eq. 2 dominates the photoconductvity, and describes how the measurement is interpreted in each case.

2. METAL CONTACTS

Most metals form blocking contacts on undoped a–Si:H [16]. As a result there is a space charge layer at the contact, with a field $E_0(x)$ whose shape is determined by the built in potential and the bulk density of states. This internal field can be studied by transient photoconductivity [4]. Undoped a–Si:H is used, so that the low conductivity and blocking contacts eliminate the second two terms in eq. 2, and the only contribution to the photoconductivity is dQ/dt. The conventional Time–of–Flight experiment is performed under these conditions, with additionally the voltage V applied just before the light pulse. This results in a field given by

$$E(x) = V/d + E_0(x) \qquad (3)$$

In practice the built–in potential of the Schottky contact to a–Si:H is < 0.5 V and the field extends at most 1–2 μm into the sample, so that a sample thickness of 5 μm and an applied voltage >1 V ensures a good approximation to a uniform field. Under these conditions it is well known that the drift mobility μ can be obtained when the carriers of fully collected (Q = ne), and $\mu\tau$ is obtained if carrier trapping results in incomplete collection.

Under conditions of zero bias, or with a d.c. bias, the internal field will be non–uniform and will be determined by the depletion layer near the contact. Figure 1 shows a comparison of the transient photoconductivity with a d.c. bias and with a voltage pulse, in which the difference between a uniform and a non–uniform field is clearly seen.

Fig. 1. A comparison of the electron response for pulsed and d.c. voltages in undoped a–Si:H. Also shown is a schematic diagram of the spatial distribution of the internal field in the two cases. With the pulsed field, the drift mobility μ is obtained from the transit time. With the d.c. bias, the internal field of the contact is measured.

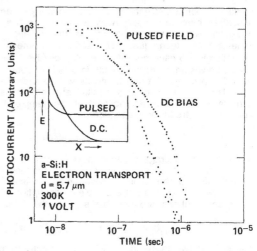

Fig. 2. Examples of the internal field profile for Cr and Pt contacts to a–Si:H. The times indicated correspond to those of the transient response. The values of w are obtained from charge collection. The density of states is derived from the exponential region of the field profile.

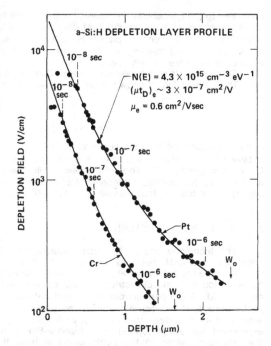

For the case of a non–uniform internal field,

$$I_{pc} = ne\mu\, E(t) \qquad (4)$$

where E(t) is the field experienced by the charge packet at time t after the excitation pulse. Provided both n and μ are independent of time, I_{pc} is directly proportional to E(t). The distance moved by the charge packet after time t is

$$x(t) = \int_0^t \mu E(T)dT \qquad (5)$$

From eqs. 4 and 5, E(x) can be obtained, and examples of the field profile at zero bias are shown in Fig. 2 for Cr and Pt electrodes. The photocurrent is carried by electrons, since these metals form electron depletion layers. Electrons are non–dispersive in a–Si:H at room temperature (see Fig. 1) so that the assumption of constant μ is valid. In addition, the pulsed voltage data gives values for μ from the transit time, ne from the saturation value of the charge collection Q, and $\mu\tau$ from the voltage dependence of Q. The first two values are needed to calculate E(x) from eqs. 3 and 4, and from $\mu\tau$ it is found that the carrier lifetime is ~10^{-6} sec [17]. The measurement times, indicated in Fig. 2, show that virtually the complete field profile is mapped out within 10^{-6} sec. confirming the assumption that n is constant. However, the field profile can also be evaluated if n and μ are not constant, provided that their time–dependence is known.

The charge collection Q directly measures the average distance moved by the carriers and is expected to be the sum of the depletion layer width plus an appropriate diffusion length in the field–free region. The contribution from diffusion is generally small since it is equivalent to carrier drift within a band bending of kT ~ 0.025 eV, compared to the typical built–in potential of 0.2–0.5 eV for the metal contact. In fact the diffusion term is reduced by the field distortion generated by the carriers. The field distortion across the charge packet is $ne/\varepsilon\varepsilon_0$ and acts to prevent the charge from diffusing in the field–free region. An accurate measurement of $E(x)$ to low values of E requires some care in the choice of n.

Further details of the measurements of Schottky contacts are described elsewhere [4]. The data of Fig. 2 show the larger barrier of Pt compared to Cr. The built–in potential is given by

$$V_{BI} = \int E(x)\, dx \qquad (6)$$

and can be obtained from the field profile. In practice, a more accurate value of V_{BI} is obtained from measurements of Q as a function of bias voltage in forward bias, since $Q \to 0$ when $V \to V_{BI}$. The slope of the field profile gives the bulk density of states in the a–Si:H. The approximately exponential $E(x)$ seen in Fig. 2 corresponds to a uniform density of states for that sample of $\sim 4 \times 10^{15}$ cm^{-3} eV^{-1}.

3. DIELECTRIC INTERFACES

The technique applied to metal contacts can also be used to study dielectric interfaces to a–Si:H, such as oxides or nitrides [5,10.11,13]. If the dielectric is transparent, as is generally the case, the excitation light is absorbed in the a–Si:H near the interface with the dielectric, and the carriers will move depending on the local electric field. Again one can perform pulsed voltage measurements to obtain ne, μ and $\mu\tau$ for the a–Si:H film, and d.c. voltage measurements to obtain the space charge layer width and its bias dependence. Just as for the experiments on metal contacts, the experiment must be performed on undoped a–Si:H with non–injecting contacts, so that the only contribution to the photocurrent is from the drift of carriers.

Figure 3 shows the bias dependence of Q for two samples in which the dielectric is plasma deposited Si_3N_4 [5]. The samples consist of a thick film (5 μm) of a–Si:H, with a thin nitride layer ($\sim$3000 Å) sandwiched between semitransparent electrodes. The two samples differ in that one has the nitride deposited on top of a–Si:H (top nitride) and the other has the nitride underneath (bottom nitride). Apart from this difference, the thicknesses and deposition conditions are identical to the two samples. However, the results show that the electronic properties of the interfaces are very different.

Fig. 3 Charge collection as a function of bias voltage for the top and bottom nitride samples. Q is given as a percentage of the total charge collection across the 5 μm sample.

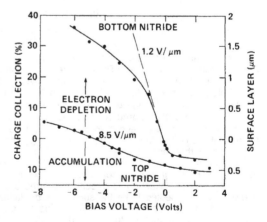

A great deal of information can be obtained from the data in Fig. 3. At zero bias, the sign of the photocurrent corresponds to an electron accumulation layer at the interface, showing that there is positive charge at or near the interface, that is screened out by negative charge in the a–Si:H. The voltage V_{FB} at which $Q = 0$ corresponds to flat bands in the a–Si:H, and $C_N V_{FB}$ is the magnitude of the interface charge that is present at zero bias, where C_N is the capacitance of the nitride. For this particular pair of samples, the top nitride has an interface charge of 5×10^{11} cm^{-2}, and the bottom nitride has about an order of magnitude less. In addition the slope of the Q(V) data gives a measure of the density of interface states, which is also deduced to be an order of magnitude larger for the top nitride compared with the bottom nitride.

The interface states can also be observed from the current response to a voltage pulse, without the light excitation. At the interface the Fermi energy is nearer to the conduction band than in the a–Si:H bulk, due to the accumulation layer. Hence the release time of electrons from interface states is much less than the thermal generation of bulk carriers, so that the transient measurement will be dominated by the interface states. Fig. 4 shows this emission of interface electrons observed as a current transient after the voltage is applied. A comparison of top and bottom nitrides confirms that the current is indeed from the interface. The measurement gives the total charge in interface states near the Fermi enery, and from the temperature dependence also gives the position of the Fermi energy at the interface. For this sample the interface Fermi energy is 0.25 eV from the conduction band edge, from which a zero bias band bending of ~0.5 eV is deduced.

Similar measurements made with silicon oxide as the dielectric find an electron depletion layer at the interface, showing an interface charge of the opposite sign to that of the nitride [11]. The origin of the band bending and the reason for the different sign are not understood. The influence of the native surface oxide can also be measured by the same technique. The only different requirement is that the upper elecrode must be close to, but not touching the surface of the sample. If the spacing is more than a few microns, the charge collection is very small and difficult to measure. The surface oxide is also found to give a depletion layer, with a band bending estimated to be ~0.2 eV, in agreement with other data [18].

Fig. 4 Examples of the transient current arising from the sweep out of carriers from the nitride/a–Si:H interface, at different temperatures.

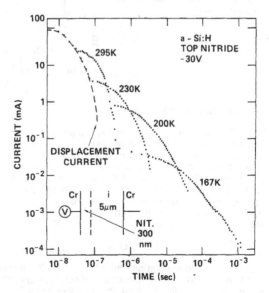

4. INDIRECT CONTRIBUTIONS TO THE PHOTOCURRENT

a) Optically Induced Injection

Eq. 2 for the photocurrent contains two indirect contributions to I_{pc} which arise from the field distortion due to the transport of charge. One possible consequence of the field distortion is a change in the current injected at one of the contacts, which will occur if there is space charge trapped near the contact, and if the contact is injecting. Fig. 5 shows such a situation, in which the transient photoconductivity response of an undoped a-Si:H sample that has n+ contacts, is measured at an applied voltage that results in a space charge limited current [4]. The photocurrent is completely different from that observed in a sample with blocking contacts, having a decay time of $\sim$ 0.1 sec and a total charge collection $Q \sim 10^4 ne$. Our interpretation of the response is that photoexcited holes are trapped near the injecting n+ contact, with the result that the field near the contact is increased, so that a larger space charge limited current results. The additional current persists until the field at the contact returns to its equilibrium value, which is evidently a very slow process. Thus neither the decay time nor the high value of Q are related to the bulk recombination properties of a-Si:H. This explanation is confirmed by measurements in which the excitation is at the non-injecting contact [10]. In this case the transient photoconductivity again has a large charge collection and slow decay. However, there is also a delay in the onset of the photocurrent that corresponds to the transit time of holes, showing that the effect only occurs when holes are trapped near the injecting contact.

Fig. 5 Transient photo-conductivity response under conditions of forward bias. Note the long response time and the very large charge collection. The inset shows the expected band diagram in forward bias.

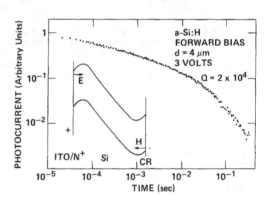

b) Bulk Conductivity

The third term in eq. 2 is the contribution to the photocurrent from bulk mobile carriers, influenced by the field distortion. In a typical experiment the induced field change is $\sim 10^2$ V/cm. Undoped a-Si:H has a bulk conductivity of $\sim 10^{-9}$ Ω^{-1} cm^{-1} giving a current 10^{-7} Amp/cm^2, which is negligible compared to the typical value of dQ/dt of $\sim 10^{-3}$ Amp/cm^2. However, in n-type a-Si:H the conductivity reaches 10^{-2} Ω^{-1} cm^{-1} and so the indirect contribution to I_{pc} can easily dominate. For an n-type sample with a metal contact, at low voltage there will be a depletion layer near the negative electrode with the bulk being essentially at zero field. The charge collection will contain two terms

$$Q = q_{drift} + q_{bulk\ screening} \tag{7}$$

corresponding to the first and third term in eq. 2. It is easy to see that the bulk current that flows due to the field distortion will continue until all the space charge is completely screened out. Hence the total charge collection will be $Q = ne$. On the other hand the contribution to Q from the drift of the photoexcited carriers is new/d, where w is the width of the depletion layer. In doped a–Si:H, w is much less than in undoped, because of the larger density of deep states. Thus for a sample thickness of about 5 μm, q_{drift} is small, and almost all the photocurrent is from the screening term.

There are several experimental tests that can distinguish between the screening and drift contribution to I_{pc} [12]. The clearest measurement is of the transient current after a voltage is applied, without light excitation. The voltage results in an increase in the depletion layer thickness, and as a result space charge is swept out of the sample. Because the sweep out is effectively the screening of the space charge layer, a similar transient response of the two experiments is expected. For our samples this result is confirmed for doping levels even as low as 1 ppm PH₃, as is described in more detail elsewhere [12].

In the voltage pulse experiment, the charge collection increases with applied voltage, provided the sample is not fully depleted, and is given by

$$Q^2(V) = 2\varepsilon\varepsilon_0 n_e eA^2 \left[(V - V'') - \sqrt{4(V-V')(V_0-V')} \right] \qquad (8)$$

where V_0 is the zero field built–in potential, and V' and V'' are correction terms. The density of mobile carriers, n_e, can be determined by the measurement as shown in the example of Fig. 6 [12]. Linearity of Q^2 versus V is observed, as expected, since the square root term in eq. 8 is small at high voltages. A density of 6×10^{14} cm⁻³ is found for a 1 ppm PH₃ doped sample, in agreement with other measurements of this quantity [19], and is interpreted as the occupied conduction band tail states.

Fig. 6 A plot of charge collection squared versus voltage for voltage pulse data from an n–type a–Si:H sample. The linear relation gives the density of shallow conducting electrons, as described in the text.

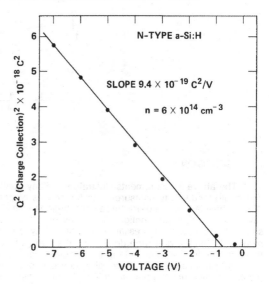

Related information is obtained from the decay time of the transient photoconductivity, which for the screening term is given approximately by

$$\tau \approx RC \approx RQ/V \qquad (9)$$

where C is the appropriate depletion layer capacitance and R is the series resistance of the bulk. Substitution of eq. 8 leads to an equivalent method of obtaining n_e from the dependence of τ^{-2} on V,. In addition the experiment allows one to observe the slow changes in the depletion layer due to release of carriers from deep dangling bond states. As the depletion layer shrinks, the capacitance increases and so also does τ.

Fig. 7 shows the voltage dependence of τ^{-2} just after the voltage is applied, for a d.c. bias, and at variable delay times. The shrinking of the depletion layer is easily seen, and covers a time period of ~ 1 sec at room temperature, which is consistent with the release times observed by DLTS [20].

Fig. 7 A plot of inverse time–constant squared versus voltage for pulsed and d.c. voltages, and for variable delay times, showing the shrinking of the depletion layer.

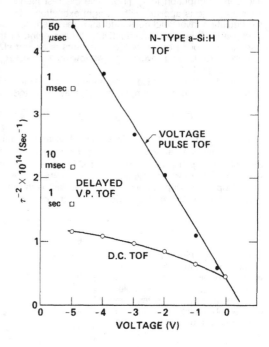

5. DISCUSSION

The above measurements illustrate the many applications of transient conductivity and photoconductivity measurements for studies of contacts and interfaces to a–Si:H. We emphasize that there are three contributions to the photocurrent as given by eq. 2. Given different sample configurations, any of these terms can dominate the response, so that in general it is necessary to identify the type of response before interpreting the results. When the drift term dominates, the information obtained is μ and $\mu\tau$ provided there is a uniform electric field in the sample. Non–uniform fields, that originate from a d.c. rather than a pulsed voltage, give information about internal field profiles, and as a result also measure the built–in potential, bulk density of states, and interface states. If on the other hand, the drift term is negligible compared to the other contributions, then entirely different information results, for example the measurement of the mobile carrier density.

It is of interest to discuss the transient photoconductivity of coplanar electrode structures, in view of the above discussion. Fig. 8 shows an example of the decay in n–type a–Si:H, with a doping level of 1 ppm PH_3 [12], and is typical of other reported measurements [15]. The decay time τ is ~ 10^{-4} sec, with an extended response evident at high voltages that persists for >1 sec. In order to evaluate the response, it is necessary to identify which type of photoconductivity dominates the response. Thus if drift dominates, as is widely assumed, then τ is interpreted as the bulk recombination time, whereas if bulk screening dominates, then τ is the effective RC time constant of the contact space charge. In fact the evidence is very strong that in this particular case the response at 10^{-4} sec is indeed the screening term, and that the decay at longer times is from the contact injection. The bulk recombination is presumably quicker than 10^{-4} sec and is not directly observed. The evidence is described elsewhere, and is largely from the results of voltage pulse data that show the same time constant of about 10^{-4} sec. In addition, the same decay is observed at zero bias, with one electrode shielded from the light, as shown in Fig. 8. Such a response can only originate from the internal field near the contact.

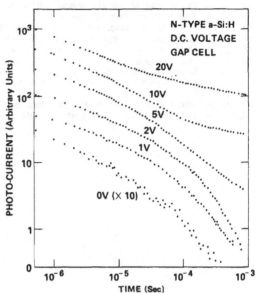

Fig. 8 Examples of the photo-current decay in an n-type sample at different d.c. voltages. The results at zero bias are with one electrode shielded from the light.

Finally we discuss a comparison of transient photoconductivity and capacitance measurements. The capacitance C also is a measure of the depletion layer width, and in fact, C-V data are equivalent to the Q-V data shown for example in Fig. 3. However in order to perform the capacitance measurement, the sample must be sufficiently conducting to allow an a.c. signal of an appropriate frequency. The C-V measurement is therefore most easily performed on doped material, and can only be made at very low frequency in undoped a-Si:H. On the other hand, the transient photoconductivity experiment on undoped samples readily gives information about the depletion layer. However in doped a-Si:H, the high conductivity that is essential for the capacitance measurement, leads to the indirect photocurrent described earlier in this paper, so that little can be learnt about the depletion layer. The two experiments therefore tend to give complementary sets of data.

ACKNOWLEDGMENTS

This research is supported by the Solar Energy Research Institute.

REFERENCES

1. H. Fritzsche, Semiconductors and Semimetals 21, Part C (1984).
2. M. Tanelian, Philos. Mag. B 45, 435 (1982).
3. M. J. Thompson, N. M. Johnson, R. J. Nemanich, and C. C. Tsai, Appl. Phys. Lett. 39, 274 (1981).

4. R. A. Street, Phys. Rev. B 27, 4924 (1983).
5. R. A. Street and M. J. Thompson, Appl. Phys. Lett. 45, 769 (1984).
6. T. Tiedje, and B. Abeles, Appl. Phys. Lett. 45, 179 (1984).
7. D. V. Lang, J. D. Cohen, and J. P. Harbison, Phys. Rev. B 25, 5285 (1982).
8. A. Madan, P. G. LeComber and W. E. Spear, J. non–Cryst. Solids 20, 239 (1976).
9. T. Datta and M. Silver, Appl. Phys. Lett. 38, 905 (1981).
10. R. A. Street, M. J. Thompson and N. M. Johnson, Philos. Mag. B 51, 1 (1985).
11. R. A. Street and M. J. Thompson, AIP Conf. Proc. 120, 410 (1984).
12. R. A. Street, Phys. Rev., in press.
13. P. B. Kirby, D. W. MacLeod and W. Paul, Philos. Mag., in press.
14. T. Tiedje, Topics in Applied Physics 56, 261 (1984).
15. J. M. Hvam and M. H. Brodsky, Phys. Rev. Lett. 46, 371 (1981).
16. C. R. Wronski and D. E. Carlson, Solid State Commun. 23, 421 (1977).
17. R. A. Street, J. Zesch and M. J. Thompson, Appl. Phys. Lett. 43, 672 (1983).
18. B. Aker, S–Q. Peng, S. Cai and H. Fritzsche, J. non–Cryst. Solids 59 & 60, 509 (1983).
19. R. A. Street and J. Zesch, Philos. Mag. B 50, L19 (1984).
20. N. M. Johnson, Appl. Phys. Lett. 42, 981 (1983).

RECOMBINATION UNDER DOUBLE INJECTION CONDITIONS IN a-Si:H BASED DIODES.

R. KÖNENKAMP* and ARUN MADAN**

*   Hahn-Meitner-Institut für Kernforschung Berlin, 1 Berlin 39, FRG
**   Solar Energy Research Institute, Golden, CO 80401, USA

ABSTRACT

The junction recovery technique is used to investigate recombination parameters and kinetics in thin film a-Si:H diodes. It is found that double injection occurs in most of the p-i-n type samples at sufficient forward bias and that the recovery currents are dominated by the extraction of trapped holes from the valence bandtail. The mobility-lifetime product of the recovered carriers is seen to be strongly dependent on the dopant concentration in the active layers of the devices.

INTRODUCTION

Analyses of a-Si:H p-i-n type solar cells have so far strongly relied upon steady state experiments, such as photovoltaic and I-V measurements, thereby allowing only for a rather indirect and model dependent determination of transport and recombination parameters (1). We have recently shown that the junction recovery technique (JR) can be applied successfully to amorphous thin film diodes (2) and that it presents a sensitive probe for the investigation of transport parameters in these devices. Further, we have pointed out the qualitative consistency of JR results with fast photoconductivity measurements such as time-of-flight experiments (3), a point which had previously been the subject of considerable controversy (4,5). In this paper we will try to outline the relevance of the method in the investigation of recombination kinetics in thin (~5000 Å) a-Si:H$_2$ diodes by addressing the changes of mobility-lifetime products ($\mu\tau$) in a range of operational conditions and for a variety of device structures. We will begin with a short review of the experimental method which has been described in more detail previously (2).

For forward voltages in excess of 1 V, i.e. in the injecting regime, a-Si:H p-i-n type diodes usually exhibit forward current densities 5 to 10 times larger than n-i-n or Me-i-n (Schottky barrier) type configurations. This can be taken as an indication that double injection prevails in the former devices and that both types of carriers effectively fill the device. Upon a fast reversal of the applied voltage into reverse bias the stored excess carriers are extracted from the center region giving rise to a measurable reverse current transient before the steady state reverse conditions with low currents are obtained. We have recently shown that in a-Si:H based diodes of the p-i-n type most of the charge which is recovered within the first 10 µs is due to holes originating from the valence bandtail (2), which is believed to comprise considerably more localized trap states than the conduction bandtail. When the reverse bias is varied, while the forward bias remains the same in consecutive experiments a change in the amount of recovered charge is observed. This behavior can be attributed to recombination losses during the extraction process and can be analyzed in terms of an equation for the recovered cahrge, given by

$$Q_r = Q_o \frac{\tau_R}{\tau_T} \left(1 - \exp\left(-\frac{\tau_T}{\tau_R}\right)\right) \qquad (1)$$

Here Q_0 is the amount of charge that would be recovered without the influence of recombination, τ_R is the recombination lifetime of the carriers and t_T their transit time.

Our previous results with a-Si:H p-i-n type diodes indicate that the predominant part of the recovered charge is due to holes originating from the valence bandtail. Hence, τ_R in Eq. (1) can be associated with the recombination time for holes. It should be emphasized that the JR experiment differs from charge collection or delayed field experiments in that trap states are filled before the extracting voltage pulse is applied, so that losses due to deep trapping are unlikely to be of importance in JR.

EXPERIMENTAL DETAILS AND SAMPLE PREPARATION

The experiment utilized a pulse generator (Wavetek 145), providing a steady state forward voltage to the sample. The switching into reverse bias typically occurred in less than 100 ns, the RC time of the circuitry. The recovery current transients were recorded with a Tektronix 7612D digitizer and the recovered charge Q_r was determined by a digital integration of the current transients over $t \simeq 10$ μs. Since the switching gave rise to strong displacement transients, it was necessary to subtract these out by a procedure which we have outlined in Ref (2). Samples were deposited by the glow-discharge technique and were typically 5000 Å in thickness with an area of ~0.02 cm^2. For the p$^+$ and n$^+$ layers (approximately 200 Å thick) 1% B$_2$H$_6$ or PH$_3$ was added to the SiH$_4$ gas flow, respectively. Junctions of the type p$^+$-p$^-$-n$^+$ and p$^+$-n$^-$-n$^+$ were fabricated with dopant/silane gas mixtures in the range of 1 to 100 vppm as indicated in the text. Schottky barrier type samples used Pd evaporated contacts.

RESULTS AND DISCUSSION

To illustrate the particularities encountered with a-Si:H based devices, we show in Fig. 1 results obtained with a crystalline Si diode (a), an amorphous p$^+$-p$^-$-n$^+$ type diode (b) and an amorphous p-i-n structure. The recovery of the crystalline device is characterized by constancy of the recovered charge and reverse transients that scale linearly with reverse bias. This behavior can be explained in terms of trap-free transport and long recombination times (6) as expected for this material. Transients measured for a-Si:H p$^+$-p$^-$-n$^+$ type samples are shown in Fig. (1b). These show a capacitive contribution to the current which is due to the switching of the bias from the forward to the reverse direction. This contribution can be eliminated (3) and is shown as a dotted curve for one of the transients in Fig. (1b). After subtraction of the capacitive contribution a linear relationship for the reverse peak current and the reverse bias is found, and the recovered charge Q_r is seen to be constant for the cases shown. The value of Q_r corresponds to approximately 10 times the value CV$_f$ (where C is the capacitance of the sample and V$_f$ the forward voltage), which is taken as additional evidence for the double injecting properties of the device. A somewhat different situation applies, however, for the optimized p-i-n-type devices with undoped center regions which are typically used as solar cell devices. In these devices the recovery current amplitudes scale sublinearly with reverse voltage V$_r$ and a voltage independent tailing is observed for all transients as shown in Fig. (1c). Further, the recovered charge increases with reverse bias saturating at V$_r \simeq 3$V.

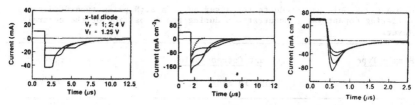

Fig (1a) Recovery transients for a crystalline Si diode obtained at a forward bias of 1.25V and reverse bias of 1; 2, and 4V.

Fig (1b) Recovery transients for an a-Si:H p-p-n diode (no. 4 of Table (1)) obtained at a forward bias of 2V and reverse bias of 1; 2 and 4 V. The dotted line shows the displacement contribution due to the switching from forward into reverse bias.

Fig (1c) Recovery transients for an a-Si:H p-i-n diode (no. 3 of Table (1)) obtained at a forward bias of 1.2V and reverse bias of 0.4; 0.8; 2 and 3V.

Apparently in the p-i-n devices the recovery process is limited by trap emission, while for the case of Fig. (1b), the rate-limiting step is the drift process in the reverse field. As a possible explanation for this difference we suggest that the valence bandtail in the center region of the p^+-p^--n^+ device comprises more states than that of the p-i-n device. This would lead to an increase in the stored hole charge in agreement with Figs. (1b) and (1c) and in turn would ensure that the trap emission rate exceeds the extraction rate so that the recovery process remains drift-limited. Such an explanation is in qualitative agreement with results from optical measurements (7), which show that doping can lead to an increase in the density of localized states in the vicinity of the valence bandtail.

The relevant recombination parameter, which can be determined from Eq. (1) is the recombination lifetime τ_R of the free carriers. This presupposes knowledge of the transit time t_T which we express by

$$t_T = \frac{L^2}{\mu V} \qquad (2)$$

where μ is the carrier mobility and V/L is the reverse electric field across the device. Using (1) and (2) one can as well express the experimental results in terms of the mobility - lifetime product $\mu\tau_R$, thus avoiding any ambiguities concerning the carrier mobility, which is not easily measured in the thin sample structures used. In Table (1) we list the results obtained for a number of devices of the type p-i-n in which the center regions were lightly doped p- or n-type. We note a strong increase of the $\mu\tau_R$ - products with p-type doping, leading to a value of 6×10^{-8} cm^2/V for the p^+- p^-- n^+ sample. This doping dependence of the hole $\mu\tau_R$ product is in qualitative agreement with the results from other measurements, such as charge collection (8), delayed field experiments (9) and the recent solar cell investigation by Okamoto et al. (1). The doping dependence can consistently be explained with a change in the density of doubly occupied dangling bond states, which are believed to be the preferred recombination sites for holes.

It should be noted that the $\mu\tau_R$ values in Table (1) are somewhat smaller than those reported from other experimental techniques as well as from theoretical considerations. For example, using the surface photovoltage method we have previously measured the hole diffusion length in intrinsic a:Si:H material to be about 0.4 - 0.5 μm (10), which is a factor 3 larger than the corresponding value for the p-i-n sample in this study.

Table (1). Hole $\mu\tau_R$ products for a series of a-Si:H p-i-n type diodes with differing dopant gas concentrations during the deposition of the central layer.

Sample Type	Dopant Concentration		$\mu\tau_R$
	(vppm in SiH_4)		(cm^2/Vs)
p n n	100	PH_3	3×10^{-9}
n n p	50	PH_3	5×10^{-9}
p i n	0		1×10^{-8}
p p n	4	B_2H_6	6×10^{-8}

Okamoto et al. (1), in an extensive experimental study have determined the hole $\mu\tau_R$ product to be approximately 10^{-7} cm^2/V in efficient p-i-n type solar cells. Since the p-i-n sample of Table (1) represents such a cell with a conversion efficiency of $\sim$ 10% (global AM1), we are led to the conclusion that the JR results underestimate the correct values. A possible explanation for this could be given in terms of a non-uniform voltage distribution across the device during the extraction process, due to screening by deeply trapped carriers or due to a faster depletion of the contact regions. Both effects would result in a diminished field across the bulk layer and thus lead to an underestimation of t_T in Eq. (2) and hence τ_R in Eq. (1). We are presently investigating the relevance of these effects in more detail and in the following will contend to point out more qualitative results of the experiment.

In Fig. (2a) we show the dependence of the recovered charge on the doping level in the active center region for p-i-n type and Schottky barrier cells. All the data were taken at approximately equal forward current densities of $I_f \simeq$ (60 mA/cm^2). It is interesting to note that the electron-dominated Schottky barrier cells, in particular the n^+- n^-- Pd sample, do not show any detectable recovery transients, although a high amount of stored charge is found for the n^+- n^-- p^+ configuration. This confirms our previous conclusions from studies on near intrinsic material (2) that electron trapping in the conduction bandtail is relatively small compared to the storage of holes in the valence bandtail. The systematic difference between the n-i-p and p-i-n samples in fig (2a) are likely to be due to carry-over of dopant from the deposition of the initial contact layer to the bulk layer. This has recently been investigated and confirmed in an independent study (11).

Further, from Fig. (2a) it is apparent that undoped center layers yield the least amount of recovered charge, which leads us to conclude that the p-i-n samples with undoped bulk layers have a comparably small amount of hole trap states. On the other hand, p-type as well as n-type center layers give rise to strong charge storage. At dopant concentrations in excess of 50 vppm PH_3 or 4 vppm B_2H_6 in the gas phase the recovery transients tend to be no longer emission limited, i.e. they become similar to those displayed in Fig. (1b), indicating that both, doping with PH_3 or with B_2H_6 gives rise to additional localized states for holes.

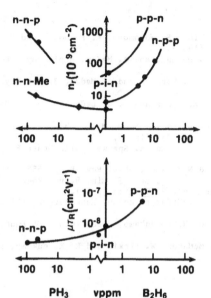

Fig (2a) Number of recovered charge carriers, n_r, in dependence of gas dopant concentration for the deposition of the center layer. (● p-i-n type diodes, ◆ Schottky barrier type diodes).

Fig (2b) Dependence of $\mu\tau_R$ for holes on the gas dopant concentration for the deposition of the center layer.

For completeness we have indicated in Fig. (2b) the behavior of the hole $\mu\tau_R$ product for some of the samples of Fig. (2a). The best performing solar cell in Fig. (2) is the p-i-n sample with an undoped bulk layer. It is interesting to note that the hole $\mu\tau_R$ product is not a maximum for this device, but yet a factor of 5 short of the $\mu\tau_R$-product for the p^+-p^--n^- sample. It might be speculated that this is due to a decrease in the electron lifetime with p-type doping (9).

CONCLUSIONS

The junction recovery technique has been applied to a variety of a-Si:H based diodes. It was shown that the amount of recovered charge as well as the hole $\mu\tau_R$ products depend sensitively upon the doping level in the active layer of the devices. We found that the amount of recovered charge increases with phosphorus <u>and</u> with boron doping, and attribute this to an increase of hole trapping states. We found that the hole $\mu\tau_R$ products increase with p-type doping in a similar way as recently reported from other experiments. The best solar cell performance can be associated with samples which show small recovery transients and intermediate $\mu\tau_R$ products for holes.

ACKNOWLEDGEMENTS

We gratefully acknowledge discussions with Prof. P.G. LeComber and Dr. T.J. McMahon. This work was partly supported under DOE Contract No. DE-AC02-83CH10093.

REFERENCES

(1) H. Okamoto, H. Kida, S. Nonomura, K. Fukumoto and Y. Hamakawa, J. Appl. Phys. 54, 3236 (1983).
(2) R. Könenkamp, A. Hermann and A. Madan, Appl. Phys. Lett. 46, 405 (1985).
(3) R. Könenkamp, A. M. Hermann and A. Madan, J. Non-Cryst. Sol. 66, 249 (1984).
(4) A. J. Snell, W. E. Spear and P. G. LeComber, Phil. Mag. B 43, 407 (1981).
(5) M. Silver, N. C. Giles, E. Snow, M. P. Shaw, V. Canella and D. Adler, Appl. Phys Lett. 41, 937 (1982).
(6) H. Benda, A. Hoffmann and E. Spenke, Solid State Electron. 8, 887 (1965).
(7) W.B. Jackson and N.M. Amer, Phys. Rev. B 25, 5559 (1982).
(8) P. B. Kirby, W. Paul, C. Lee, S. Lin, B. von Roedern and R. L. Weisfield, Phys. Rev. B 28, 3635 (1983).
(9) W. E. Spear, H. L. Steamers and H. Mannsperger, Phil. Mag. B 48, L49 (1983).
(10) T. J. McMahon and R. Könenkamp, Proc. 16th IEEE Photvolt. Spec. Conf, 1389 (1982).
(11) A. Madan, T.J. McMahon, W. Pickin and R. Könenkamp, submitted for publication.

PHOTOEMISSION STUDIES OF AMORPHOUS SILICON/GERMANIUM HETEROJUNCTIONS

F. EVANGELISTI*, S. MODESTI+*, F. BOSCHERINI*, P. FIORINI*, C.QUARESIMA**,
M. CAPOZI*** and P. PERFETTI**
* Dipartimento di Fisica, Universita' "La Sapienza", 00185 Roma, Italy
** Istituto di Struttura della Materia, 00044 Frascati, Italy
*** PULS, Laboratori Nazionali di Frascati, 00044 Frascati, Italy

ABSTRACT

The heterostructures obtained by growing a-Ge on a-Si:H and a-Si have been investigated by synchrotron radiation photoemission. We measured valence band and core level spectra on the heterostructures grown in situ under ultrahigh-vacuum conditions. A step-by-step monitoring of possible band-bending changes during the interface formation enabled us to determine unambiguously the band discontinuities. The measured values of the valence band discontinuity were 0.2 ± 0.1 eV for a-Si:H/a-Ge and 0.0 ± 0.1 eV for a-Si/a-Ge, respectively. Evidence was found for the formation of abrupt interfaces without interdiffusion.

INTRODUCTION

In the crystalline case the interfaces between Si and Ge have been extensively investigated [1] and represent one of the test systems [2] for the theories on the heterojunction formation. The study of their amorphous counterpart can provide, besides new information on these disordered systems, new insight on the driving force that determines the band alignement, diffusion potential, etc. Moreover, this study has an immediate technological interest, since efficient photovoltaic devices have been realized [3] with amorphous hydrogenated silicon-germanium alloys used as absorbing layers with an optical gap smaller than that of a-Si:H. The band discontinuities of the Si/Ge heterojunctions provides an upper limit to the discontinuities that can occur at the interface between silicon and Si-Ge alloys.

EXPERIMENTAL

The experiments were performed in ultrahigh vacuum conditions (operating pressure 5×10^{-11} - 5×10^{-10} Torr). The heterostructures were grown in situ by thermal evaporation of a-Ge on a-Si and a-Si:H. Amorphous silicon samples were prepared in situ by electron-gun evaporation. The electron gun was a homemade, miniature source in which electrons with ~ 3 KeV energy were directed against a Si single crystal. By using this source it was possible to evaporate the films keeping the chamber pressure in the high 10^{-10} Torr range and thus to avoid contamination. The hydrogenated silicon samples were grown by r.f. glow-discharge in a capacitive reactor from pure silane and cleaned in the measurement chamber by Ar ion sputtering. Auger spectroscopy was used to monitor the surface cleanliness.

The photoemission measurements were performed at the Frascati Synchrotron Radiation Facility at two different beam lines, one equipped with a toroidal grating monochromator and the other with a grasshopper monochromator. The experimental approach used consists in growing step-by-step a-Ge on the substrate and following the evolution of the valence band and core levels at each step. When the valence band discontinuity is small, this technique allows distinguishing unambiguously between the shift of the valence band maximum E_v due to the discontinuity setting in and the shift caused by possible band-bending effects.

RESULTS AND DISCUSSION

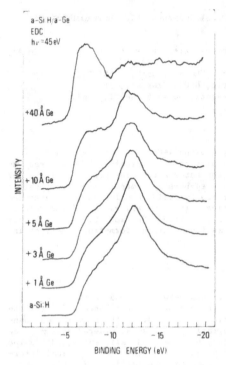

Fig. 1 Evolution of the valence-band energy distribution curves during the a-Si:H/a-Ge heterojunction formation.

As an example, we report in Fig.1 the photoemission energy distribution curves (EDC's) of the valence band of clean a-Si:H and that at different a-Ge coverages. The dominant feature in the a-Si:H spectrum is the hydrogen-induced structure at ~ 6.5 eV below the valence edge. The higher energy shoulder is due to the Si 3p states. The prominence of the hydrogen induced peak is a cross section effect [4]. The top spectrum in Fig.1 is that of amorphous germanium. Its dominant feature near the valence band edge is due to the 4p states. The broad structure extending from ~ 4 eV to ~ 12 eV below the band edge is due to the 4p-s and 4s derived states that, in the crystalline case, give rise to two well resolved peaks. The intermediate spectra in Fig.1, corresponding to lower a-Ge coverages, are a superposition of two contributions: that coming from the a-Si:H underneath and that of the a-Ge overlayer. This behavior is the one expected from an abrupt junction without interdiffusion (see discussion below). At the first two coverages any difference with the clean a-Si:H EDC is hardly appreciable. The presence of Ge is shown by the Ge 3d core level photoemission, however.

The valence band maximum E_V is defined as the linear extrapolation to zero of the EDC leading edge. The error in this determination is evaluated to be ~ 0.1 eV. The values of E_V as a distance from the Fermi level are reported in Fig.2 as a function of a-Ge thickness. A shift of 0.2 eV upon deposition of a fraction of monolayer is clearly visible. In order to ascertain whether this shift is due to a band discontinuity or is the result of a band-bending change upon chemisorption of the first Ge atoms, we have measured the Si 2p and the Ge 3d core-level EDC's. Their position are also shown in Fig.2. It is evident that no displacement in energy of the core levels is occurring. We conclude, therefore, that the shift of E_V is due to a valence band discontinuity between the a-Si:H and a-Ge of 0.2 ± 0.1 eV. Since the optical gap of the a-Si:H we use is 1.7-1.8 eV and that of a-Ge evaporated at room temperature is 0.7 eV [5], it follows that a large conduction band discontinuity of 0.8-0.9 eV must be present in the conduction band at the interface.

As mentioned previously, from the intensity dependence of the core lines as a function of overlayer thickness, it is possible to extract information on the type of the heterojunction. In particular, if the interface is abrupt and the overlayer is uniform we have:

$$(1 + \alpha \ I_{OL}/I_{SU}) = \exp(d/\lambda) \qquad (1)$$

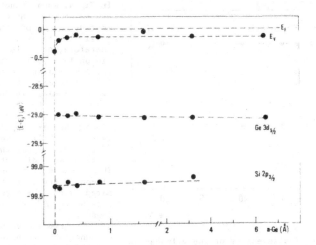

Fig. 2 Position in energy of the valence-band maximum, of the Si 2p and Ge 3d core levels as a function of a-Ge coverages for the a-Si:H/a-Ge heterojunction.

where I_{OL} and I_{SU} are the core intensities of the overlayer and substrate, respectively, α is a factor which accounts for the different photoemission cross-sections, λ and d are the overlayer escape depth and thickness. The results obtained using the Si 2p and Ge 3d core levels are shown in Fig.3 . The exponential behavior is strictly obeyed, suggesting that the growth at room temperature of a-Ge on a-Si:H proceeds without interdiffusion, although for a final proof a more precise determination of the overlayer thickness than that presently available would be necessary [6]. This abrupt behavior is in agreement with the results of the deposition of Ge on crystalline sili-con.

Following the same experimental procedure, we have also studied the for-mation of the heterojunction between a-Si and a-Ge. The results on the ener-gy position of the valence band maximum and Ge 3d core level are shown

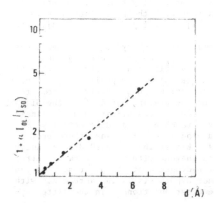

Fig. 3 Ratio of the core-level in-tensities as a function of overla-yer thickness.

98

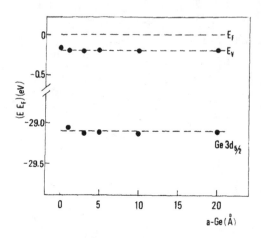

Fig. 4 Position in energy of the valence-band maximum and of Ge 3d core level as a function of a-Ge coverage for the a-Si/a-Ge hetero-junction.

in Fig.4, as a function of Ge coverage. No displace-ment of E_v and of the core level takes place. Therefore, no valence band discontinuity is present at the a-Si/a-Ge interface. The difference between the two pseudogaps is entirely accomodated by the conduc-tion band discontinuity. Since the a-Si optical gap is 1.26 eV [7] we have ΔE_c = 0.5 eV for this con-duction band discontinuity. The absence of valence band discontinuity is to be com-pared with the crystalline case where ΔE_v = 0.17 eV is found.

The two amorphous het-erojunctions we have inves-tigated in the present work exhibit two different va-lence-band discontinuities that, moreover, differ from the value obtained by grow-ing Ge on crystalline Si.

Although the effect is experimentally significant, it is small compared to the average accuracy [2] of the theories on heterojunction formation and this makes its understanding difficult. Few comments can be made in this short communication. For the a-Si/a-Ge system the absence of discontinuity can be the result of the smearing of the valence band maximum due to the localized states induced by disorder. The width of the tail of these states, resulting mainly from fluctuations of the dihedral angle, is expected to be of the order of 0.2-0.3 eV in a-Si [8]. On the other hand, our previous data [9] are compatible with a upward shift of 0.1 eV of E_v going from c-Si to a-Si while data by Reichardt et al.[10] suggest 0.3 eV. Since the Ge grown on crystalline Si is likely to be amorphous, it seems that the zero discontinui-ty a-Si/a-Ge interface compared to ΔE_v =0.17 eV in the crystalline case is a result of a disorder effect in silicon. It is worth pointing out that this kind of reasoning implicitly assumes the validity of one of the "linear" dis-continuity models [2] proposed to predict theoretically band alignement. It can be shown that the present result agrees with the prediction of some of these theories, in particular Harrison's [11], Frensley-Kroemer's [12], and Tersoff's [13].

As for the a-Si:H/a-Ge system, we remark that the hydrogenated amorphous silicon samples were deposited in growing conditions that usually introduce 5-10 at% of hydrogen. It was shown [9] by investigation of the a-Si:H/a-Si interface that such hydrogen content leaves the energy position of the va-lence band maximum unchanged. The same conclusion can be inferred from re-cently published data [10,14], by comparing the distance of E_v from the Si 2p core level in a-Si:H and a-Si. As a consequence the application of a linear discontinuity model would predict zero valence band discontinuity for a-Si:H/a-Ge, too. The 0.2 eV value found experimentally leads, therefore, to one of the following two possibilities. Either the "linear" discontinuity models are not able to predict the band alignement with an accuracy better than 0.2 eV or that we had a much larger hydrogen content in our samples (an unlikely possibility, however). The former hypothesis is in agreement with the findings of Ref.2 on the crystalline heterojunctions. In the latter hypothesis a displacement of E_v towards higher binding energy can take place [10] and that would naturally explain the discontinuity in the framework of the "linear" discontinuity models.

We have not yet investigated the a-Si:H/a-Ge:H interface. However, by considering previous results on the valence-band/core level distance in a-Ge and a-Ge:H [15] and the valence band alignement in the a-Si/a-Si:H inteface at low hydrogen content ($< \sim 10$ at%) [9] we can safely guess that also at a-Si:H/a-Ge:H interface the valence band discontinuity will be small ($< \sim 0.2$ eV). We conclude, therefore, that heterojunctions between silicon and germanium, hydrogenated or not, amorphous or not, exhibit a small ΔE_v and larger ΔE_c and that the same happens when a-Si$_x$Ge$_{1-x}$:H is employed instead of germanium. In a p-i-n photovoltaic cell this band alignement is very advantageous if the material with large pseudo-gap is used as a p-doped layer at the p-i interface while it would be a handicap at the i-n interface. In fact, in the former case the conduction band discontinuity gives rise to a barrier which hinders the electron backdiffusion and then reduces the collection losses [16,17]. Quite to the contrary, in the latter case ΔE_c would work against the electron collection. This different behavior must be kept in mind in the design of heterojunction solar cells.

REFERENCES

+ Present address: Istituto di Fisica, Universita' di Trieste, Trieste, Italy

1) G.Margaritondo, N.G.Stoffel, A.D.Katnani and F.Patella, Solid State Commun. 36, 215 (1980); P.Perfetti, N.G.Stoffel, A.D.Katnani, G.Margaritondo, C.Quaresima, F.Patella, A.Savoia, C.M.Bertoni, C.Calandra, and F.Manghi, Phys. Rev. B24, 6174 (1981).

2) A.D.Katnani and G.Margaritondo, Phys. Rev. B28, 1944 (1983).

3) G.Nakamura, K.Sato, T.Ishihara, M.Usui, K.Okaniwa and Y.Yukimoto, J. Non-Cryst. Solids 59-60, 1111 (1983).

4) L.Ley "The Physics of Hydrogenated Amorphous Silicon II", Ed. Joannopoulos and Lucovsky, Springer Verlag 1984, p.82.

5) G.A.N.Connel, R.J.Temkin and W.Paul, Adv. Phys. 22, 643 (1973).

6) R.S.Bauer and J.C.McMenamin, J. Vac. Sci. Technol. 15, 1444 (1978).

7) M.H.Brodsky, R.S.Title, K.Weiser, and G.D.Pettit, Phys. Rev. B1, 2632 (1970).

8) F.Yonezawa and M.H.Cohen, "Fundamental Physics of Amorphous Semiconductors", Ed. F.Yonezawa, Springer-Verlag 1981, p.119.

9) F.Patella, F.Evangelisti, P.Fiorini, P.Perfetti, C.Quaresima, M.K.Kelly, R.A.Riedel and G.Margaritondo, "Optical Effects in Amorphous Semiconductors", Ed. P.C.Taylor and S.G.Bishop, AIP Conference Proc. No.120 (1984) p.402.

10) J.Reichardt, L.Ley and R.L.Johnson, J. Non-Cryst. Solids 59-60, 329 (1983).

11) W.Harrison, J. Vac. Sci. Technol. 14, 1016 (1977).

12) W.R.Frensley and H.Kroemer, Phys. Rev. B16, 2642 (1977).

13) J.Tersoff, Phys. Rev. B30, 4874 (1984).

14) D.Wesner and W.Eberhardt, Phys. Rev. B28, 7087 (1983).

15) F.Patella, F.Sette, P.Perfetti, C.Quaresima, M.Capozi, A.Savoia and F.Evangelisti, Solid State Commun. 49, 749 (1984).

16) H.Okamoto, H.Kida, K.Fukumoto, S.Nonomura and Y.Hamakawa. J. Non-Cryst. Solids 59-60, 1103 (1983).

17) F.Evangelisti, P.Fiorini, C.Giovannella, F.Patella, P.Perfetti, C.Quaresima and M.Capozi, Appl. Phys. Lett. 44, 764 (1984).

MINORITY CARRIER INJECTION AND SERIES RESISTANCE EFFECTS IN HYDROGENATED AMORPHOUS SILICON SCHOTTKY BARRIER DIODES

Jerzy Kanicki

IBM Thomas J. Watson Research Center

P.O.Box 218, Yorktown Heights, New York 10598, U.S.A.

The minority-carrier injection and series resistance effects on the electrical properties of a-Si:H Schottky barrier diodes are described. The conductivity modulation was observed, for the first time, in metal/HOMOCVD a-Si:H contacts. Its effect on capacitance-voltage characteristics are discussed. The minority-carrier injection ratio is estimated from current-voltage characteristics as a function of total forward current for different metals. It is shown that these effects cannot be neglected in the interpretation of the AC and DC measurements. The caution, therefore, must be taken when using a-Si:H diodes structures to obtain the fundamental physical parameters characterizing either the interface or bulk properties of amorphous semiconductors.

INTRODUCTION

Much interest has been shown in hydrogenated amorphous silicon (a-Si:H) and related compounds for possible application to various amorphous semiconductors devices [1]. The critical aspect in the devices such as solar cells or flat panel displays is the metal or insulator/a-Si:H interfaces. So, the full understanding of these interfaces is of crucial importance. Furthermore, the behavior of a-Si:H interfaces is very different from that observed in crystalline silicon. Since, most of a-Si:H devices operate under non-equilibrium condition (e.g., the relationship $np = n_i^2$ is violated) their performance is very much related to their tendency to return to equilibrium via relaxation-recombination process. Some experimental and theoretical investigation have been carried out on a-Si:H Schottky contacts [1]. However, the series resistance effects on the capacitance-voltage characteristics and minority-carrier injection in a-Si:H have not been investigated until now. In this paper the effect of minority-carrier injection on electrical characteristics of metal/a-Si:H Schottky diodes is reported for the first time. The minority-carrier injection ratio has been determined as a function of the total forward current.

EXPERIMENT

Diode Structure

Metal/a-Si:H Schottky diodes structure were used for the investigation of series resistance effect and minority-carrier injection phenomenon in a-Si:H films. The intrinsically n-type a-Si:H films of comparable electronic quality (comparable electrical and optical properties) were grown by homogeneous chemical vapor deposition (HOMOCVD) technique [2] on glass substrate coated with molybdenum (Mo). The film thickness (t=0.6 μm) was such that the width of the quasi-neutral region (t-W) is much less than the diffusion length of holes ($L_p \approx 0.5$ μm). The width of the depletion region (W) being of the order of 0.4 μm. An heavily doped n^+ – layer was used to ensure back low contact resistance to the Mo. A 0.15 μm thick various metal front contact were deposited by vacuum evaporation (10^{-7} Torr) using an electron-beam gun source on a chemically etched surface (buffered HF). However, even under this condition the metal and a-Si:H are not in intimate contact. There inevitably exists a very thin (<5 Å) interfacial oxide layer. The contact areas ($A_c \cong 1.824 \times 10^{-2}$ cm^2) were defined by stainless steel masks. The current-voltage and capacitance-voltage characteristics were measured by the usual techniques.

Experimental Results

Typical forward current density (J_F) - applied voltage (V_F) characteristics of metal/HOMOCVD a-Si:H diodes made from the comparable electronic quality films are shown in Fig. 1.

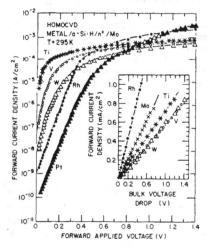
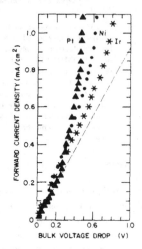

FIGURE 1. Forward characteristics, at room temperature, of different Schottky barrier diodes. The solid lines represent the theoretical curves. The insert shows the variation of J_F with V_B.

FIGURE 2. Forward characteristics of quasi-neutral region of several metal/HOMOCVD a-Si:H/ n^+ /Mo diodes exhibiting the modulation effect.

An interesting feature in Fig. 1, at higher forward current density, is a cross-over of low-(W,V or Ti) by high-(Pt) Schottky barrier characteristics. After the cross-over point the forward current density for a higher barriers is observed to be larger than bulk current under equilibrium condition. This excess current is due to the minority-carrier (holes) injection from the metal front contact into the quasi-neutral region of a-Si:H layer; the metal serves only for charge replacement at the interface. After the injection the minority carriers will flow away from the barrier (front contact) by the diffusion process, the rate of which is determined by their gradients concentrations. At the same time about the same concentration of majority carriers (electrons) will be injected by the back Mo contact in the direction to the front contact in order to neutralize the excess space-charge induced by the injected minority carriers. The injection from the back contact is necessary to maintain the overall bulk quasi-equilibrium (electronic neutrality) of a-Si:H device. The rearrangement of electrons and holes distribution, till a new quasi-equilibrium is establish, will happen during about the relaxation time of the semiconductor.

The barrier height for minority carriers (from the metal contact to the a-Si:H) decreases as the barrier height for majority carrier (from the a-Si:H to the metal) increases. The injection of minority-carrier , therefore, becomes increasingly important as the barrier height for majority carriers increases. Criteria for the transition barrier height that roughly separates high-barrier from low-barrier behavior can be given by: $\phi_n \geq E_g - E_{th}$ (E_{th} is thermal activation energy). The lower limit for the transition barrier height for the undoped device quality a-Si:H layer seems to be $\phi_n \geq 0.9$ eV. This value, however, is very much dependent on the hydrogen content in the HOMOCVD a-Si:H films and the injection phenomenon takes place, generally, if the barriers are substantially higher than about half the band gap of a-Si:H; this is in order to bring the minority carrier band-edge into the vicinity of the Fermi level. As a result of the minority-carrier injection the barrier profile will be no longer parabolic and an inversion layer (p>n) can be produced at the interface. When this is the case, the region of the n-type a-Si:H adjacent to the metal contains a high concentration of holes and thus becomes p-type. Consequently, this p-type region becomes more conductive, by virtue of the high hole concentration, than one might expect under bulk equilibrium condition. This process is known in crystalline silicon devices as conductivity modulation [3,4] and plays an important role in reducing the diodes series resistance (R_S). At the same time this leads to an improved rectification ratio. Indeed, the decrease of R_S (e.g., increase of conductivity), at higher forward current density, is clearly visible in Figs. 2 and 4 for the high work function metals (Pt,Ir,Ni,etc.). For that reason the a-Si:H film becomes more "heavily doped " as injection level is increased with increasing forward bias. This experimental finding contradict the simple Schottky diode equation (Eq. (3)), even, modified for an constant series resistance, where the minority carriers influence are neglected.

The measured capacitance (C_m) - applied voltage (V_a) characteristics, at different frequencies, of W/a-Si:H and Pt/a-Si:H diodes are presented in Fig. 3.

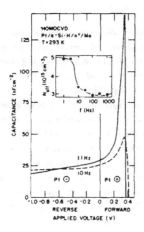 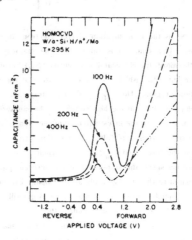

FIGURE 3. Capacitance characteristics, at room temperature, of Pt and W/a-Si:H/ n$^+$/Mo Schottky barrier diodes; the insert shows the variation of effective density with the frequency of the test signal at room temperature.

These curves represent two typical behavior encountered in metal/HOMOCVD a-Si:H contacts. In general, the increase in capacitance with the reverse and low forward bias, $V_F \leq V_{FB}$ (V_{FB} is the flat band voltage equivalent to the contact potential), should mainly be due to the depletion capacitance of the junction. Therefore, in this bias domain, in order to measure the true space-charge (depletion) capacitance (C_{SC}) of Schottky contact the measurements should be done at such angular frequency (ω) that $(\omega C_{SC})^{-1} > R_S$ and under such static bias voltage that $R_d > R_S$ (R_d is the diode differential resistance) [5]. These conditions are usually satisfied for a-Si:H diodes, which have a very large series resistance ($R_S = 10^5 - 10^6$ Ω), under a reverse bias or at low forward voltage, $V_F \leq V_{FB}$ and at measuring frequency, $f_m \leq 100$ Hz. However, even under the above conditions the $C_m^{-2} - V_a$ plot is usually not linear over a wide potential range, which makes difficult the determination of reliable values of diffusion potential (V_d) and dopant concentration (N_D) from experimental data. The experimentally observed deviations can be associated, in the first approximation, with the presence of deep gap states in the depletion and bulk region of the semiconductor [1,6,7]. The influence of these states on capacitance measurements was already discussed extensively in the literature [1,6,7] and there is not need to repeat it in this paper. To sum up, the presence of deep traps in the depletion region of the Schottky barrier makes the junction capacitance a complicated function of the bias voltage, measuring frequency and temperature. Therefore, the expectation to obtain a comparable experimental data for metal/a-Si:H contacts as were observed for metal/c-Si diodes will be rather a pure coincidence than the typical behavior of amorphous semiconductors. So, the classical equation developed for metal/c-Si devices [8] has to be modified and can be eventually expressed as:

$$C_m^{-2} \cong \frac{V_d + V_R}{2 \, q \, \epsilon_s \, N_{eff}} \tag{1}$$

where

$$N_{eff} \cong N_D^+ \left[1 + \frac{N_t \, e_n^2}{N_F \, (e_n^2 + \omega^2)} \right]^2 \tag{2}$$

The N_F is the concentration of free electrons in the bulk and e_n the electron emission rate of deep traps. This equation is valid when the majority carriers dominate the current, for $\phi_n \leq 0.9$ eV. Furthermore, the classical donor density (N_D) becomes, now, the effective density (N_{eff}) of electrically active impurities which are the combination of the ionized donor (N_D^+) and deep traps (N_t) present in depletion region. The $N_{eff} \cong N_D$ if the traps time constant is so large that they are not able to respond to changes either in the bias voltage or the measuring signal or temperature. This is shown in insert of Fig. 3, where, in the investigated frequency range, the N_{eff} makes a transition from a state in which all deep states are able to follow the test signal ($e_n > \omega$) to another state in which most of the traps cannot follow the test signal ($e_n < \omega$). Furthermore, when the traps can follow the bias variation but not the measuring signal the situation is even more complicated. In such situation if the N_t is not negligible compared to the N_D (specially at low doping level or for intrinsic materials) then the impurity distribution obtained from $C_m - V_a$ measurements gets distorted due to the presence of traps. Thus the great care should be taken in determining the impurity profiles from capacitance measurements. However, even if N_{eff} varies smoothly in x-direction it remains constant in planes parallel to the metal contact, such that $N_{eff}^+(x) = N_{eff}(x)$ for all x. Hence, the local variation of $N_{eff}(x)$ can be obtained and used for determination of N_t and N_D from Eq. (2), if the other parameters are known. This will be discussed in the separate paper. Therefore, the presence of deep gap states can, in certain bias region, explain the observed deviation of capacitance data. Eventually, even if the observed decrease of C_m at higher forward voltage can be predicted by some theoretical models including deep gap states [6,7] the theoretical calculations cannot predict why the capacitance becomes negative, at higher forward voltage, for the high work function metal contacts such as Pt,Ir, Ni, etc. Furthermore, on the one hand, while the contribution of the deep gap states at $V_F > V_{FB}$ is negligible, on the other hand, the influence of the interfacial layer and the series resistance is not. Indeed, the experimental and theoretical works developed for the crystalline silicon devices clearly demonstrated that the decrease in capacitance at forward voltage should be rather associated with either a high diode series resistance [9,10] or an interfacial layer [11] and not necessarily only with the presence of deep gap states in the a–Si:H films.

DISCUSSION

The J_F-V_F characteristics of an metal/HOMOCVD a–Si:H Schottky barrier diodes obey an exponential law only if the whole applied forward voltage remains concentrated on the potential barrier of the space-charge layer. This is the case in the low forward voltage region ,$V_F < 0.4$ V. At higher bias, as a–Si:H layer is the highly resistive material ($\rho_B \cong 10^8 - 10^9$ Ω cm), the voltage drop across the quasi-neutral bulk region will become an ineligible part of applied voltage. Therefore, the forward characteristics will deviate increasingly with forward bias from the straight line (Figs.1 and 2). Now, if the R_S remain constant, in the investigated voltage domain, the J_F-V_F characteristics, even, at high forward current densities, will take the usual series resistance limited shape and can be fitted by:

$$J_F \cong J_0 \exp \left[\frac{q (V_F - V_B)}{n k T} \right] \{ 1 - \exp \left[\frac{q (V_F - V_B)}{n k T} \right] \} + \frac{(V_F - V_B)}{R_{sh}} . \qquad (3)$$

The J_0 is the saturation current density at $V_F = 0$ V, V_B the bulk voltage drop ($V_B = R_S I_F$) across the quasi-neutral region and n the ideality factor. In general, the exponential index (n) is slightly higher than unity and the observed deviation is associated with the carrier diffusion-recombination process via deep gap states present in both depletion and bulk region of the semiconductor. The q, k and T symbols have the usual meaning. This equation does not take into account the effect of conductivity modulation and may differ significantly from actual J_F-V_F characteristics under high-injection condition especially for the high-barrier Schottky diodes, for $\phi_n \geq 0.9$ eV. In such situation the multiplier in front of the exponential may not be connected directly with the true saturation current density and the exponential index will indicate what fraction of the applied voltage is across the depletion region itself; both values can be very different from their corresponding magnitudes at low levels of injection. The theoretical-experimental curves fitting using Eq. (3) is only true for the Schottky contacts with low-barrier heights, for $\phi_n \leq 0.9$ eV. In fact, as it can be seen from Fig. 1, for such metals as Mo, W, V, Ti and Rh the forward current density vary linearly with the voltage drop across the bulk region and the influence of the minority-carrier injection is negligible. The R_S is current or voltage independent and Eq. (3) describes well the J_F-V_F curves over large bias range; the R_S in some cases can be greater than the normal ohmic resistance of the semiconductor. To sum up, for $\phi_n \leq 0.9$ eV, at small forward bias, the diode current is dominate by recombination process in the depletion region; at moderate forward voltage, it is determinant by diffusion rates through the diode bulk; and at larger bias the bulk series resistance becomes evident. For the high-barrier heights, $\phi_n \geq 0.9$ eV, in which the conductivity modulation (decrease of the series resistance) due to the minority-carrier injection was observed, Figs. 2 and 3, the exact solution must be used; and the experimental curves can be fitted using the Finite Element Device Analysis Program (FIELDAY) specially developed at IBM for c–Si devices [12]. The conductivity modulation, in the present case, for such metals as Pt, Ni, Ir, Pd and Au is expressed by a curved line surpassing the straight line (ohmic regime) at lower forward current density, Figs. 2 and 4. This observed phenomenon can be better appreciated by developing the correlation between forward conductivity (σ_F) and forward current density:

$$\sigma_F \cong \frac{I_F (t - W)}{V_B A_c} . \qquad (4)$$

The results for Pt/a–Si:H diode, at different temperature, are plotted in a bilogarithmic plot in Fig. 4.

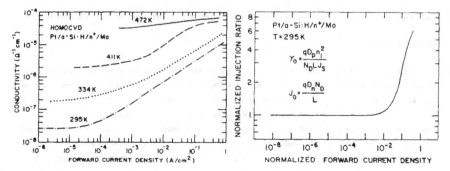

FIGURE 4. Forward conductivity increase in the quasi-neutral region of Pt/a-Si:H Schottky barrier diode exibiting the modulation effect, at different temperatures.

FIGURE 5. Normalized injection ratio (γ/γ_0) as a function of normalized forward current density (J_F/J_0) for Pt/a-Si:H / n^+/ Mo diode.

In this figure the dark forward conductivity is plotted as a function of forward current density. These curves can be obtained after a simple manipulation of the experimental data. If σ_F is constant with increasing J_F the σ_0 represents the unmodulated DC conductivity of the HOMOCVD a-Si:H layer. The deviation from σ_0 represents, here, the local modulated conductivity. The intensity of this modulation increases at lower temperature, e.g., at lower conductivity. Consequently, the current density marking the beginning of modulation is obviously bulk conductivity and Schottky barrier height dependent. The similar behavior has been observed for other high work function (Ir, Ni, Pd, etc.) metals.

The phenomenon of conductivity modulation can be explained by a high density of free holes in the n-type diode neutral zone. This holes density may rise to be about the same as of the electrons, and the both will increase about same amount with increasing current density for reason of quasi-neutrality. However, the conductivity modulation is not directly due, as it was already pointed out, to the holes density but to an extra internal field (E_{in}) associated with minority-carrier injection. This extra internal field which will compensate for the electron diffusion can be given by:

$$E_{in} \cong - \frac{k\,T\,(b-1)}{q\,(b\,n+p)} \; \nabla\, \delta n \; . \tag{5}$$

where b is μ_n/μ_p. It is evident from this equation that $E_{in} \cong 0$ if either there is no gradient concentration ($\delta n \cong \delta p$) of the carriers or if $\mu_n \cong \mu_p$ (or $D_p \cong D_n$). For the a-Si:H layer, where $\mu_n \neq \mu_p$, E_{in} will contributed to the total electric field present at the junction. An extra field in the bulk region is equivalent to a local series resistance decrease. Therefore, the E_{in} is due to the fact that the diffusion constant of electrons $(D_n \leq 1.3 \times 10^{-2}$ cm^2/ sec) and holes $(D_p \geq 2.6 \times 10^{-5}$ cm^2/ sec) are very different from each other. The electrons will diffuse faster than the holes and when the effective separation takes place between both types charge carriers, the E_{in} will appear in addition to the applied field in order to compensate for the electron diffusion; because the current density has to be constant everywhere. The E_{in} can become eventually so strong that the both charge carriers will diffuse together. This phenomenon is called ambipolar diffusion which plays important role in the Schottky barrier diodes. Because of increase of the intensity of modulation with minority-carrier injection, consequently, the modulation conductivity is favored by large-barrier heights, by low doping level and by low mobility materials. This is in full accordance with our experimental observations.

The injection ratio of minority-carrier (γ_p) (defined as the ratio of minority carrier current to total measured current) in the low bias forward voltage, initially developed for c-Si devices [13 – 15], can be given by the following relationship:

$$\gamma_p \cong \frac{N_v L_p}{t N_{eff}} \exp \left[q(\phi_n - E_g)/k T \right] . \tag{6}$$

This equation takes into account only the diffusion component of the hole current density. Therefore, below the critical current density value of $8 \times 10^{-2} A/cm^2$ the γ_p is virtually constant for different metals, Fig. 5. Above this value for larger forward bias the γ_p follows approximatevily the forward current density (or electric field) according to:

$$\gamma_p \cong \frac{N_v L_p^2 J_F}{q b t D_p N_{eff}^2} \exp \left[q(\phi_n - E_g)/k T \right] . \tag{7}$$

The increase of γ_p ,at larger bias voltage, is due to the existence of the internal electric-field in the quasi-neutral region, which gives rise to a significantly larger drift-field than the diffusion current component observed at lower bias voltage. The both regimes for Pt/a-Si:H diode are shown in Fig. 5, where the normalized γ_p is plotted as a function of the normalized diode current density. The normalization has been done in a way similar as for the c-Si Schottky diode [13]. In a-Si:H devices the γ_p can be very large for $\phi_n \geq 0.9$ eV and the minority-carrier injection can be the major contribution to the dark current under forward bias voltage. This is because the minority-carrier injection is greatly enhanced over the c-Si as a result of reduction in carriers mobilities. Therefore, to reduce the minority-carrier injection ratio or modulation conductivity one must use a metal/a-Si:H system with high doping level (corresponding to high conductivity material), large saturation current (corresponding to low-barrier height, ϕ_n must be lower than 0.9 eV) and small n_i (corresponding to large band gap).

In general, for metal/HOMOCVD a-Si:H contacts the space-charge capacitance modified for the deep gap states predominates in the reverse and low forward polarization displaying a depletion mode characteristics. Indeed, in this bias region the minority-carriers cannot be generated sufficiently rapidly to maintain an inversion layer. However, at higher forward voltage the situation can be different. The capacitance can break-off at certain voltage from depletion mode behavior and reverts again to the normal behavior at higher voltage. For example for W/a-Si:H the break-off happens around 0.3 V and reversion at 1.2 V; and the maximum in capacitance is achieved at flat-band voltage. The decrease in capacitance, which follows, is due to the fact that with increasing V_F a larger and larger portion of applied voltage appears across the bulk quasi-neutral region of device. If minority-carrier effects are neglected, for $\phi_n \leq 0.9$ eV, the C_m will approached the diode geometrical capacitance. Eventually, the capacitance can increase again at higher forward bias due to the presence of very thin interfacial native oxide layer or the influence of the interface states. The similar situation was observed for metal/c-Si diodes [11]. Therefore, the shape of the humps and its position will depend on the bulk resistivity of a-Si:H, on the nature of the metal contacts and the oxide layer thickness. In fact, the amplitude of the humps increases with decreasing oxide thickness layer and decreases with increasing metal work function. The region between 0.2 and 1.2 V corresponds to the transition from semiconductor-limited to tunnel-limited behavior. Now, if the minority carriers effects can not be neglected and when the injected carrier concentration is not small compared to the equilibrium concentration, for example for $\phi_n \geq 0.9$ eV, the drift and diffusion processes interact, and the diode capacitance becomes negative, e.g., an inductance is developed in the bulk region at higher forward voltage. The inductance can be represented, here, as an equivalent capacity to: $C_m = (\omega^2 L)^{-1}$. The inductive contribution observed under positive polarization can be associated with the conductivity modulation of series resistance caused by the hole injection; the inductive susceptance is proportional to the rate at which the bulk conductivity will change under injection condition. Indeed, the modulation of the bulk conductivity requires some time and, hence, it is delayed with respect to the variation of the applied voltage. Therefore, the impedance of the a-Si:H diode has to be inductive as it is experimentaly demonstrated. Consequently, in order to get inductance the diode must be made in such a way that a part of the applied voltage appears across the bulk and a part across the barrier region. The similar effect can be also observed in P–N and P–I–N a-Si:H junctions, but does not become important until the $p \cong N_D$ or $n \cong N_A$ in the bulk regions of the device. For a-Si:H Schottky diodes the inductive effect can be dominant at minority carrier densities which can be several orders of magnitude smaller. The inversion voltage (V_{inv}, $V_F = V_{inv}$ when $C_m = 0$) at which C_m becomes inductive is temperature and frequency dependent; V_{inv} decreases linearly with increasing temperature and increases logarithmically with increasing frequency. At the same time the maxima in the measured capacitance decreases with increasing frequency or increases with increasing temperature. Consequently, the inductive effect can be eliminated at higher frequencies or lower temperatures, Fig. 6.

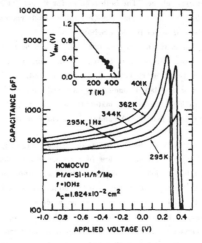

FIGURE 6. Capacitance characteristics, at different temperatures, of Pt/a-Si:H/ n[+] / Mo contact; the insert shows the variation of inversion voltage with the temperature.

So, the inductive contribution seems to be strongly influenced by the density of deep gap recombination centers at the edge of the space-charge region. The lower density of these centers will be, e.g., at higher temperatures or under forward polarization, the larger positive capacitance peak will be and the lower the bias at which the device will turn inductive, Fig. 6. In addition, because of the presence of the inversion layer at the interface introduced by minority-carrier injection the $C_m^{-2} - V_a$ plot cannot be used to obtain the absolute values of barrier heights. In fact, the measured barrier height will be less than the actual barrier height; the largest calculated barrier heights from $C_m^{-2} - V_a$ plot are not obtained when the actual barrier are largest. This can also explain why ϕ_n is not very sensitive to the variation of the metal work function [16]. The additional factor that will also influence the AC characteristics is the high diode series resistance. Indeed, it has been demonstrated for c-Si devices [10,17] that the capacitance maximum can be observed when the differential resistance approaches the diode series resistance . This is exactly what happens for HOMOCVD a-Si:H Schottky contacts at about the flat band voltage [5]. This is why, even, if the minority-carrier injection can be neglected in some cases the presence of high R_S causes the maximum to occur in the total measured capacitance.

In conclusion, the influence of both or either series resistance and minority-carrier injection cannot be neglected in AC and DC measurements of metal/a-Si:H contacts properties. Specially, if the accurate values of fundamental physical parameters are needed in order to fabricate the reliable, predictable and reproducible a-Si:H devices.

REFERENCES

1. Hydrogenated Amorphous Silicon, in Semiconductors and Semimetals, vol. 21, Part A-D, ed. J. I. Pankove, Academic Press Inc., New York, 1984.

2. B. A. Scott, R. M. Plecenick and E. E. Simonyi, Appl. Phys. Lett. 39 , 73 (1981).

3. Y. Kanai, J. Phys. Soc. Japan 10 , 718 (1955).

4. T. Misawa, J. Phys. Soc. Japan 12 , 882 (1957).

5. J. Kanicki, M. Osama Aboelfotoh and W. Bauhofer, Proc. Int. Conf. Phys. Sem., San Francisco, USA, August 6-10, 1984.

6. I. Chen and S. Lee, Appl. Phys. Lett. 40 , 487 (1982).

7. I. W. Archibald and R. A. Abram, Phil. Mag. B 48 , 111 (1984).

8. E. H. Rhoderick, Metal-Semiconductor Contacts, Clarendon Press, Oxford, 1980.

9. J. R. MacDonald, Solid-St. Electron. 5 , 11 (1962).

10. O. V. Konstantinov and O. A. Mezrin, Sov. Phys. Semicond. 17 , 193 (1983).

11. M. A. Green and J. Shewchum, J. Appl. Phys. 46 , 5185 (1975).

12. E. M. Buturla and P. E. Cottrell, FIELDAY-Finite Element Device Analysis Program, IBM TR 19.0356 (1975).

13. D. L. Scharfetter, Solid-St. Electron. 8 , 299 (1965).

14. A. Y. C. Yu and E. Snow, Solid-St. Electron. 12 , 155 (1969).

15. J. C. Manifacier and H. K. Henish, Phys. Rev. B 17 , 2640 (1978).

16. J. Kanicki, C. M. Ramson, W. Bauhofer, T. I. Chappell and B. A. Scott, J. Non-Cryst. Solids 66 , 51 (1984).

17. V. I. Stafeev, Sov. Phys.-Tech. Phys. 3 , 1502 (1958).

PHOTOELECTRONIC PROPERTIES OF AMORPHOUS SILICON/SILICON OXIDE HETEROSTRUCTURES

F. Carasco, J. Mort, F. Jansen and S. Grammatica
Xerox Webster Research Center, Webster, NY 14580

A glow-discharge deposited a-Si:H/insulator heterostructure has been characterized by a range of measurements including optical absorption, internal photoemission, xerographic discharge and spectral dependence of photoconductivity. Efficient injection of photocarriers from a-Si:H into, and transport through, films of $SiO_x:N:H$ up to 10 μm thick has been achieved. Unlike the conventional thermal oxide on Si, no significant energy barrier to injection is found in the plasma deposited heterostructure. The use of the structure as a potential xerographic device is demonstrated. A mobility lifetime product as high as 6 x 10^{-10} cm^2/volt is found for electrons in the $SiO_x:N:H$.

INTRODUCTION

There is immense current interest in glow-discharge deposited hydrogenated amorphous silicon (a-Si:H), because of the ability to efficiently dope it [1], for possible applications in a range of technologies such as solar photovoltaics, thin film electronics and electrophotography [2]. For some of these applications, heterolayers of a-Si:H and wide bandgap insulators such as Si_3N_4, SiO_2, etc., are required. For optimum efficiency of fabrication, it is clearly desirable to produce these materials by the glow discharge process. Films produced by the glow-discharge deposition of mixtures of silane and ammonia or nitrous oxide are properly described as $SiN_x:H$ or $SiO_x:N:H$ respectively and as such can be expected to differ significantly from Si_3N_4 or SiO_2 produced by CVD or thermal oxidation. In general, of course, they do share common properties with their stoichiometric counterparts, if the NH_3/SiH_4 or N_2O/SiH_4 ratios are made sufficiently high. Under such circumstances, highly insulating materials with wide bandgaps are obtained. Significant studies on these glow discharge materials have been reported and they have been shown to function adequately in thin film transistors [2]. Little in detail, however, is known about the charge transport properties of these materials particularly as a function of composition.

One of the most widely used techniques in the study of the electronic structure and transport properties in wide gap materials is that of internal photoemission [3,4]. In SiO_2, for example, the band gap between extended states of ~8 eV precludes the use of intrinsic photoconductivity. Instead, the injection of photoexcited carriers in metals or photogenerated carriers in the contiguous crystalline silicon has been used to determine the electronic energy levels and subsequent charge transport in SiO_2.

Studies of a-Si:H/$SiO_x:N:H$ heterolayers have been recently described [5]. The work had several motivations. The primary one was to explore the potential value of such layered structures as electrophotographic devices. Hydrogenated amorphous silicon itself is being extensively studied for such applications [6]. However, a bulk a-Si:H photoreceptor, which must be ~20 μm thick, possesses intrinsic limitations. The first is the fact that the thermal generation of carriers in the dark of such devices is higher than desirable for applications at lower process speeds. This is compounded, because of the required smaller bandgap material, if spectral extension to solid state GaAs laser wavelengths is contemplated. Therefore, significant advantages would accrue to a much higher resistivity device. Such a device could in principle comprise of a relatively thick highly insulating wide bandgap material such as SiO_x:N:H in conjunction with a thin spectrally sensitizing layer such as a-Si:H or a-Si:Ge:H. Thus, the resistivity of the device would be determined by the thicker SiO_x:N:H layer, but the photoelectronic properties would be determined by the sensitizing layer. For such a structure to work as a useful photoreceptor, the carriers photogenerated in the sensitizing layer must be injected with high efficiency into and then traverse the SiO_x:N:H layer without significant loss due to deep trapping. Based on the conventional view of the relative positions of the conduction bands and valence bands of the component layers, the possibility of achieving unity injection efficiency would be zero. This dilemma can be conceptually resolved as shown in Figure 1. Here it is assumed, in the simplest picture, that a discrete level of deep lying states lies isoenergetically with, e.g., the conduction band of a-Si:H. High injection efficiency and transport could then occur if the density of these states were, or could be made, high enough to produce sufficient wave function overlap for defect hopping transport of the injected carriers to occur. In concept, this is similar to the molecular doping of polymer matrices to render otherwise inactive polymers electronically useful by using isolated molecules as localized states [7]. In general, the probability of a single discrete level lying isoenergetic to either conduction or valence band in a-Si:H is remote. A more realistic picture is the presence of a continuum of states as indicated by the dashed line in Figure 1.

In this paper we report the progress to date in exploring this concept. After describing the experimental procedure, results are presented on optical absorption, internal photoemission, xerographic discharge measurements and spectral dependence of photoconductivity. Finally, a discussion of these results in terms of the concept just outlined and the implications for the optimum functioning of devices, such as xerographic photoreceptors, is presented.

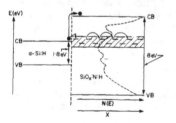

Fig. 1. Schematic diagram of the injection process from a-Si:H into a discrete defect level in a-SiO_x:N:H or a distribution of defect states within the gap. The horizontal axis denotes either spatial extent or density of states distribution as a function of energy.

EXPERIMENTAL DETAILS

Sample preparation

Films of a-SiO_x:N:H and dual layers of a-Si:H and a-SiO_x:N:H were prepared in a plasma deposition system in which the rf power (13.56 MHz) was capacitively coupled to one electrode. The compositional variation of the oxide was achieved by controlling the ratio of the flow rates of N_2O and SiH_4. Samples with gas ratios ranging from 5:1 to 32:1 were studied. The notation used corresponds to the ratio of N_2O/SiH_4 gas flow rates. The total pressure during deposition was maintained at 500 mTorr, the substrate temperature at 250 °C and the power density at 0.02 W/cm^2 to yield deposition rates of about 2 μm/hr. Film were typically 0.5 to 10 μm thick and were deposited on a variety of substrates including UV grade quartz, aluminum and stainless steel.

The study of the heterolayers was carried out primarily on samples in which the $SiO_x:N:H$ was overlayed with a-Si:H of 0.5 μm thickness. The electroded measurements required thin thermally evaporated Au (150A) films to be deposited on the samples.

Measurement techniques

The photoconductivity/photoemission studies employed light from a 1000 W xenon lamp which was passed through a ISA HR-320 grating monochromator controlled by a computer via an ISA Spectra Link. Appropriate grating and filters were used, depending on the energy range being investigated. The dc measurements were made with a HP 4140B pA/dc voltage source, with sufficient time being allowed for a quasi-steady state dark level to be reached before commencing a wavelength scan. The data were taken point by point under steady illumination with sufficient time being allowed for stabilization at each photon energy.

The xerographic discharge measurements were made by conventional methods previously described [8]. The films were charged to an initial voltage of appropriate polarity using a corotron charging device. The surface potential was monitored with a thin wire loop probe coupled directly to an electrometer whose output was fed to a high speed recorder. The films were illuminated through the loop with monochromatic light of known intensity to produce the photoinduced discharge. Light of 4000A which is strongly absorbed in a-Si:H was mainly employed.

Figure 2 indicates the basic features of the xerographic discharge technique. The surface of sample thickness L, Figure 2(a), is uniformly exposed to light of constant intensity, I_0, the absorption depth being much less than L. In Figure 2(b), the film is charged to an initial voltage and in the absence of illumination exhibits dark decay which may be determined either by bulk-generated dark carriers or, in the case of dual layer structures, by injection of carriers from the contiguous layer. On exposure, photodischarge occurs from the initial voltage, V_0. For an emission-limited discharge, a condition satisfied at low light intensities, with negligible trapping, i.e., $\mu\tau E \gg L$, the initial rate of discharge is given by

$$| \, dV/dt \, |_{t=0} = eF\eta/C \qquad (1)$$

where e is the elemental charge, F is the absorbed photon flux, C is the capacitance of the photoreceptor, and η is the supply of carriers out of the excitation region per absorbed photon, i.e., xerographic gain. The latter may include any limitations on the carrier injection between the generator and the transport layer. Thus, it is possible to determine η absolutely without any specific knowledge about the carrier transport. As the discharge progresses, the field decreases and $\mu\tau E$ can become comparable or even smaller than L. Then the accumulation of trapped charge becomes significant enough to diminish the discharge rate, resulting in a finite residual voltage V_r. It has been shown by Kanazawa and Batra [9] that the ratio of this residual voltage to the initial voltage, V_r/V_0, is independent of the injection rate and is solely determined by the ratio of the initial drift length to the thickness of the transport layer, $\mu\tau V_0/L^2$. Therefore, using the universal curve relating V_r/V_0 to $\mu\tau V_0/L^2$ the carrier range $\mu\tau$ can be determined from the observed residual voltage.

EXPERIMENTAL RESULTS

In Figure 3 the optical absorption of samples made with N_2O/SiH_4 gas ratios of 5:1 and 12:1 are shown. No measurable absorption was observed in the same energy range for a sample of the same thickness (4.5 μm) prepared with a gas ratio of 20:1. Given the observed absorption edges are fairly broad and featureless, it is only possible to get a rough estimate of the band gap. However, it is clear that for SiO_x:N:H made with gas ratios >5:1 they are very large. By comparing this data to that of Holzenkampfer et al., [10] who investigated the variation of the absorption coefficient with oxygen content of SiO_x:N:H films, one can estimate that x is within 10% of the value 2 for these samples. This was also confirmed by Auger spectroscopy.

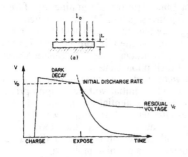

Fig. 2. Basic features of the xerographic discharge technique. a) Sample of thickness L exposed to a light flux of intensity of I_o. b) Schematic discharge curves with and without range limitations.

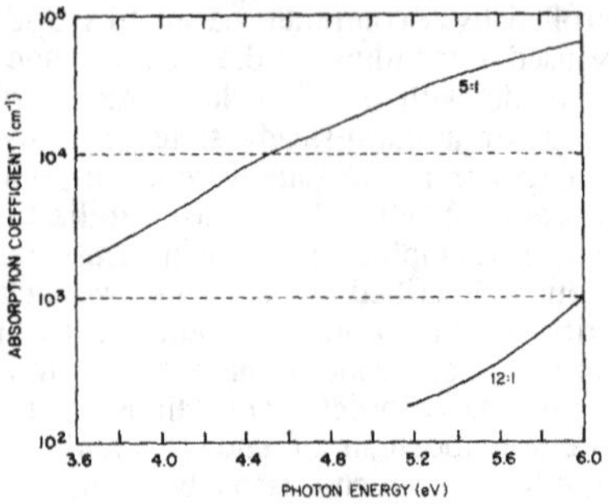

Fig. 3. Plot of absorption coefficient vs. energy for SiO_x:N:H prepared with 5:1 and 12:1 N_2O/SiH_4 gas ratios.

Figure 4 shows the square root of photoresponse against light energy for the structure Au/SiO_x:N:H/Al with the Au biased negatively for 20:1 gas ratio SiO_x:N:H. Also shown in the same figure is the result with Au biased positively. The choice of plotting the square root of photoresponse is based on a conventional Fowler theory as described by Goodman [4]. It is clear that for Au biased negatively there is an additional photoresponse to the background. This is tentatively attributed to internal photoemission of electrons from the gold to the conduction band of SiO_x:N:H. It was found that the relative energy position of ~ 4.7eV, between the Fermi level of gold and the conduction band of the SiO_x:N:H, measured by this method does not change significantly with gas ratio. This is so even though it is known from the absorption measurements that the optical gap is changing.

Figure 5 shows a typical xerographic discharge for a heterolayer with 12:1 SiO_x:N:H. As described earlier, the ratio of the residual to the initial voltage, V_r/V_o, can be used to determine the carrier range $\mu\tau$. The variations of V_r/V_o and $\mu\tau$ for samples of different gas ratios are shown in Figure 6. All these samples had 0.5 μm of a-Si:H on 1.0 μm of SiO_x:N:H. (Note the initial voltage refers to that across the oxide layer only). In terms of the minimum V_r/V_o and the maximum $\mu\tau$, the optimum gas ratio is 12:1. The maximum $\mu\tau$ value is 6 x 10^{-10} cm^2/volt. A self consistency check was done by deducing the $\mu\tau$ of samples with the same SiO_x:N:H compositions but with thickness of 1, 5 and 10 μm. The values were found to be within a factor 3. The xerographic discharge data shown in Figure 5 indicate a significant photodischarge effect. Following the discussion earlier, Equation 1 can be used to obtain a value of η, which is the supply of carriers out of a-Si:H. It was found that for light of wavelength 4000 Å, at a field of 8 x 10^5 volt/cm and for negative charging, $\eta > 0.6$ was routinely achievable on structures with one micron of $SiO_x:N:H$.

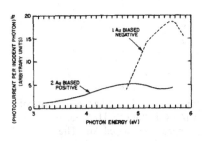

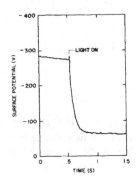

Fig. 4. Square root of normalized photo-response of Au/SiO_x:N:H/Al vs. photon energy. For curve 1, the Au is biased negatively while for curve 2 it is biased positively. The gas ratio for the SiO_x:N:H is 20:1.

Fig. 5. A typical discharge curve for a 0.5 μm a-Si:H on 5.0 μm SiO_x:N:H with negative charging. Light of wavelength 4000 Å was used.

Figure 7 shows the spectral dependence of injection of Au/a-Si:H/SiO_x:N:H/Nesa for the Au, and therefore a-Si:H, biased negatively with a field of 7 x 10^5 volt/cm. It is clear that the structure has the same spectral photoresponse as a-Si:H. This is in sharp contrast with the crystalline analog or a-Si:H with thermally grown oxide where one observes a threshold of about 4.0 eV. Further, the efficiency of photoresponse which was of the order of 0.1 for the sample shown in Figure 7 is at least several orders higher than that obtained with a thermally grown oxide.

DISCUSSION

Yamamoto et. al. [11] showed that for thermally grown SiO_2 (250 A) on a-Si:H, there is a energy threshold of about 4.0 eV below which photoresponse is not observed. In the crystalline system with thermal and /or CVD SiO_2/Si, over which a metal or metal/CVD Si rich SiO_2 [12] is deposited, one always observes a similar threshold. The widely accepted explanation is that photoresponse is only observed when the photon energy is larger than the energy difference between the metal Fermi level and the CB of the SiO_2 (metal biased negative) or the VB of the semiconductor and the CB of the SiO_2 (metal biased positive). In all situations the barrier was never less than 3.0 eV. Figure 7 shows that a similar threshold is not

116

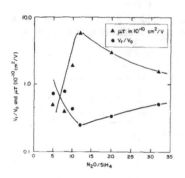

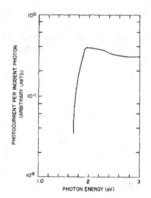

Fig. 6. The variation of V_r/V_0 and $\mu\tau$ for different gas ratios used during the preparations of the SiO_x:N:H in the heterolayers for negative charging.

Fig. 7. Plot of the spectral response of Au/a-Si:H/SiO$_x$:N:H/Nesa, for negatively biased Au. The SiO$_x$:N:H was made with 10:1 gas ratio.

observed for the heterostructures in this study. In fact, the spectral response is essentially the same as that of the semiconductor (a-Si:H). This observation is consistent with the high injection efficiencies of 0.6, as compared with injection efficiencies of $<10^{-4}$ for systems involving crystalline silicon and thermally grown oxides. Further, while the Au/SiO$_x$:N:H/metal internal photoemission experiments are not a direct analog to the photoemission experiments on SiO$_2$/Si they are interpretable in a similar fashion. The photoemission photon energy threshold observed, taken together with the optical absorption measurements, indicate that the SiO$_x$:N:H is basically similar in its main band structure to thermal or CVD SiO$_2$. Based on the 4.7 eV energy barrier between the Fermi level of gold and the extended states of SiO$_x$:N:H, injection efficiencies of ~0.6 for injection of electrons from a-Si:H into a-SiO$_x$:N:H are inconceivable with a process involving the extended states of SiO$_x$:N:H. Therefore, it is supposed that the steady state photoresponse and xerographic discharge in the heterolayers is due to injection of electrons into a transport path in the SiO$_x$:N:H lying below the CB. This transport occurs via hopping between localized states in the band gap of a-SiO$_x$:N:H whose density, and therefore wavefunction overlap, is controlled by the N$_2$O/SiH$_4$ gas ratio.

The results discussed have a number of technological ramifications. The primary observation is that high injection efficiency of carriers from a panchromatic photoconductor into and transport through a transparent, highly insulating, wide band gap material can be achieved by utilizing hopping states in the gap. This enlarges the range of materials and device properties available for potential applications as amorphous tetrahedral xerographic photoreceptors. Thus, extended infrared sensitivity, lower dark decay and utilization of thinner device structures, giving improved device performance, increased process latitude and decreased production costs may be envisaged. For practical device applications, further improvement of the $\mu\tau$ products of the injected carriers within the insulator is desirable. Such improvements may be sought by improved defect control, impurity doping, process variations or by alternative insulator materials. Realization of these objectives obviously require increased knowledge and understanding of the transport properties in materials such as SiO$_x$:N:H produced by glow discharge.

References

1. W.E. Spear and P.G. LeComber, Phil. Mag., **33**, 735 (1976).
2. See for example: Proc. of 10th Int. Conf. on Amor. and Liq. Sem., K. Tanaka and T. Shimizu, eds., J. of Non-Cryst. Solids, **59-60**, 1 (1983).
3. R. Williams, Phys. Rev. **140**, A569 (1965).
4. A.M. Goodman, Phys. Rev. **144**, 588 (1966).
5. F. Carasco, J. Mort, F. Jansen and S. Grammatica, J. Appl. Phys., May(1985)
6. J. Mort, S. Grammatica, J.C. Knights and R. Lujan, Photo. Sc. and Eng., **24**, 241 (1980).
7. J. Mort, G. Pfister and S. Grammatica, Solid State Comm., **18**, 693 (1976).
8. J. Mort and I. Chen, Appl. Solid State Science, **5**, 69(1975).
9. K.K. Kanazawa and I.P. Batra, J. Appl. Phys., **43**, 1845 (1972).
10. E. Holzenkämpfer, F.W. Richter, J. Stuke and U. Voget-Grote, J. of Non-Cryst. Solids, **32**, 327 (1979).
11. T. Yamamoto, Y. Mishima, M. Hirose and Y. Osaka, Jap. J. Appl. Phys., Supplement **20-2**, 185 (1981).
12. D.J. DiMaria and D.W. Dong, J. Appl. Phys. **51**, 2722 (1980).

Multilayered Structures

AMORPHOUS SEMICONDUCTOR MULTILAYER STRUCTURES:
INTERFACE AND LAYER THICKNESS EFFECTS IN PHOTOLUMINESCENCE

T. Tiedje
Corporate Research Laboratory
Exxon Research and Engineering Co.
Annandale, N.J. 08801

BACKGROUND

A number of new developments have occurred recently in research on the synthesis and properties of amorphous semiconductor multilayer structures ("amorphous superlattices") since the discovery of this class of materials in 1983.[1] This and more recent work have shown that tetrahedrally bonded amorphous semiconductors can be fabricated in the form of multilayer structures, with highly uniform layers and atomically abrupt interfaces. The remarkably high degree of structural perfection in these materials on the length scale of the superlattice period (> 5A) has been demonstrated by transmission electron microscopy.[2,3]

The class of materials that can be fabricated into superlattice structures has been expanded[4,5,6] from the original a-SiN:H/a-Si:H materials to a-Si:H/a-Ge:H, a-SiO:H/a-Si:H and a-SiC:H/a-Si:H, as well as to alternately doped a-Si:H multilayers.[7,8] The abruptness of the interfaces and the uniformity of the layers make it possible to observe a number of interesting phenomena not normally observed in conventional homogeneous amorphous semiconductors.

For example, the blue shift in the optical gap originally observed in single ultrathin layers of a-Si:H and attributed to quantum size effects,[9] has now been reported in a variety of compositionally modulated structures in which the thickness of the small band gap sublayer is less than about 40A.[4,6] Other effects that have been observed include transfer doping of a-Si:H by gap states in a-SiN:H,[10] persistent photoconductivity in modulated doping structures,[7] large internal electric fields generated by asymmetric interface charges,[11] and enhanced photoluminescence.[12]

PHOTOLUMINESCENCE EFFICIENCY

In this paper, as an illustration of some of the dimensional and interface-related effects that are observed in these materials, we describe the sublayer thickness dependence of the low temperature photoluminescence (PL) in two different types of multilayered structures, namely a-Si:H layers alternating with wide bandgap layers (a-SiN:H) and a-Si:H alternating with a smaller bandgap material (a-Ge:H). The latter class of materials can be studied from the point of view of a-Ge:H encapsulated with a-Si:H or as a-Si:H encapsulated with a-Ge:H. In other words the thickness dependence of the PL can be studied for the two layers independently in the same series of samples since the emission spectra for the two different sublayers are well separated in energy.[12]

The thickness dependence of the PL efficiency measured at 10K on a series of a-Si:H/a-Ge:H multilayer samples is shown in Fig. 1 for the a-

Ge:H and the a-Si:H layers independently. The PL efficiency for the thick a-Si:H layers in Fig. 1 is low compared to the values normally observed in high quality bulk material due to absorption of both the excitation light and the emitted light in the a-Ge;H layers. Notice also that the efficiency drops rapidly to zero for a-Si:H layers thinner than about 200A. Qualitatively this last result is not surprising since photoexcited electron-hole pairs will have a tendency to escape into the neighboring lower bandgap a-Ge:H layers if the a-Si:H layer is too thin.

On the same graph in Fig. 1 we also show the corresponding thickness dependence of the efficiency for the a-Ge:H layers in the same series of samples. In this case, however, the efficiency increases dramatically with decreasing layer thickness.

Finally in Fig. 2 we show the thickness dependence of the PL efficiency for a-Si:H layers in a-Si:H/a-SiN:H multilayer structures. In this case, beginning at a thickness of about 100A the PL efficiency drops by about an order of magnitude with decreasing layer thickness and saturates at the lower value with further reduction in the layer thickness.

DISCUSSION

The thickness dependence of the efficiency in Figs. 1 and 2 can be understood with reference to Fig. 3. In this figure we show schematically an a-Ge:H layer sandwiched between a-Si:H layers, a-Si:H between a-SiN:H layers and a-Si:H between a-Ge:H layers. The x's in these figures are meant to represent dangling bonds or other non-radiative centers. The shaded areas represent regions of zero quantum efficiency and unshaded areas regions of unit quantum efficiency.

For example, in Fig. 3a the continuous shaded bands in the silicon layer next to the germanium interfaces represent the regions where the electron-hole pairs are more likely to escape into the germanium than recombine radiatively in the silicon layer. If these have a thickness R_c then the net a-Si:H layer PL efficiency is,

$$\eta = \eta_0 \ \frac{L-2R_c}{t}$$ (1)

where L is the a-Si:H layer thickness. The broken line through the a-Si:H data in Fig. 1 is a fit of Eq. (1) to the data with $R_c = 80A$. In the a-Si:H/a-Ge:H multilayers the entire interface represents a non-radiative center as far as the a-Si:H layers are concerned.

In the silicon/silicon nitride system only isolated defects at the interfaces act as non-radiative centers as indicated schematically in Fig. 2b. In this case if the layer thickness L is less than the capture radius R_c of the non-radiative centers then the PL efficiency is

$$\eta = \eta_0 \ \exp(-\pi R_c^2 N_s)$$ (2)

where η_0 is the efficiency of bulk a-Si:H and N_s is the density of non-radiative centers at the interfaces. The solid line in Fig. 2 is the efficiency calculated numerically from a more general version of Eq. (2) valid without restriction on the relative value of L and R_c.[13] The fit shown by the solid line in Fig. 2 corresponds to $R_c = 70A$, $N_s = 1.5 \times 10^{12}$ cm^{-2} and assumes that all of the non-radiative centers are at one of the interfaces.

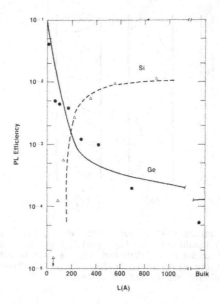

Fig. 1   Photoluminescence efficiency for
a-Si:H and a-Ge:H sublayers in a-Si:H/
a-Ge:H multilayer structures as a function
of the individual layer thicknesses. The
solid and broken lines are theoretical
fits as discussed in the text.

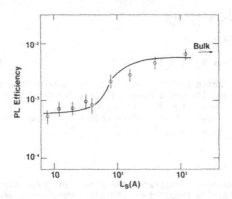

Fig. 2   Photoluminescence efficiency
for a-Si:H/a-SiN:H superlattices as a
function of the a-Si:H sublayer thickness.

124

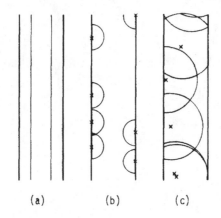

(a)   (b)   (c)

Fig. 3   Schematic diagrams shwoing non-radiative regions (shaded)
and radiative regions (unshaded) for (a) a-Si:H layer bounded by
a-Ge:H (b) a-Si:H bounded by a-SiN:H and (c) a-Ge:H bounded by
a-Si:H. The diagrams are drawn to scale for a layer thickness
of 260A. The x's indicate non-radiative centers.

The PL efficiency in the a-Ge:H layers is limited by bulk defects
within the layer and not by interface effects, as illustrated in Fig.
3c. It follows that the efficiency of thin layers is greater than for
bulk material because of a geometrical phase space effect when L < 2 R_C
. Many of the centers that would cause non-radiative recombination at a
given position in bulk material are located in positions that would be
outside the layer in a thin slice of material and hence are excluded in
the thin layer. It follows that the PL efficiency increases from

$$\eta = \eta_0 \exp \left(-\frac{4}{3} \pi R_c^3 N_b\right) \qquad (3)$$

for bulk material[14] to,

$$\eta = \eta_0 \exp \left(- \pi R_c^2 L N_b\right) \qquad (4)$$

for thin layers in which L < 2R_c. A fit to the a-Ge:H PL efficiency
based on Eqs. (3) and (4) and a numerical calculation[12] of the
intermediate regime where L ~ 2R_c, is shown in Fig. 1. The solid line
in this figure corresponds to a capture radius R_c = 200A, and a bulk
density of defects N_b = 2.2 x10^{17} cm-3.

SUMMARY

In summary, we have illustrated the effect of three different types
of interfaces on the low temperature photoluminescence in amorphous
semiconductor superlattice structures. The three different types of
interfaces, namely collecting, defective and reflecting, lead to three
different forms for the thickness dependence of the PL efficiency, and
in particular to a large increase in the efficiency for a-Ge:H layers
capped with a-Si:H. These PL results are one example of the wealth of
phenomena and the new opportunities for understanding the basic physics
of disordered semiconductors that are available in amorphous
semiconductor superlattice structures.

REFERENCES

1. B. Abeles, T. Tiedje, Phys. Rev. Lett. 51, 2003 (1983).
2. H. W. Deckman, J. H. Dunsmuir, B. Abeles, Appl. Phys. Lett. 46, 171 (1985).
3. R. Cheng, S. Wen, J. Feng, M. Fritzsche Appl. Phys. Lett. 46, 592 (1985).
4. T. Tiedje, B. Abeles, P. D. Persans, B. G. Brooks, G. D. Cody, J. Non-Cryst. Solids 66, 345 (1984).
5. B. Abeles, T. Tiedje, K. Liang, H. W. Deckman, H. E. Stasiewski, J. C. Scanlon, P. M. Eisenberger, J. Non Cryst. Solids 66, 351 (1984).
6. M. Hirose, S. Miyazaki, J. Non Cryst. Solids 66, 327 (1984).
7. J. Kakalios, H. Fritzsche, Phys. Rev. Lett. 53, 1602 (1984).
8. M. Hundhausen, L. Ley, R. Carius, Phys. Rev. Lett. 53, 1598 (1984).
9. H. Munekata, H. Kukimoto Jpn. J. Appl. Phys. 22, L544 (1983).
10. T. Tiedje, B. Abeles, Appl. Phys. Lett. 45, 179 (1984).
11. C. B. Roxlo, B. Abeles, T. Tiedje, Phys. Rev. Lett. 52, 1994 (1984).
12. T. Tiedje, B. Abeles, B. G. Brooks (to be published).
13. T. Tiedje, B. Abeles, B. G. Brooks, AIP Conf. Proc. 120, 417 (1984).
14. R. A. Street, Semiconductors and Semimetals vol. 21B ed. J. Pankove, p. 197 (1984).

PHOTO-INDUCED EXCESS CONDUCTIVITY IN DOPING MODULATED AMORPHOUS SEMICONDUCTORS

J. KAKALIOS and H. FRITZSCHE
The Department of Physics and the James Franck Institute
The University of Chicago, Chicago, IL 60637

ABSTRACT

The metastable excess conductivity $\sigma(E)$ observed in hydrogenated amorphous silicon (a-Si:H), that is alternately doped n- and p- type, is compared with the Staebler-Wronski effect and other metastable conductivity changes observed in compensated a-Si:H and in oxidized p- type a-Si:H respectively. We find that Döhler's model of electron-hole pair separation in the pn-junction fields cannot account for the long life of $\sigma(E)$ near and above 300°K. A defect complex associated with boron having a large configurational relaxation after releasing an electron by photoexcitation is considered as an explanation for $\sigma(E)$.

INTRODUCTION

Recently the synthesis of doping modulated amorphous semiconductors consisting of hydrogenated amorphous silicon (a-Si:H) that is alternately doped n- type and p- type has been reported.[1-6] These multilayers are the amorphous analogs of modulation-doped superlattices such as GaAs nipi crystals.[7] One of the new properties of our doping modulated a-Si:H films[1-4] is a photo-induced enhancement of up to two orders of magnitude in the in-plane dark conductivity which decays on a time scale of days at room temperature. This photo-induced excess conductivity is reversible upon annealing above 400°K.

In this paper we compare and contrast this phenomenon with other photo-induced metastable conductivity changes observed in a-Si:H. These include the Staebler-Wronski effect[8] which is attributed to a photo-induced creation of dangling bond defects, the photo-induced excess conductivity observed in compensated a-Si:H,[9] and the photo-induced enhancement of the space charge conductance[10] near the oxidized surface of p- type a-Si:H. We also discuss Döhler's model[7] which predicted the presence of a long-lived excess conductivity in doping modulated semiconductors.

EXPERIMENTAL DETAILS AND RESULTS

The doping modulated a-Si:H films were grown[1-4] on 2.5X1.25 cm² large Corning 7059 glass substrates which were clamped on a flat section of a stainless steel ball that was kept at 220°C. This ball could be rotated so that the substrates faced alternately a plasma reactor containing 0.05 Torr of SiH_4 and 0.20 Torr of Ar with 100 ppm PH_3 for growing n- type a-Si:H via plasma enhanced chemical vapor deposition, or a second plasma reactor containing the same ratio of SiH_4 and Ar with 100 ppm B_2H_6 for growing the p- type layers. Stationary substrates mounted in either chamber provide us with reference samples of pure p- or n- type a-Si:H. These reference samples are also layered because the plasma was extinguished before the ball was rotated to its new position.

We present here a few new experimental results which we need, in addition to those published earlier, for the discussion presented below. Figure 1 shows as a function of accumulated exposure time the increase in excess dark conductivity $\sigma(E)$, relative to its value $\sigma(A)$ after annealing, measured 10 min after the light is turned off. We used heat-filtered white light of

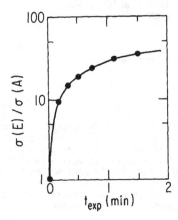

Fig. 1. Ratio of excess conductivity to annealed state conductivity as a function of exposure to $F=50mW/cm^2$ light. Sample has 41 layers, each $d_n=d_p=150Å$ thick.

intensity $F=50mW/cm^2$. The point to note is that saturation is essentially reached after about 1-2 min. During this short illumination time the photoconductivity remains constant. Smaller light intensity yields a lower saturation value of $\sigma(E)$ without substantially affecting the saturation time. This sample had 41 layers, each $d_n=d_p=150Å$ thick. These were scratched with a diamond scribe to assure that the carbon paint electrodes contacted each layer. The I-V characteristic was ohmic.

Figure 2 shows the in-plane conductivity at 300°K of a sample, having 11 layers and $d_n=d_p=500Å$, before light exposure (A), during a $t_e=60$ min exposure to light intensity F, and after the light is switched off (state E). One notices a decrease in the steady state photoconductivity which is characteristic of the Staebler-Wronski effect.[8] It is attributed to a decrease in recombination lifetime as new dangling bonds are created by light. Figure 3 shows the corresponding results obtained with an n- type reference sample that consists of 6 n- type layers and that was prepared and measured simultaneously with the modulation doped sample of Fig. 2. This reference sample has a high conductivity in the annealed state A. Its photoconductivity is a factor 3 smaller than that of the modulation doped film but the relative decrease of σ_p during light exposure is approximately the same. After the light is turned off the dark conductivity $\sigma(B)$ is much less than $\sigma(A)$. This is normally found in doped a-Si:H because the neutrality energy of the new dangling bonds created by the Staebler-Wronski effect lies only slightly above the gap center.[11] $\sigma(B)$ is of the same order of magnitude as $\sigma(E)$ of the modulation doped film.

FIELD SEPARATION MODEL

Very long recombination lifetimes that lead to the appearance of light-induced excess conductivities have been predicted for doping modulated semiconductors by Döhler.[7] The photo-excited electron hole pairs are expected to get separated by the built-in pn- junction fields with the electrons going to the n- layers and the holes to the p- layers. Here they occupy mostly localized gap states. The corresponding rise in trap quasi Fermi levels can then lead to an increase in the coplanar conductance. Several experimental results are difficult to reconcile with this model. As shown in Fig. 1, saturation of the excess conductivity requires about 1 min. This time is much longer than the photoconductivity rise time during which a steady

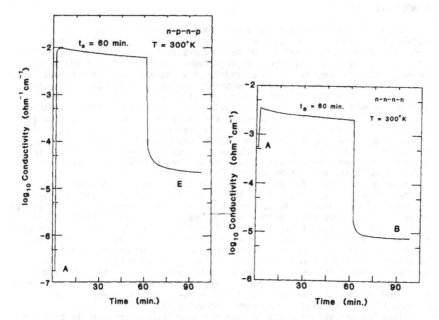

Fig. 2. Coplanar conductivity of a-Si:H having 11 layers, each $d_n=d_p=500$Å thick. A=annealed state; $t_e=60$ min exposure to F=50mW/cm^2 light. E=excess dark conductivity.

Fig. 3. Coplanar conductivity of reference sample consisting of 6 n- type layers (500Å thick) before, during, and after light exposure.

state is established between the drift and diffusion currents across the junctions as well as between the localized gap states and the extended transport states. Moreover, there are problems with explaining the long decay time at and above room temperature. The series of pn- junctions becomes forward biased by the illumination. After the light is turned off, the dark equilibrium between drift and diffusion currents should be reestablished within the dielectric relaxation time of the resistive region of the junction. This is less than 1s at 300°K. Equilibrium between deep traps and the transport states on the other hand may be slowed down by a very small emission rate of deep gap states

$$\nu_e = \nu_o \exp(-\Delta E/kT)$$

With deep traps having $\Delta E \approx 0.8$eV in the intrinsic layer of the pn- junction and a small but not unreasonable attempt-to-escape frequency $\nu_o=10^{10}s^{-1}$ one obtains $\nu_e=3 \times 10^{-5}s^{-1}$ or an equilibration time $\nu_e^{-1} \sim 10$h. This is close to the time needed for $\sigma(E)$ to decrease by a factor 2 at 300°K. Nevertheless this explanation, which invokes deep gap states with a small capture cross section, must be ruled out because such states should yield similarly long photoconductivity decay times in highly resistive homogeneous a-Si:H. Since this is not the case, we cannot explain our results with Döhler's model but must consider processes or defects that do not occur in singly doped or intrinsic a-Si:H.

COMPENSATION MODEL

The light-induced metastable excess conductivities observed by us in modulation doped a-Si:H multilayers are in many respects similar to those observed by Mell and Beyer[9] in compensated a-Si:H films: the dark conductance is increased by about two orders of magnitude, the decay time of the excess conductivity, defined as the time needed for a decrease in excess conductivity by a factor of two, is of similar magnitude and is activated by about 0.55eV. The excess conductivity of the doping modulated multilayers and of the compensated samples disappears upon annealing near 420°K about 50°K lower than the anneal temperature of the Staebler-Wronski effect. The same is true for the light-induced excess space charge conductance observed by Aker et al.[10] in boron-doped p- type a-Si:H. The presence of boron appears to be essential for these metastable excess conductance effects; we were unable so far to observe the excess conductivity effect in nini multilayer films consisting of alternating layers of n- type and intrinsic a-Si:H.

However, a major difference between the three excess conductivity effects is the following. With the same light intensity, the excess conductivity in doping modulated films needs an exposure of only one minute whereas exposures of one or more hours are required in compensated films or for the oxidized p- type films of Aker et al. However, this difference in efficiency does not necessarily rule out a common origin of these effects because the efficiency might depend on the presence of competing recombination processes which may be affected by the internal pn- junction fields in the multilayers. The absence of excess conductivities in the reference samples, that were grown simultaneously with the multilayers, suggests that defect complexes containing boron and perhaps phosphorus located at the pn- junction interfaces are responsible for the long lived excess conductivity.

Since the activation energy of the characteristic decay time is only about 0.5eV it is unlikely that the long equilibration time is due to deep traps having a small capture cross section. The long, nonexponential decay of $\sigma(E)$ may instead be caused by an atomic relaxation of the defect complex which acts as a repulsive barrier against the recapture of an electron after its release by optical excitation similar to the proposed DX center in crystalline compound semiconductors.[12] About $10^{12} cm^{-2}$ of such defects are needed to account for the magnitude of the observed $\sigma(E)$. If such defect complexes are responsible then they should act either as donors or as acceptors. Mell and Beyer indeed find a large asymmetry between n- and p- type compensated samples. On the basis of this argument we would predict that dominant p- type modulation doped multilayers will not exhibit a light-induced excess conductivity. Experiments are in progress to check this point.

STAEBLER-WRONSKI EFFECT

The most extensively explored metastable effect in a-Si:H is the light-induced creation of dangling bond defects.[8,13] At these light intensities this effect approaches saturation within an hour or two. It manifests itself in a decrease of the steady state photoconductivity because the dangling bond defects act as recombination centers. This decrease is of the same relative magnitude in the modulation doped film of Fig. 2 and the reference sample of Fig. 3. However, the magnitude of the excess conductivity $\sigma(E)$, which is established after an exposure of a few minutes, remains essentially unaffected by the creation of Staebler-Wronski defects. From this we conclude that the two effects are unrelated.

SUMMARY AND CONCLUSIONS

The photo-induced excess conductivity found in doping modulated a-Si:H[1-4]

is in many ways similar to that observed in compensated n- type a-Si:H films.[9]
The 100 times shorter exposure times needed for saturating the effect in doping
modulated films may be due to an action of the pn- junction field on the
detailed kinetics of the process. A defect complex associated with boron
having a large configurational relaxation after the release of an electron
may be responsible for the long-lived excess conductivity. Döhler's model[7]
of field separation of photo-excited electron-hole pairs may account for the
long recombination lifetimes at low temperatures [5,6] but not for the long
lived excess conductivity at and above room temperature.

This work was supported by Energy Conversion Devices and NSF DMR8009225.
We profitted very much from the materials preparation facility of the Materials
Research Laboratory supported at the University of Chicago by NSF. We grate-
fully acknowledge many clarifying discussions with Huseyin Ugur, S. Guha, M.
Hack and S. Agarwal. One of us (J.K.) is an A.T.&T. Bell Labs. Ph.D. Scholar.

REFERENCES

1. J. Kakalios and H. Fritzsche, Phys. Rev. Lett. 53, 1602 (1984).
2. J. Kakalios, H. Fritzsche and K. L. Narasimhan, AIP Conf. Proc.
   120, 425 (1984).
3. J. Kakalios and H. Fritzsche, Proc. of the 17th Intl. Conf. on the
   Physics of Semiconductors, San Francisco, CA (1984), Springer Verlag
   (in press).
4. J. Kakalios and H. Fritzsche, Proc. of the 1984 Fall Meeting of the
   Materials Research Society, Nov. 26-30, Boston, MA.
5. M. Hundhausen, L. Ley and R. Carius, Phys. Rev. Lett. 53, 1598 (1984).
6. M. Hundhausen, J. Wagner and L. Ley, Proc. of the 17th Intl. Conf. on the
   Physics of Semiconductors, San Francisco, CA (1984), Springer Verlag
   (in press).
7. G. H. Döhler, in "Advances in Solid State Physics: Festkörperprobleme",
   Vol. 23, ed. by P. Grosse (Heyden, Philadelphia, 1983) p.207.
8. D. L. Staebler and C. R. Wronski, J. Appl. Phys. 51, 3262 (1980).
9. H. Mell and W. Beyer, J. Non-Cryst. Solids 59/60, 405 (1983).
10. B. Aker and H. Fritzsche, J. Appl. Phys. 54, 6628 (1983).
11. M. H. Tanielian, N. B. Goodman and H. Fritzsche, J. de Physique (Paris)
    42, C4-375 (1981).
12. D. V. Lang and R. A. Logan, Phys. Rev. Lett. 39, 635 (1977).
13. C. R. Wronski, in "Semiconductors and Semimetals" Vol. 21C, ed. by
    J. I. Pankove, (Academic Press, N.Y. 1984) p.347.

PART V

Si-Ge Alloys

LOCAL BONDING IN A-Si,Ge ALLOY FILMS

G. LUCOVSKY, R.A. RUDDER, J.W. COOK, Jr. and S.Y. LIN
Department of Physics, North Carolina State University
Raleigh, North Carolina 27695-8202

ABSTRACT

This paper discusses the bonding of hydrogen in a-Si,Ge:H alloy films prepared by reactive magnetron sputtering (RMS). We compare our results for H atom bonding with films produced by: (a) the glow discharge decomposition (GDD) of silane and germane mixtures, and (b) reactive diode sputtering (RDS). We discuss the energy states associated with Si and Ge atom neutral dangling bonds in the context of an empirical tight-binding model. The model places the Ge atom dangling bond state deeper in the gap than the corresponding Si atom defect state. The differences between the electronic properties of GDD and RDS films, and RMS films are explained in terms of the degree of H compensation of Si and Ge atom dangling bonds.

INTRODUCTION

There has been considerable interest in a-Si,Ge:H alloys for photovoltaic applications. This derives from the fact that the effective optical band-gap can be varied between about 1.1 eV and 1.8, the values in the respective end-members. The properties of the end-member binary alloys vary considerably with deposition techniques and conditions. However, photovoltaic quality intrinsic a-Si:H films have been produced by all three of the techniques mentioned above; the GDD decomposition of silane [1], and the RDS [2] and RMS [3] sputtering of Si in a hydrogen ambient. These photovoltaic quality a-Si:H alloys display similar dark- and photo-conductivities, with bonded H in predominantly monohydride configurations and in the composition range of 10 to 15 atomic percent. In contrast, the properties of a-Ge:H films show a wider range of behavior. Intrinsic a-Ge:H films produced by the GDD [4,5] and RDS [6,7] methods have displayed photoelectronic properties that are considerably poorer than those reported for films produced by RMS [8,9]. These difference correlate with H atom bonding.

The initial attempts to make a-Si,Ge:H alloys were based on extensions of the procedures used to make photovoltaic a-Si:H films, the GDD decomposition of silane/germane mixtures [10,11], and the RDS sputtering of Si and Ge from composite targets [12]. The films produced displayed a similar degradation in photoelectronic properties as the Ge concentration was increased. As the Ge concentration was increased: (a) the total bonded H concentration decreased [10,11,13]; (b) the attachment of H was preferentially to Si rather than Ge atom sites [10-12], (c) the majority carrier photoconductivity decreased [10-12], (d) the activation for dark conductivity increased [14], and (e) an ESR signal associated with Ge atom neutral dangling bonds also increased [15]. There have recently been attempts to deposit a-Si,Ge:H,F alloys via the GPD process using gas mixtures that have included SiF_4, GeF_4 and H_2 [16]. These have yielded films with improved photoelectronic properties. There has also been progress in the deposition of a-Si,Ge:H films from silane/germane discharges, and depositions at temperatures in excess of $280°C$ have yielded films with improved photoelectronic properties [17].

The second section of this paper deals with properties of Si and Ge atom neutral dangling bonds (D_g centers) as they are derived from model calculations based on the empirical tight-binding method applied to structural models using cluster Bethe Lattices [18-21]. We include these calculations to introduce a theoretical framework for discussing the defect states in a-Si,Ge:H alloy films. The model reflects trends that are related to the chemistry of the constituent Si and Ge atoms, and as such we expect that refinements based on more rigorous theoretical approaches will not change the basic qualitative conclusion of the calculation, namely that the Si atom neutral dangling bond state is always higher in energy in the pseudo-gap than the corresponding neutral Ge atom dangling bond state.

DEPOSITION OF a-Si,Ge:H ALLOYS BY RMS

The a-Si,Ge:H alloy films were deposited in a dual magnetron sputtering system employing separate Si and Ge targets. The details of this system have been discussed elsewhere [3,8,9]. In this paper, we highlight the important features of the system that play a role in the quality of the deposited films. These are: (a) the deposition chamber is UHV compatible with a base pressure of less than 5×10^{-9} torr; (b) substrates are introduced through a load lock arrangement; (c) the Si and Ge targets have been modified with: (i) a c-Si washer used as the aperture of the Si target front ground shield, and (ii) a tubular structure used as the Ge target front ground shield; (d) the target to substrate distance is 18 cm, and about the same as the mean free path for collisions under the sputtering conditions, 0.4 to 4.0 mTorr of hydrogen and 2.0 mTorr of argon; and (e) the plasma regions in front of the Si and Ge targets are separated from each other, and neither extends to the substrate. Table I gives the deposition conditions for optimized a-Si:H and a-Ge:H binary alloys and an optimized a-Si,Ge:H ternary alloy film with approximately 50 atomic percent Ge.

Table I: Deposition Conditions for Optimized* Alloy Films [9-11]

| Material | Gas Pressure | | RF Power | | Substrate Temperature |
	Hydrogen (mTorr)	Argon (mTorr)	Si Target (Watts)	Ge Target (Watts)	T_s (degrees C)
a-Si:H	0.4	2.0	200	---	240
a-Ge:H	4.0	2.0	---	70	180
a-Si,GeH	1.4	2.0	200	100-150	180

* Optimization = highest ratio of photo- to dark-conductivity.

The important things to note are: (a) the hydrogen partial pressures for optimized a-Si:H and a-Ge:H films differ significantly; (b) a lower T_s is required for the a-Ge:H films; and (c) the ternary alloy is deposited at an intermediate hydrogen partial pressure, but at the same T_s as the a-Ge:H alloy.

PROPERTIES OF DEPOSITED FILMS

Alloy Composition Series

These films were prepared at a constant partial pressure of H (0.4 mTorr), a constant RF power to the Si target (200W), a fixed T_s (180°), and a variable RF power input to the Ge target. Fig. 1 of Ref. 9 gives the IR absorption. This is dominated by monohydride bonding groups associated with the Si and Ge atoms, with stretching mode absorptions at 2000 and 1880 cm^{-1}, respectively. A stronger feature near 600 cm^{-1} has contributions from both Si-H and Ge-H bending vibrations. The alloy films also display weaker absorption between 800 and 900 cm^{-1} which is due to isolated Si and Ge dihydride bonding groups. None of the films displays any features (e.g., the 845/890 cm^{-1} or 760/825 cm^{-1} doublets) due to polymerized dihydride groups, either polysilane or polygermane. The most important aspects of H

Fig. 1 Relative concentrations of bonded hydrogen in alloy composition series. The SiH/Si, GeH/Ge and $SiH_2/(SiH_2 + SiH)$ ratios are plotted as functions of the relative Ge concentration (Ge + Si = 1.0).

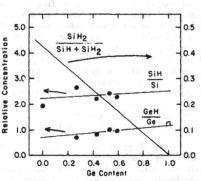

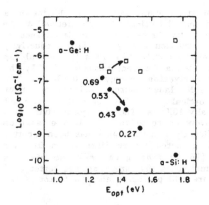

Fig. 2 Dark conductivity (solid circles) and photoconductivity (open squares) for alloy composition series. Data are plotted as a function of the optical band-gap. Ge compositions are indicated. The photoconductivity is for a flux of 10^{16} photons/cm^2-s from a He-Ne laser. The arrows and additional points are for an optimized alloy with 53 % Ge. The photoconductivity of the a-Ge:H film is approximately equal to the dark conductivity for a flux of 10^{15} photons/cm^2-s.

Fig. 3 IR absorption spectra: (a) optimized a-Si:H alloy (0.4 mTorr H); (b) optimized a-Ge:H alloy (4.0 mTorr H); (c) optimized a-Si,Ge:H alloy with 53 % Ge (1.4 mTorr H); and (d) a-Si:H film (1.4 mTorr H).

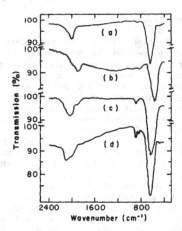

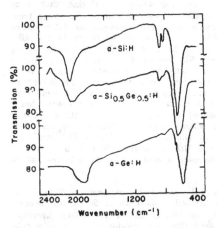

Fig. 4 IR absorption for films deposited at 80°C.

atom bonding are displayed in Fig. 1, where we show the number relative numbers of SiH and GeH bonds as a function of alloy composition. The preferential attachment ratio, used by the Harvard Group [10,11] is given by SiH/Si divided by GeH/Ge. There are three points to be made: (a) the relative concentrations of SiH and GeH bonds remain approximately constant over the entire alloy range; (b) there is a small preferential attachment for H to Si rather than Ge, of the order of two; and (c) the fraction of SiH_2 bonds decreases as the Ge concentration is increased. The total bonded hydrogen concentration varies between about 20 and 13 atomic % as the Ge concentration is increased. The bonding of H in these films is quanitatively different from what has been reported in GDD and RDS films, wherein: (a) the amount of bonded hydrogen decreases from about 10 to 5 atomic % as the Ge concentration is increased, and (b) the preferential attachment ratio rises with increasing Ge concentration and approaches values in excess of ten [10-12].

Figure 2 gives a summary of conductivity data. Instead of plotting data against alloy composition, we have used the optical band-gap (Tauc's method) as the x-axis variable. The log of the dark conductivity displays a linear dependence on the activation energy, consistent with the activation energy for dark conductivity being a fixed fraction of the optical gap, approximately 0.45 times the optical gap over the entire alloy range. For an alloy with a Ge concentration of 50%, the optical band gap is 1.43 eV and the activation energy is 0.63 eV. The photoconductivity is obtained using a He-Ne laser source, and has not been corrected for the relative values of the optical absorption constant at 1.96 eV (about 100 A) and the film thickness, nominally 10,000A. The photoconductivity falls off in the mid-alloy range, and then recovers en route to a-Ge:H. The maximum degradation in the mid-alloy range is significantly less than what has been reported initially for GDD and RDS films [10-12,14]. As noted earlier, films grown from gas mixtures containing SiF_4, GeF_4 and H_2 [16] or from germane/silane mixtures deposited at high T_s [17], display photoconductivities that are somewhat better than those quoted in the earlier GDD films, but are still poorer by about a factor of ten than what we have found in the RMS films.

Optimization Series

We have optimized the photoelectronic properties of alloys with approximately 50 atomic percent Ge by varying the hydrogen partial pressure and keeping all other deposition parameters fixed. This choice derives from the fact that optimized end-member binary alloy films have been grown at very different hydrogen partial pressures, 0.4 mTorr for a-Si:H and 4.0 mTorr for a-Ge:H. We find that alloy properties improve with increasing partial pressure of hydrogen over the range from about 0.4 to 1.4 mTorr, and then show degradation as the hydrogen partial pressure is further increased (see Fig. 2 of Ref. 8). The amount of hydrogen in the film increases as the hydrogen pressure is increased, with the hydrogen going predominantly to Si sites (see Fig.3 of Ref. 9). The GeH/Ge ratio shows only a small increase and the preferential attachment ratio in an optimized film is about 4.5. The values of the photoconductivity and dark conductivity in the optimized films are also shown in Fig. 2. The photoconductivity is about 7 to 10 times lower than in photovoltaic grade a-Si:H [2,3]. Sub band-tail absorption and space charge limited dark currents have also been measured in these optimized films, by the groups at ECD and SERI respectively. Preliminary measures indicate defect state densities within the pseudo-gap that are in the range of $2-3 \times 10^{16}$/cm^3 and therefore higher than those in photovoltaic a-Si:H by similar factors of 7 to 10. This means that the photoconductivity may be used a measure of the relative defect state density.

Figure 3 gives the IR absorption in: (a) optimized end-member films a-Si:H and a-Ge:H; (b) an optimized a-Si,Ge:H ternary alloy with about 50 atomic percent Ge; and (c) an a-Si:H film grown at a T_s of 180°C and a 1.4 mTorr hydrogen partial pressure. We note that: (a) all optimized films are dominated by monohydride bonding; but (b) they also display weaker absorption due to a small fraction of isolated dihydride groups; (c) optimized films do not display any absorption due to polysilane or polygermane groups; and (d) the a-Si:H alloy grown at $T_s = 180^{\circ}$C and with a hydrogen pressure of 1.4 mTorr displays relatively strong absorption associated with polysilane groups. This last observation means that the bonding of H to Si is qualitatively different in a ternary alloy film with about 50 atomic percent Ge, and in a binary a-Si:H alloy grown under the same conditions. In order to gain more insight into this phenomenon, we have studied the IR absorption in films deposited at still lower T_s.

Low Temperature Series

Figure 4 gives the IR absorption for three films deposited with $T_s = 80^{\circ}$C and a hydrogen partial pressure of 1.4 mTorr. The IR of the a-Si:H film is dominated by polysilane absorption with: (a) the dominant absorption in the stretching regime being at 2100 cm^{-1}; and (b) the doublet absorption at 845/890 cm^{-1}. The a-Ge:H film shows a larger fraction dihydride bonding than the film grown at $T_s = 180^{\circ}$C, but does not display the doublet feature found in GDD films and characteristic of polygermane incorporation. The ternary alloy film displays increased dihydride bonding, but this is due to isolated GeH$_2$ and SiH$_2$ groups, rather than any polymerized species.

MODEL FOR FILM GROWTH

The RMS deposition system minimizes gas phase reactions between the sputtered species Si and Ge, and between each of these and H. This is achieved by the spatial isolation of the plasma regions in front of each of the sputtering targets. The plasma glow regions extend out about 10 cm in front of each of the targets, and the tubular ground shield in front of the Ge targets insures isolation of the two plasmas. The target to substrate distance is about 18 cm, so that neither plasma glow region extends to the substrate. This limits charged particle bombardment of the substrate during deposition, and in addition does not provide an energy source for stimulating surface chemistry. The partial pressures of the reactive and sputtering gases are in range of 0.5 to 5.0 mTorr, and the effective partial pressures of Si and Ge, estimated from a growth rate of about 1 A/sec and an assumed sticking probability close to unity, are in the uTorr range. Under these conditions we anticipate very few molecular collisions between the Si and Ge species en route to the substrate or between Si and H or Ge and H atomic species. There may be some H attachment to Si and Ge atoms near the respective targets, however, we believe that this will be small compared to the number of Si and/or Ge atoms liberated.

Our model for the deposition process assumes that the species incident on the substrate for the deposition of a-Si:H films are Si, H, and Ar atoms, and a small fraction of diatomic molecular species such as Si$_2$ and SiH. We believe that the species are largely neutral, but our model is not refined where we predict different behavior for neutral or charged species. Since the H and Si atoms fluxes are proportional to their respective partial pressures, we believe that the H atom incorporation is dominated by H attachment and rejection at the growth interface. Each incident Si atom is trapped on the growth surface and is decorated with H. The amount of H retained is proportional to T_s. The bonding configurations, SiH, isolated SiH$_2$ and polysilane are determined by the amount of H retained in film. Support for this model derives from a comparison between the H incorporation in our films and that in films produced by the GDD method from dilute silane plasmas [1]. At a sufficiently high hydrogen partial pressure, we H find identical IR spectra that vary in the same way with T_s. It has been established that in such GDD films, the H atom incorporation is determined by surface rather than gas phase reactions. The similarity between the RMS films and GDD films then supports the growth mechanism described above.

There are qualitative differences between the MRS a-Ge:H films, and a-Ge:H films grown by either GDD of germane or by RDS. The major difference is the incorporation of polymerized GeH_2 or polygermane groups. In GDD films, grown under a variety of conditions in which either gas phase or surface polymerization reactions dominate, there is always evidence for polygermane bonding groups in films deposited with T_s below about $200^{\circ}C$ [4,5]. In RDS films polygermane incorporation also occurs [7]. However, there is no evidence for polygermane incorporation in MRS films produced under any of the deposition conditions that we have explored to date. This is consistent with our picture of the deposition process if we assume that polygermane incorporation is only possible if there is either gas phase or surface generation of digermane and higher GeH polymers [4,5,22]. We believe that these can only be surface or gas phase generated from recombination between specific germane molecular species such as GeH_4 and GeH_2. Since these are not probable in our magnetron process, the only bonded species in RMS a-Ge:H films are GeH and isolated GeH_2 groups.

The growth model for a-Si,Ge:H films with approximately 50 atomic percent Ge and Si is an extension of the models discussed above. The a-Si,Ge:H alloy films display H atom incorporation predominantly in SiH and GeH monohydride groups with a small fraction of the bonded H in isolated dihydride groups, SiH_2 and GeH_2. These alloy films in addition show no IR features characteristic of either polysilane or polygermane groups. The absence of polygermane is expected, but the absence of polysilane must be accounted for. We believe that polysilane formation is supressed by the statistics of Si and Ge atom attachment. For example in a 50-50 alloy, the arrival rates of Si and Ge atoms are the same and there is an equal probability for a given Si or Ge atom to have either Si or Ge neighbors. This in effect means that it is unlikely for large chain of Si-Si bonds to develop. This type of pictures implies that polysilane incorporation will increase as the Ge atom concentration is decreased. This is indeed the case. Moreover since H atom rejection is simply a function of T_s, the fraction of SiH bonds in an alloy film should be independent of the Ge concentration (for a given T_s and partial pressure of H) but the fraction of H in various bonding configurations should vary. This is also the case. This means that in a low T_s ($<200^{\circ}C$), we can exceed the solubility of H in monohydride groups as found in a binary a-Si:H films. We find that about 20-25 % of the Si atoms in the optimized 50 atomic % Ge film are in monohydride bonding groups, whilst in a-Si:H alloys the limit is closer to 15 atomic %.

THEORY OF DANGLING BOND DEFECTS

We have performed theoretical studies of neutral dangling bond defects in a-Si:H alloys containing C, N ,0 and Ge impurities [19,20] and more recently of Si and Ge atom dangling bonds in a-Si,Ge alloys [21]. The calculations used the empirical tight-binding method as applied to a cluster Bethe Lattice structural model. It was found that: (a) alloy or impurity atoms such as C, N, O and Ge do not introduce new electronic states into the pseudo-gap of a-Si; but (b) that can shift the positions of dangling bond states if they are back-bonded to the Si atom with the dangling bond [19,20]. The amount of the shift, referenced to the energy of a dangling bond configuration in which the Si atom with the dangling bond has only Si nearest neighbors, is proportional to the energy difference between a Si-Si bond and the bond between Si and the impurity atom. Therefore C, N and O shift dangling bonds up the gap and Ge shifts them down [20,21].

We have also studied the energies of Si and Ge dangling bonds in a-Si,Ge alloys [21]. Fig. 5 indicates the structural model employed and Figs. 6 and 7 summarize the results. Figure 5 gives a schematic representation of the Bethe Lattice structure and the local cluster containing a dangling bond. For the alloy calculations we consider Si and Ge atom dangling bonds in a random binary alloy of Si and Ge atoms. In treating a Si atom dangling bond, we consider four discrete environments of nearest neighbors, three Si atoms, two Si and one Ge, and so forth. We attach these defect clusters to Bethe Lattices built up of virtual atoms with

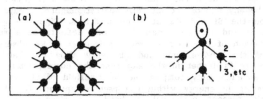

Fig. 5 Schematic representation of: (a) Si (or Ge) atom Bethe lattice; (b) Neutral dangling bond defect configuration.

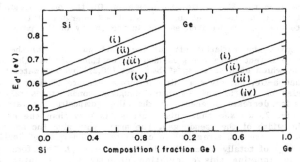

Fig. 6 Relative energies of Si and Ge atom dangling bond defects. The energy of the defect state, relative to the valence band edge has been normalized with respect to the energy gap. The numbers (i) through (iv) correspond to the four configurations of atoms that are back-bonded the the Si or Ge atom with the dangling bond: (i) three Si atoms; (ii) two Si and one Ge; (iii) one Si and two Ge; and (iv) three Ge.

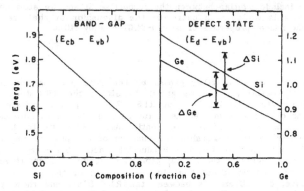

Fig. 7 (a) Calculated variation of the band-gap versus alloy composition; (b) Average defect state energies as a function of the alloy composition. The arrows indicate the widths of the defect state bands that result from the four different nearest neighbor bonding arrangements. (See Figs. 6(a) and (b).)

the average properties at a particular alloy composition [21]. We assume that the probability of the any one of the discrete nearest neighbor groups is statistical and depends only on the Si and Ge atom compositions. Figure 6 gives the relative energy of the Si and Ge atom dangling bond defects as a function of alloy composition for each of the four discrete environments. We observe the following: (a) for each of the environments and at every alloy composition the Ge atom dangling bond state is deeper in the gap than the corresponding Si atom dangling bond state; (b) at each alloy composition the manifold of Si and Ge atom dangling bond states overlap in energy within the pseudo-gap; and (c) the relative defect state energy ,with respect to the valence band edge, rises relative the width of the pseudo-gap as the Ge concentration is increased. The trends with alloy composition, and energy differences between the Si and Ge defect states reflect "chemical differences" between the Si and Ge atomic states. On the other hand, there is an uncertainty of about 0.2 to 0.3 eV in the calculated energies of the defect states with respect to the valence band edges, and in the energy of the pseudo-gap.

Fig. 7 gives the variation of: (a) the band gap and (b) the statistically weighted average defect state energies also as functions of the alloy composition. The calculation, which employs an sp^3s^* basis set [20,21] overestimates the band-gap of both a-Si and a-Ge, but gives a fairly accurate measure of their difference. The average energies of the Si and Ge atom dangling bonds states relative to the valence band edge decrease in energy as the alloy composition is varied; however, referring to Fig. 7, we see that this decrease is less than the corresponding decrease in the width of the pseudo-gap. The arrow-bars give the spread of defect state energies. We have not attempted to calculate the energies of doubly occupied defects states (D_-) of totally unoccupied defects states (D_+). Therefore caution must be observed in applying this calculation directly to the interpretation of experimental data. We apply this model in the discussion section, but qualify the extent to which we believe the model is applicable. At this point, we also note that Stutzmann, Nemanich and Stuke [23] have studied ESR in boron-doped p-type a-Si,Ge:H alloys and have proposed a model for defects states in which the neutral Ge atom dangling bond is higher in the pseudo-gap than the neutral Si atom dangling bond. Their model has the distributions of these Ge and Si defect states separated rather than overlapping in energy. The Harvard group [17] has modified the model to include overlap between the respective Si and Ge atom defect states, and has used this model as a basis for the discussion of their results on the electronic and optical properties of GDD a-Si,Ge:H alloy films.

DISCUSSION

The a-Si,Ge:H alloy films grown by the RMS process are qualitatively different than similar alloy films grown by either the GDD decomposition of silane/germane mixtures [10,11,17] or by RDS [12]. The primary differences are: (a) in the amount of bonded hydrogen as a function of the alloys composition; (b) in the composition dependence of the preferential attachment ratio for H bonded to Si and Ge atoms; (c) in the composition dependence of the photoconductivity; and (d) in the relative magnitude of the dark conductivity activation energy, i.e, the ratio of this activation energy to the optical gap. We shall now consider these points one at a time, and then point out that in the context of the defect state model discussed above, the differences between the RMS films, and those produced by GDD and RDS process can be explained in terms of Fermi level pinning by a small number of Si atom dangling bonds in the first case, and a larger number of Ge atom dangling bonds in the second.

We will base the majority of our comparisons on the data given in Ref. 17. The properties of these films are similar to those discussed in Refs. 10 and 11, but with considerable improvement in the photoconductivity and the luminescence. We will focus on films with about 50 atomic percent Ge and Si. The films grown by the MRS process, and deposited with Ts = 180°, display bonded H atom concentrations in excess of about 15 atomic % H over the entire alloy range. Optimized alloys with 50 atomic percent Ge have even higher bonded H concentrations, approaching 25 atomic percent. In contrast optimized alloys grown by the GDD process have bonded H atom concentrations of about 5 atomic percent H. The problem is the GDD films relates to polysilane incorporation. Films grown at 230° show substantial polysilane incorporation. Polysilane is not present in a-Si:H films grown at the same T_{s} and RF power, but is present in films grown from silane/germane mixtures. The photoconductivity in the GDD films displays a strong temperature dependence [17] and increases dramatically above a T_{s} of about 280°. The increase in photoconductivity correlates with the disappearance of polysilane groups in the deposited films. In addition to having relatively low bonded H atom concentrations, the GDD and RDS films also display larger preferential attachment ratios. These are at least twice the nominal values of 4 to 5 in the optimized RMS films [10,11].

The variation of majority carrier photoconductivity with alloy composition is very different in the RMS, and GDD and RDS films. As shown in Fig. 2, the RMS films show a drop of about two orders of magnitude in going from a-Si:H to a 50 atomic percent Ge alloy, and recover almost completely in returning to a-Ge:H. In contrast, the optimized GDD films, see Fig. 9(b) of Ref. 17, display a drop of at least 4 orders of magnitude with a minimum in photoconductivity at about 70 to 80 atomic percent Ge. They do not recover in a-Ge:H to the any where near the same degree as the RMS films. GDD films with about 50 atomic percent Ge display photoconductivities that are about an order of magnitude poorer than the MRS films. The data in Ref. 17 has been adjusted for the ratio of sample thickness to penetration depth of light, whilst the data in Fig. 2 of this paper has not been adjusted.

Finally, the dark conductivity activation energy in the RMS films is found to be about 45% of the Tauc's optical gap energy. The activation energy in the GDD films is generally more than 50% of the E04 optical gap [17]. The Tauc's gap is always greater than E04 so that care must be exercised in comparisons of this sort. In any case, typical values of the activation energy for GDD films with 50 atomic percent Ge range from 0.75 eV [17] to 0.85 eV [14] in GDD films, but are only about 0.63 eV in the RMS films.

We interpret these results in terms of the defect picture we have developed using the tight-binding model. All of these results are in accord with a model in which the Fermi level is pinned by Si atom dangling bond defects in the RMS films and by Ge atom dangling bonds in the GDD and RDS films. This is based on the Ge neutral dangling bond state being deeper in the gap than the Si atom neutral dangling bond state. Since the Si state is higher in the gap, essentially all of the Ge atom dangling bonds are doubly occupied in the RMS material, and the activation energy for dark conductivity is pinned by the Si neutral dangling bonds. The measurements of the photoconductivity and sub-band tail absorption on the RMS films yield estimated defect densities of 2-3 X 10^{16}/cm^{3}. For the GDD films, there is a smaller concentration of bonded hydrogen, and a higher preferential attachment ratio. This, coupled with ESR measurements on similar films to those discussed above [15], yields a situation in which the Fermi level in the GDD and RDS films is pinned by neutral Ge atom dangling bonds. The ESR signal corresponds to more than 10^{17}/cm^{3} [15], consistent with the number of neutral Ge atom dangling bonds exceeding the nominal concentration of Si defect states. The observation of a smaller Si atom ESR signal [15] is also consistent with the relative energies of the neutral atom defect states as predicted by the tight-binding model. Finally, the activation energies are also in accord with the Ge defect state being deeper in the gap than the Si defect state.

There is clearly much experimental and theoretical work to be done on the alloy films. More systematic and complete experimental studies of GDD films produced (a) using the F substituted gases, and (b) at higher deposition temperatures from silane/germane and hydrogen mixtures are necessary. There have been alternative models proposed for defect states in a-Si,Ge:H alloys in which the Ge neutral and doubly occupied states are higher in energy than the corresponding Si defect states [17,23]. This proposal is not accord with differences in the energies of the atomic states of Si and Ge, or with the model in Ref 21. In order to resolve the differences between the two defect state models, one must consider structural relaxation at the defects sites, and determine the energies of the associated charged defect states.

ACKNOWLEDGEMENTS

This work has been supported under SERI Subcontract XB-2-02065-1 and ONR Contract N00014-79-C-0133.

REFERENCES

1. G. Lucovsky, R.J. Nemanich and J.C. Knights, Phys. Rev. B 19, 2064 (1978).
2. T.D. Moustakas, C.R. Wronski and T. Tiedje, Appl. Phys. Lett. 39, 721 (1981).
3. R.A. Rudder, J.W. Cook,Jr., J.F. Schetzina and G. Lucovsky, J. Vac. Sci. Tech. A 2, 326 (1984).
4 G. Lucovsky, S.S. Chao, J. Yang, J.E. Tyler and W. Czubatyj, J. Non-Cryst. Solids 66, 99 (1984).
5. G.Lucovsky, S.S. Chao, J. Yang, J.E. Tyler, R.C. Ross and W.Czubatyj, Phys. Rev. B 31, 2190 (1985).
6. G.A.N. Connell and J.R. Pawlick, Phys. Rev. B 13, 787 (1976).
7. D. Bermejo and M. Cardona, J. Non-Cryst. Solids 32, 421 (1979).
8. R.A. Rudder, J.W. Cook,Jr. and G. Lucovsky, Appl.Phys. Lett. 45, 887 (1984).
9. R.A. Rudder, J.W. Cook,Jr. and G. Lucovsky, J. Vac. Sci. A 3 (1985), in press.
10. D.K. Paul, B. von Roedern, S. Oguz, J. Blake and W. Paul, Jpn. J. Appl. Phys. Suppl. A 49, 1261 (1980).
11. W. Paul, D.K. Paul, B. von Roedern, J. Blake and S. Oguz, Phys. Rev. Lett. 46, 1016 (1981).
12. S.Z. Weirz, M. Gomez, J.A. Muir, O. Resto, R. Perez, Y. Goldstein and B. Abeles, Appl. Phys. Lett. 44, 634 (1984).
13. L. Chahed, C. Senemaud, M.L. Theye, J. Bullet, M. Galin, M. Gauthier and B. Bourdon, Solid State Commun. 45, 649 (1983).
14. W. Paul in U.S. Department of Energy Report, SERI/CP-211-2167 (1983).
15. A. Morimoto, T. Muira, M. Kumeda and T. Shimuzu, Jpn. J. Appl. Phys. 20, L833 (1981).
16. K. Nozawa, Y.Yamaguchi, J. Hanna and L Shimuzu, J. Non-Cryst. Solids 59 & 60, 533 (1983).
17. K.D. Mackenzie, J.R. Eggert, D.J. Leopold, Y.M. Li, S. Lin and W. Paul, Phys. Rev. B 31, 2198 (1985).
18. D.C. Allan and J.D. Joannopoulos, in The Physics of Hydrogenated Amorphous Silicon II, Ed. by J.D. Joannopoulos and G. Lucovsky (Springer-Verlag, Berlin, 1984), p. 5.
19. S.Y. Lin, G. Lucovsky and W.B. Pollard, J. Non-Cryst. Solids 66, 291 (1984).
20. G. Lucovsky and S.Y. Lin, AIP Conf. Proc. 120, 55 (1984).
21. G. Lucovsky and S.Y. Lin, to be published.
22. R.C. Ross, G. Lucovsky, S.S. Chao, J.E. Tyler and W. Czubatyj, J. Vac. Sci. Tech. A 3 (1985) (in press).
23. M. Stutzmann, R.J. Nemanich and J. Stuke, Phys. Rev. B 30, 3595 (1984).

ELECTRONIC AND OPTICAL PROPERTIES OF a-(Si, Ge):H ALLOYS

VIKRAM DALAL, JAMES F. BOOKER AND MARK LEONARD

Spire Corporation
Patriots Park
Bedford, MA   01730

ABSTRACT

We describe the preparation and electronic and optical properties of amorphous (Si, Ge) alloys. A-(Si, Ge):H alloys were prepared by glow discharge decomposition of SiH_4 and GeH_4. The bandgap was varied between 1.78 and 1.42 eV by changing the $GeH_4:SiH_4$ ratio in the gas phase. We find a distinct influence of growth temperature on electronic properties. Films grown at low temperatures (200-250C) tend to have much lower photo conductivity than films grown at higher temperatures (300-325C). The electron $(\mu\tau)$ products of high temperature films are generally $> 1X10^{-7}$ cm^2/V. We also obtain very sharp valence band tails in a-(Si, Ge):H alloys, with slopes of ~ 40 meV. The hole $(\mu\tau)$ product is generally $\sim 1-2X10^{-8}$ cm^2/V. All these properties suffer a catastrophic decline when bandgap is reduced below about 1.5 eV.

Introduction

Alloys of a-Si:H and a-Ge:H have potentially useful electronic properties, particularly in solar energy conversion. The bandgap of such alloys can be varied over a range of 1.1 - 1.75 eV[1], and as such, these alloys are very promising as low gap solar cells in a tandem-junction cell structure.[2-4] However, not much is known about the basic electronic properties of these alloys, such as tail-state densities, mid-gap defects, or electron and hole $(\mu\tau)$ products, where μ is the mobility and τ is the lifetime. In this paper, we report some results on these basic electronic properties of a-(Si, Ge):H alloys, and show that the material remains very promising for use in tandem-junction solar cell structures.

Material Preparation

a-(Si, Ge):H alloys were grown by plasma decomposition of SiH_4 and GeH_4 on 7059 glass substrates. GeH_4 used was diluted to 10% in H_2. Typical growth conditions are listed in Table I.

Table I
Growth Conditions

pressure	=:	200-300 μm
T	=:	200-325°C
Flow rate	:	100 SCCM
(SiH_4 + GeH_4)/H_2	:	1/6 to 1/7
RF Power	:	50-70 mW/cm^2
Bias on Substrate	:	0 to -100 V
Electrode Geometry	:	Triode. Substrate isolated from plasma.

For certain measurements, p-i-n devices were made on SnO_2 coated glass. The configuration was: $SnO_2/p/i/n/Al$, with light incident from p side. SnO_2 was typically textured with a transmission loss of ~20% in 400-500 nm range. The pin devices had a p-type a-(Si, C):H layer, followed by i a-(Si, Ge):H and n+ a-(Si, Ge):H. n+ layer was grown by doping with PH_3 at 1% concentration in the gas phase.

We observed a significant influence of substrate temperature on growth kinetics. Typically, higher substrate temperatures lead to films with high stress, and films which show evidence of etching during growth.[5] Increasing power and substrate negative bias tend to reduce the growth rate, a strong indication of etching, probably by H ions. Further work on etched films is in progress.

Optical Properties

The bandgap of the films was varied from 1.78 eV to 1.42 eV by varying GeH_4/SiH_4 ratio. We decided to stay in this regime because the previous work indicated a catastrophic reduction in electronic properties for Eg < 1.45 eV.[6-8]

In Figure 1, we show a typical $\sqrt{\alpha}$ E vs. E (tauc curve) for an a-(Si, Ge):H film with Eg = 1.57 eV. The bandgap quoted is the intercept of the Tauc curve.

We measured dark (σ_D) and photoconductivity(σ_{pc}) of these films. All films were first annealed at 150° for > 1 hr. to eliminate any variations due to Stabler-Wronski effect. A photoconductivity was measured both under an ELH lamp at 100 mW/cm^2, and under illumination from a He-Ne laser at 1.96 eV. We also measured σ_{pc} as a function of light intensity from 5 mW/cm^2 to 100 mW/cm^2 of ELH lamp illumination.

In Figure 2, we show σ_D and pc of films grown at 250°C as a function of bandgap. We note that σ_D is very low, in the range of 10^{-11} S-cm^{-1}, and does not seem to vary significantly with Eg. σ_{pc}, on the other hand, shows a continuous decrease with decreasing Eg.

Quite a different behavior is observed for films grown in 300-325°C range. Figure 3 shows σ_{pc} and σ_D as functions of bandgap. In contrast to Figure 2, we see a systematic increase in σ_D with decreasing Eg., and a relatively flat σ_{pc} curve, until about 1.52 eV, at which time σ_{pc} drops precipitously. Note that even at 1.5 eV, $\sigma_{pc} > 10^{-5}$ S-cm^{-1}, a value high enough for efficient photovoltaic devices.

An important point to note from Figure 2 and 3 is the very high ratio of σ_{pc}/σ_D. This ratio is an indication of material quality, and a high ratio indicates a high electron ($\mu\tau$) product and a low Fermi level.[9] In a-Si:H, we obtain ratios > 10^6, and in a-(Si, Ge):H, ratio > 5×10^4 for Eg. > 1.5 eV.

In Figure 4, we show the electron ($\mu\tau$) products for 300°C films calculated from σ_{pc} under He-Ne laser illumination. From Figure 4, we see that electron ($\mu\tau$) decreases from ~1×10^{-6} cm^2/V for a-Si:H to ~1×10^{-7} cm^2/V for a 1.5 eV a-(Si, Ge):H. However, this latter value is quite adequate for a photovoltaic device, since it translates into a field-free diffusion length of ~0.5 μm, and a range ($\mu\tau$E) of ~5 μm, assuming an region field of 5×10^3 V/cm. Thus, electron transport in photovoltaic devices made from a-(Si- Ge):H for Eg. > 1.5 eV should not be a problem. Of course, the precipitous decrease in electron ($\mu\tau$) for Eg. < 1.5 eV will make efficient devices in such materials difficult.

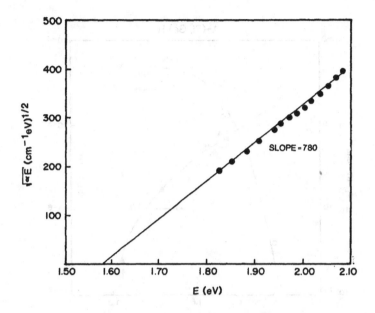

FIGURE 1.   ABSORPTION CURVE FOR A-(Si,Ge):H GROWN AT 250°C

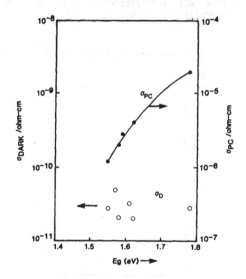

FIGURE 2.   DARK AND PHOTOCONDUCTIVITY FOR A-(Si,Ge):H FILMS
             DEPOSITED AT 250°C

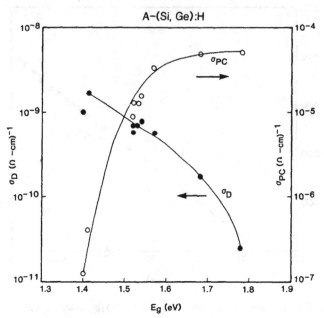

FIGURE 3.   DARK AND PHOTOCONDUCTIVITY OF A-(Si,Ge) FILMS
            DEPOSITED AT 300°C

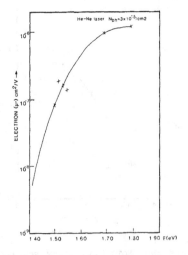

FIGURE 4.   ELECTRON $(\mu\tau)$ PRODUCT FOR a-(Si,Ge):H FILMS DEPOSITED AT
            300°C

We believe that one of the reasons why higher temperatures lead to better a-(Si, Ge):H films is that Ge may introduce structural disorder in the material.[10] As temperature increases, the disorder decreases. However, at higher temperatures ($\sim$325°C), H begins to evolve rapidly from the films, and thus, there may be an optimum temperature of $\sim$300°C for growth of high quality films.

Measurement of Valence Band Tail States

To measure valence band tail states, and hole ($\mu\tau$) products, we made pin devices as described in Section 2 above. The technique for measuring valence band tail states was to measure the tail of quantum efficiency (QE) of the device under reverse bias. When the device is under reverse bias, all electron-hole pairs generated by light are collected. Therefore, quantum efficiency is directly proportional to αt, where α is the absorption coefficient, and t is the thickness of the i layer assuming Xt < 1, as is the case for small α.[11]

Thus, by measuring the tail of QE vs. photon energy for low energy, (sub-bandgap) photons, we can obtain the tail absorption coefficient (Uhrbach tail). Since, generally speaking, sub-bandgap absorption occurs from valence band tail states to conduction band,[12] (see Figure 5), $\alpha \sim \exp (E/E_o)$, where E is the photon energy, and E_o is a characteristic slope of the uhrbach tail.[13] E_o is a measure of disorder,[13] and a small E_o means a sharp valence band trail, and low mid-gap defects.

In Figure 6, we plot log QE vs. E, for an a-(Si, Ge):H device of $\sim$1.55 eV i layer, and also show a similar plot for an a-Si:H device with 1.8 eV i layer. Both devices were made at 250°C from gas mixtures diluted at least 5:1 with H_2. We note that both a-Si:H and a-(Si, Ge):H can have sharp valence band tails, indicating low valence band disorder.

Hole ($\mu\tau$) Product

Hole ($\mu\tau$) product is an important parameter for devices. A high ($\mu\tau$) indicates a high diffusion length. We use the technique developed by Dalal and Alvarez (11) to measure hole ($\mu\tau$). The technique consists of measuring QE of a device under forward and reverse bias and matching the experimental points to a theoretical curve where ($\mu\tau$) is the only unknown. In Figure 7, we plot the experimental points for QE of a 1.6 eV device vs. voltage and also show the analytical curve. From the fit, $\mu\tau \sim 1.5 \times 10^{-8}$ cm^2/V, or a diffusion length of 0.2 μm. Further work on measuring ($\mu\tau$) products for different a-(Si, Ge):H alloys is in progress.

Conclusions

In conclusion, we have shown that a-(Si, Ge):H with high electron and hole ($\mu\tau$) products can be made, and that these alloys have very sharp valence band tails. A catastrophic decline in electronic quality sets in at Eg. < 1.5 eV, which may be due to increased structural disorder due to increased Ge concentration in the alloy.

Acknowledgments

We thank C. Botts and F. Berry for their technical assistance during the course of this work. This work was partially supported by SERI under subcontract ZB-4-03055-1, and by Polaroid.

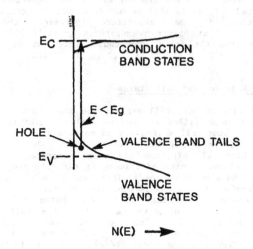

FIGURE 5. A MODEL FOR ABSORPTION FROM VALENCE BAND TAILS

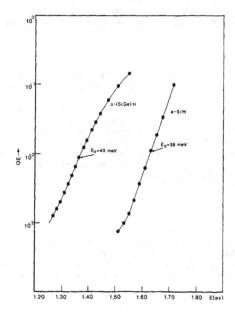

FIGURE 6. PLOT OF LOG OF QE OF PIN DEVICES VS. PHOTON ENERGY

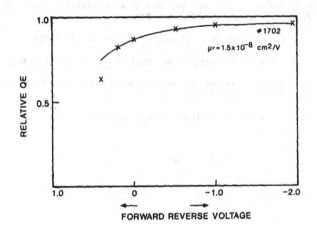

FIGURE 7. PLOT OF QE VS. APPLIED VOLTAGE IN a-(Si,Ge) PIN DEVICE

References

1. J. Chevallier, H. Wieder, A. Onton and C.R. Guarnieri, Solid State Comm. 24, 867 (1977)

2. Y. Marfaing, 2nd European Photovoltaic Energy Conf. (Berlin, 1979) p. 287

3. V.L. Dalal and E.A. Fagen, Proc. 14th IEEE Photovolt. Conf. (San Diego, CA, 1980) p. 1066

4. V.L. Dalal, Proc. 17th IEEE Photovolt. Conf. (Orlando, FLa., 1984) p. 86

5. V.L. Dalal, C.M. Fortmann and E. Eser, AIP Proc. #73, p. 15 (1981)

6. G. Nakamura, K. Sato, Y. Yukimoto, Proc. 15th IEEE Photovolt. Conf. (San Diego, CA, 1982) p. 1331

7. G. Nakamura et al. J. Non-Cryst. Solids, 59-60, IIII (1983)

8. G. Nakamura et al. Proc. of 3rd European PV Solar Energy Conf. (Cannes, France, 1980) p. 835

152

9. P.E. Vanier, A.E. Delahoy and R.W. Griffith, Proc. AIP Conf. #73, (Carefree, AZ, 1981) p. 227

10. W. Paul & J. McKenzie, Phys. Rev. B, (Feb. 15, 1985)

11. V. Dalal & F. Alvarez, J. de Physique 42, C-4, 491 (1981)

12. G. Cody, in "Semiconductors and Semimetals", Ed. J. Pankove 21B, 11 (1984) (Academic Press, NY)

13. G. Cody et al. J. de Physique 42, C-4, 301 (1981)

PREPARATION OF α-(Si,Ge):H ALLOYS BY D.C. GLOW DISCHARGE DEPOSITION

D. SLOBODIN, S. ALJISHI, R. SCHWARZ, AND S. WAGNER
Department of Electrical Engineering and Computer Science
Princeton University, Princeton, NJ 08544

ABSTRACT

We report the preparation of α-(Si,Ge):H alloy films by decomposition of SiF_4, GeF_4, and H_2 in a d.c. glow discharge. Germanium is incorporated very efficiently from GeF_4. The germanium content and optical gap can be controlled by varying the GeF_4 flow while keeping the SiF_4 and H_2 flows constant. The films, all prepared at 300 °C substrate temperature, exhibit high photo- to dark-conductivity ratios for compositions of up to ~25% germanium.

INTRODUCTION

Amorphous hydrogenated silicon/germanium alloys, α-(Si,Ge):H, are potentially valuable materials for solar cell and other device applications. Their attraction stems from the possibility of tailoring their optical or mobility gap to specific applications while still maintaining the favorable electronic properties exhibited by unalloyed amorphous silicon (α-Si:H). Nozawa et al. have shown that α-(Si,Ge):H alloys with a high photo- to dark-conductivity ratio can be made for Ge concentrations up to 50% [1].

α-(Si,Ge):H films have been prepared in a variety of ways including reactive sputtering and r.f. and d.c. glow discharge decomposition of silane/germane gas mixtures [2-5]. Regardless of the deposition method, a common feature of these films is an apparently large density of states in the mobility gap [6]. These states are thought to be predominantly associated with germanium dangling bonds that remain unpassivated by hydrogen [7]. This is believed to be a direct result of the relative weakness of Ge-H bonds as compared to Si-H bonds.

Consequently, fluorine appears particularly attractive as a dangling bond passivator in these materials in light of the greater bond strength of fluorine to germanium and silicon. Nevertheless, little effort has been made to investigate the possibility of involving fluorine in the alloys. Nozawa et al. have prepared α-(Si,Ge):H:F films by r.f. glow discharge of an SiF_4, GeF_4, and H_2 gas mixture [1]. The high photo- to dark-conductivity ratios for those films appear promising. However, more research must be carried out in order to evaluate the true potential of α-(Si,Ge):H alloys made from fluorides and to understand the effect of fluorine on alloy structure.

We have prepared α-(Si,Ge):H alloy films by d.c. glow discharge deposition technique using a SiF_4, GeF_4, and H_2 gas mixture. Films with a range of germanium contents were prepared by varying the amount of GeF_4 in the gas mixture while keeping all other deposition parameters fixed. In this paper we discuss the nature of the deposition process and report the structural, optical, and electronic properties of these films.

EXPERIMENTAL

Preparation

A schematic of the deposition system is shown in Figure 1. The system is elasto-mer sealed and pumped by a 200 l/s turbomolecular pump backed by a mechanical pump. The deposition chamber is baked out with an infrared quartz heat lamp. The piping for the three gases is baked out with heat tape. Prior to each deposition, the system is allowed to reach a base pressure of below 10^{-7} Torr. The residual gases are monitored with a UTI 100 quadrupole mass spectrometer. (The major residual gases are water and nitrogen). The base pressure is monitored with a cold cathode gauge.

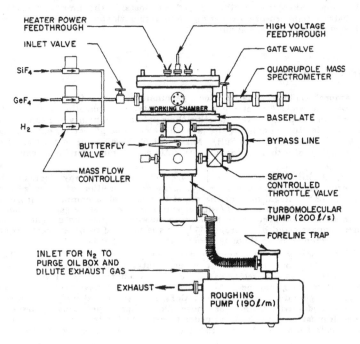

SCHEMATIC OF VACUUM SYSTEM USED FOR DEPOSITION
OF $\alpha - (Si, Ge):H$

Fig. 1

A proximity (or cathode screen) mode of deposition was used. The electrode configuration is shown in Figure 2. Three 1" X 3" substrates (or one 3" X 3" sub-strate) are held in place by a stainless steel frame with a 5 inch outside diameter. The frame clamps the substrates onto a copper block covered with silver foil provid-ing good thermal contact. The copper block is heated by ceramic resistance heaters designed for UHV use. The substrate temperature is measured with a chromel-alumel thermocouple that is precision press fitted into the copper block such that the thermocouple junction is as close to the substrate side as possible. The substrate temperature is controlled by a three-term temperature controller. The substrate electrode is held ~1 cm above the perforated stainless steel cathode screen by a high voltage feedthrough that allows the entire electrode to be set to any potential. In these experiments the substrate electrode was grounded. The cathode screen rests on a quartz ring of 4.5 inch outside diameter, 1 inch height, and 1/16 inch wall thickness. The quartz ring insulates and uniformly separates the cathode from the anode. In this configuration, the glow occurs primarily between the anode and cathode since the distance between the cathode and substrate is equal to or less than the dark space. Quartz rings of other heights may be used to vary the cathode to anode spacing. A quartz ring of height smaller than the dark space would allow for normal d.c. cathodic or anodic deposition. The anode is a screen with small perforations which makes it act as a gas diffuser to promote uniform film deposition. This electrode design can be modified and cleaned easily.

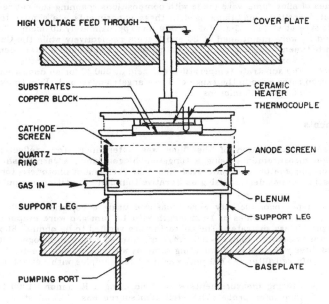

ELECTRODE CONFIGURATION FOR D.C. GLOW DISCHARGE DEPOSITION

Fig. 2

The gases are metered by individually set mass flow controllers. The gas mixture is introduced into a plenum below the anode screen. The gas passes through the anode screen, through the glow discharge region and through the cathode screen. The used gas is then pumped though a port centered on the chamber baseplate. During deposition the main butterfly valve is closed and the chamber is pumped through a bypass line. A servo-controlled throttle valve is mounted in the bypass line to control pressure during deposition. This pressure is measured by a capacitance manometer.

The semiconductor gases used were of the highest purity currently available. The silicon tetrafluoride used was Synthatron 99.999 vol. % purity (batch analysis). The germanium tetrafluoride was manufactured by Ozark-Mahoning without analysis. The gas was analyzed for major impurities by bleeding into the vacuum chamber and taking mass spectra with the quadrupole mass spectrometer. No significant impurities above the background could be detected after checking mass range from 1 to 250 amu. The hydrogen was Scientific Gas Products research grade 99.9999 vol. % analyzed.

Three type of substrates were used: 10 ohms/square tin oxide coated glass, Corning 7059 glass, and 7059 glass metallized with a semi-transparent layer of chromium. All substrates were cleaned in MOS grade methanol, blown dry in filtered dry nitrogen, rinsed in filtered deionized water, and air dried in a laminar flow hood. Prior to deposition, the substrates were glow discharge cleaned for 3 minutes in a 15% hydrogen/85% argon plasma at 0.5 Torr pressure and 40 watts power.

A series of alloy films were made with compositions spanning the entire range. The typical deposition conditions used in these experiments are similar to those used by Nozawa et al. [1]. The total gas flow was approximately 30 sccm. The flows for SiF_4 and H_2 were maintained at 20 and 9 sccm respectively while the GeF_4 flow was varied between 0 and 1.8 sccm. A total power of 62 watts and a power density of 0.52 watts/cm^2 was used throughout. The discharge was current controlled at 100 milliamperes. The substrate temperature was held at 300 °C for all deposition runs. The deposition rates were in the range of 3 to 4 angstrom/second. Film thicknesses ranged between 0.4 and 2.0 microns.

Measurements

The optical absorption of the films was determined from reflection and transmission measurements using a tungsten-halogen lamp, a Jarrell-Ash double grating monochromator, and calibrated silicon or germanium photodetectors. The optical absorption was derived using an iterative thin film calculation computer program.

Dark and photoconductivity were measured using a coplanar parallel stripe geometry. Aluminum stripes of 1 cm length with 1.6 mm gap were evaporated on films with plain glass sustrates. The contacts were verified to be ohmic. Photoconductivity was measured under ~100 mW/cm^2, filtered tungsten-halogen illumination. A red filter was used in illuminating samples with optical gap greater than 1.5 eV while an infrared filter ($\lambda > 750$ nm) was used for samples with optical gap less than 1.5 eV.

Raman scattering measurements were done using a Ramanor U1000 (Instruments S.A.) Raman microprobe. The excitation source was 5145 angstrom, 3 mW argon ion laser light. The beam was focused on a 15 μm spot. Backscattering geometry was used with no polarization selection. A 280 to 530 cm^{-1} range was examined. The measurement had ~ 3 cm^{-1} resolution.

A Physical Electronics Auger system was used for Auger electron spectroscopy (AES). The relative concentrations of silicon and germanium were measured by comparing the peak heights of the Si(KLL) and Ge(LMM) signals with those of a calibrated standard. The standard was an α-(Si,Ge):H film whose composition was accurately determined by proton elastic scattering [8]. The concentration of fluorine was estimated using the known approximate sensitivity factor. Secondary ion mass spectrometry (SIMS) measurements were performed with a Cameca IMS 3F system. A cesium ion source was used. Silicon, germanium, and hydrogen concentrations were measured using the same calibrated standard as used in AES.

The microscopic structure of the films was observed using scanning electron microscopy (SEM). The films on insulating substrates were sputter coated with gold to avoid charging effects. Film thicknesses were measured by observing the broken edge of a substrate.

RESULTS

    Germanium is incorporated very efficiently into films from GeF_4. This is seen in Figure 3 which shows film composition versus GeF_4 flow and GeF_4 flow relative to SiF_4 flow. Only a few percent of GeF_4 are needed in the gas phase to produce a germanium-rich film. This is a consequence of the unexpectedly different plasma chemistries of GeF_4 and SiF_4. This was striking in earlier experiments in which we found it nearly impossible to deposit pure α-Ge:H:F from a GeF_4/H_2 gas mixture in analogy with the deposition of α-Si:H:F from SiF_4/H_2 [9,10]. The main product of the GeF_4/H_2 runs, over a wide range of GeF_4/H_2 ratios and deposition power, was a white, crystalline, hygroscopic powder that formed on all cold surfaces in the deposition chamber. From the mass spectrum of its vapor, this substance is believed to be GeF_2, which is a stable solid at room temperature [11].

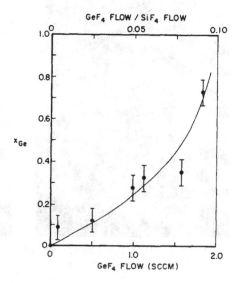

Fig. 3 Variation of germanium fraction of α-$Si_{1-x}Ge_x$:H alloy with GeF_4 flow.

    The preferential incorporation of Ge out of $GeF_4/SiF_4/H_2$ mixtures can lead to marked GeF_4-depletion of the reacting gas and thus to local composition gradients on the substrates. This effect became increasingly important at higher GeF_4 flows. For the experimental results reported here, we chose the substrate section that showed the least Ge-concentration gradient. Nevertheless, some of the scatter in our data may result from differences in the location on the film between the various measurements. It is clear that uniform gas flow is crucial to obtaining uniform film composition.

    The amorphicity of the films was verified by both Raman scattering and X-ray diffraction measurements. Samples with 0 and ~50% germanium were examined. The Raman spectra revealed that the films contained no crystalline phase. Wide angle X-ray diffraction measurements confirmed these results. No peak structure was observed over scattering angles (2Θ) ranging from 20° to 70° for both samples.

    The composition of the films was determined by using both AES and SIMS. The ratios of germanium to silicon content obtained from both techniques agreed within 5% for all samples. The concentration of fluorine was estimated by AES to be approximately 1% for all samples. No concentration trends were observable within the resolution of the Auger system. In contrast, the SIMS results suggest a trend of increasing fluorine concentration with increasing germanium fraction of the film. The absolute magnitude of the fluorine concentration could not be ascertained due to the lack of an adequate fluorine calibration standard. The hydrogen concentration ranged from 2 to 10% with no obvious trends apparent.

The film morphology was very substrate dependent. Numerous pinholes were observed in films deposited on conducting tin-oxide coated glass substrates while few pinholes were observed in films deposited on plain glass. The pinhole problem appeared to increase as the GeF₄ flow was increased. Figure 4 shows a typical alloy film deposited on a tin-oxide coated glass substrate and a bare tin-oxide coated glass slide for comparison. While it is obvious that the alloy film texture follows that of the substrate, it is more important to note that the film appears to coalesce around substrate irregularities and form islands. This may be partially responsible for pinhole formation.

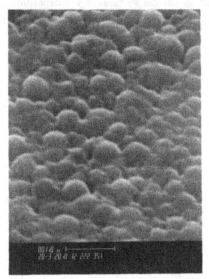 

ALLOY FILM ON TIN OXIDE GLASS SUBSTRATE

TIN OXIDE GLASS SURFACE

SCANNING ELECTRON MICROGRAPHS ×20000 MAGNIFICATION

Fig. 4

The optical gaps of the alloy films were determined using the conventional Tauc plot of $(\alpha E)^{\frac{1}{2}}$ versus E, where E is the energy of the incident photons and α is the optical absorption coefficient at that particular photon energy. For all samples, the Tauc plots were linear over a range sufficient for a meaningful optical gap to be derived. Figure 5 shows a plot of the optical gap versus film composition. The optical gap is seen to vary from 1.73 eV for pure α-Si:H to 1.0eV for a sample having a 0.8 germanium fraction. (Samples of greater germanium fraction had pinholes that disturbed the optical measurements.) The plot appears nearly linear over the range of the compositions of the samples. This is similar to that reported previously for alloys prepared from glow discharge of SiH₄/GeH₄ [3,5]. Extrapolation of the plot to $x_{Ge}=1.0$ suggests that a pure germanium film would have an optical gap of ~0.8 eV.

This asserts that the germanium-rich films contain relatively low amounts of hydrogen [12].

An upper limit of 5×10^{17} spins/cm^3 was assigned to the samples since no spin signal was observable on our calibrated electron spin resonance (ESR) machine. Figure 6 shows the photo- and dark-conductivities (room temperature) versus germanium fraction. There is approximately 3 to 4 orders of magnitude difference between photo- and dark- conductivities for alloys up to $x_{Ge} \sim 0.25$. However the difference drops abruptly for $x_{Ge} > 0.25$. The change is the result of the abrupt increase in dark conductivity. This type of behavior has been reported previously [13]. The dark conductivity was observed to be temperature activated over a range of temperatures extending from room temperature to 170 °C.

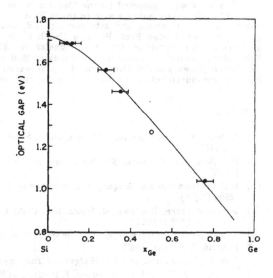

Fig. 5 Dependence of optical gap on α-Si$_{1-x}$Ge$_x$:H alloy composition. Open circle is for a film whose composition was interpolated using Fig. 3 and the known flow conditions.

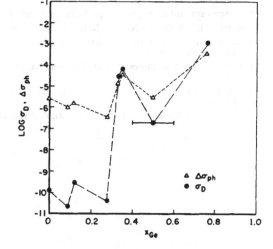

Fig. 6 Variation of room temperature dark- and photo-conductivity with α-Si$_{1-x}$Ge$_x$:H alloy composition.

CONCLUSIONS

We have shown that alloy films with a range of compositions can be prepared by varying the GeF_4 in the gas mixture. The results of this preliminary study suggest that films prepared by d.c. glow discharge of a $GeF_4/SiF_4/H_2$ mixture are quite similar in their properties to those prepared from glow discharge of SiH_4/GeH_4 mixtures. The role of fluorine in determining film properties has yet to be determined.

ACKNOWLEDGEMENTS

This work was supported by the Electric Power Research Institute under contract RP-1193-6. The authors wish to thank Prof. Antoine Kahn and Kevin Stiles for their help in performing and interpreting AES measurements. The authors also gratefully acknowledge Prof. Philippe Fauchet and Prof. Steve Lyon for Raman scattering measurements, Dr. Steve Schafer for ESR measurements, and Mark Schneider of Siemens, Princeton, NJ for making SEM micrographs. Special thanks to Dr. Steven Schwarz of Bell Communications Research in Murray Hill, NJ for carrying out SIMS measurements and aiding in their interpretation.

REFERENCES

[1] K. Nozawa, Y. Yamaguchi, J. Hanna, and I. Shimizu, J. Non- Cryst. Solids, **59-60**, 533 (1983)

[2] P.K. Banerjee, R. Dutta, S.S. Mitra, and D.K. Paul, J. Non-Cryst. Solids **50** 1-11 (1982)

[3] M.C. Cretella and J.A. Gregory, J. Electrochem. Soc. **129** 2850 (1982)

[4] B. von Roedern, D.K. Paul, J. Blake, R.W. Collins, G. Moddel, and W. Paul, Phys. Rev. B **25** 7678 (1982)

[5] Y. Yukimoto, JARECT**6**, 136-147 (1983), Y. Hamakawa, ed. Ohm-sha and North Holland, New York, 1983.

[6] C.-Y. Huang, S Guha, and S.J. Hudgens, J. Non-Cryst. Solids **66** 187-192 (1984)

[7] W. Paul, D.K. Paul, B. von Roedern, J. Blake, and S. Oguz, Phys. Rev. Lett. **46**(15) 1016 (1981)

[8] R. Schwarz and S. Wagner, Appl. Phys. Lett. **46**(6) 552 (1985)

[9] A. Madan, S.R. Ovshinsky, and E. Benn, Phil. Mag. B **40** 259 (1979)

[10] Y. Uchida, T. Ichimura, O. Nabeta, and H. Haruki, Jpn. J. Appl. Phys. **21** Suppl. 21-2, 193-198 (1982)

[11] F. Glockling, **The Chemistry of Germanium**, (Academic Press, New York, 1969) p. 32

[12] J.A. Reimer, B.A. Scott, D.J. Wolford, and J Nijs, Appl. Phys. Lett. **46**(4) 369 (1985)

[13] N.V. Dong, T.H. Danh, and J.Y. Leny, J Appl. Phys. **52**(1) 338 (1981)

AMORPHOUS SILICON-GERMANIUM DEPOSITED BY PHOTO-CVD

H. ITOZAKI, N. FUJITA AND H. HITOTSUYANAGI
Sumitomo Electric Industries, Ltd., 1-1 Koyakita 1-Chome, Itami 664, Japan

ABSTRACT

Hydrogenated amorphous silicon germanium (a-SiGe:H) films were deposited by photo-chemical vapor deposition (Photo-CVD) of SiH4 and GeH4 with mercury sensitizer. Their band gap was controlled from 0.9 eV to 1.9 eV by changing the gas ratio of SiH4 and GeH4. High quality opto-electrical properties have been obtained for the a-SiGe:H films by Photo-CVD. Hydrogen termination and microstructure of a-SiGe:H were investigated by infrared absorption and transmission electron microscopy. An a-Si:H solar cell and an a-Si:H/a-SiGe:H stacked solar cell were made, each of which has conversion efficiency 5.3% and 5.1%, respectively.

INTRODUCTION

Recently a-Si solar cells with high efficiency have been developed [1]. Multi-gap stacked solar cell is one of the most hopeful candidates to get higher efficiencies [2,3]. Materials with different band gap are necessary to make this kind of solar cells. A-SiGe:H which has a narrower band gap than a-Si:H was applied to the stacked cell [4]. A-SiGe:H deposited by glow discharge (GD) has not sufficient opto-electrical properties yet. In GD plasma, high energy charged particles exist to some extent and make damages to the film. However, in Photo-CVD this radiation-induced damage might be ignored. Therefore, high-quality films can be expected by Photo-CVD. Although this technique has been already applied to a-Si:H [5-7] , a-SiGe:H deposited by Photo-CVD has not been reported. In this paper, the growth of a-SiGe:H by Photo-CVD is introduced and its application to solar cells is also discussed.

EXPERIMENTAL

Fig.1 shows a schematic diagram of a Photo-CVD reactor. The light source is a low pressure mercury lamp emitting 184.9 nm and 253.7 nm ultra-violet lights. A quartz window is placed on reactor and its internal surface was coated with low pressure oil in order to prevent the film deposition on the window. The gas inlet system has a mercury vaporizer so that sufficient mercury vapor is introduced into the reactor chamber. Mercury vapor pressure is controlled by the temperature of mercury which was kept 60°C for this work. 100% SiH4 and 10% diluted GeH4 with H2 were used for raw gases. The detailed experimental conditions are shown in Table 1.

The structure of the deposited films was investigated by infrared absorption and transmission electron microscopy.

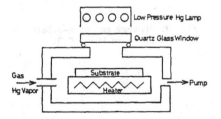

Fig. 1. Schematic diagram of Photo-CVD Reactor.

Table 1. Experimental condition

Light Source	Low Pressure Mercury Lamp
Gas	SiH4 (100%), GeH4 (10% / H2)
Gas Flow Rate	10 - 200 sccm
Pressure	5 Torr
Temp. of Hg	60 °C
Substrate Temp.	200 °C
Substrate	Glass, TCO coated Glass, Si-Single Crystal

RESULTS AND DISCUSSIONS

A-SiGe:H deposition

A-SiGe:H films with a various band gap were obtained by changing gas flow ratios of GeH4/(SiH4+GeH4). Fig.2 shows the relation between band gap (Eg) and GeH4/(SiH4+GeH4). Eg was determined by plottings of $(\alpha h\nu)^{\frac{1}{2}}$ vs. $h\nu$. Eg ranges from 0.9 eV to 1.9 eV.

Band gap energy decreases rapidly as increasing the gas ratio of GeH4/(SiH4+GeH4). Photo-CVD requires less GeH4 gas than glow-discharge to get a-SiGe:H with the same Eg. It suggests that Photo-CVD is a lower cost process for a-SiGe:H film than GD.

Opto-electrical properties

Fig.3 shows the photo conductivity (under the illumination of AM 1.5, 100 mW/cm^2) ($\Delta\sigma$ph) and dark conductivity (σd) vs. Eg. σd increases from 10^{-9} to 10^{-6} (ohm-cm)$^{-1}$ as Eg decreases from 1.7 eV to 1.3 eV. $\Delta\sigma$ph is 10^{-4} (ohm-cm)$^{-1}$ and constant for Eg of from 1.8 eV to 1.5 eV, and decreases to 10^{-5} (ohm-cm)$^{-1}$ as Eg decreases to 1.3 eV. For Eg of less than 1.5 eV, $\Delta\sigma$ph decreases with decreasing Eg. While $\Delta\sigma$ph of a-SiGe:H by GD is constant for Eg of down to 1.6 eV and decreases as Eg decreases for Eg of less than 1.6 eV. The dark conductivity increases rapidly with decreasing Eg. In contrast, in GD films, $\Delta\sigma$ph decreases rapidly with Eg, but σd does not incease as rapidly as in Photo-CVD film.

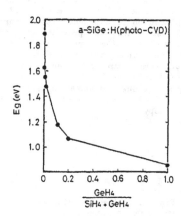

Fig. 2. Energy gap of a-SiGe:H and Gas Ratio in Photo-CVD.

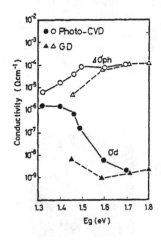

Fig. 3. Conductivity vs. Eg of
a-SiGe:H by Photo-CVD and GD method.

Structure of the film

Structure of the a-SiGe:H by Photo-CVD has been investigated using infrared absorption and transmission electron microscopy (TEM).

Firstly, Fig.4 shows the infrared absorption coefficient spectrum of a-SiGe:H (Eg = 1.4 eV) by Photo-CVD around 2000 cm^{-1}. The a-SiGe:H has absorptions at 2000 cm^{-1} and 1900 cm^{-1}, which indicates that dangling bonds of silicon and germanium are mainly terminated by one hydrogen atom (Si-H, Ge-H). The a-SiGe:H film has also the absorption at 2100 cm^{-1} that is due to Si-H2 bonds. Therefore, these Si-H2 bonds may be one of the reasons why opto-electrical properties of the a-SiGe:H by Photo-CVD degrade as Eg decreases.

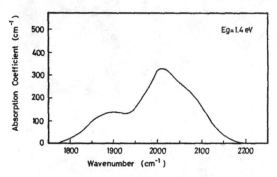

Fig. 4. Infrared Absorption of a-SiGe:H by Photo-CVD.

Fig.5 shows TEM micrographs of a-SiGe:H by Photo-CVD and by GD. Eg of
these a-SiGe:H are 1.6 eV. Electron diffraction patterns of both films show
halo patterns, which indicates these films are amorphous. Although, the
structure of a-SiGe:H by Photo-CVD is very uniform, the structure of the
a-SiGe:H film by GD is not so uniform and has tiny particles at about 50 nm
interval. These particles are microcrystalline. This microcrystalline phase
may be one of the causes of poor quality in GD a-SiGe:H films.

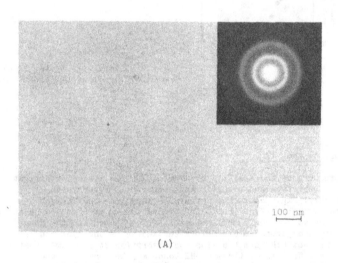

(A)

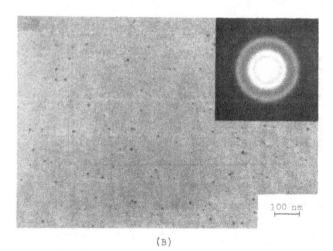

(B)

Fig. 5. TEM images and electron diffraction patterns of a-SiGe:H
by Photo-CVD (A) and GD method (B).

Application of a-SiGe:H to solar cells

An a-SiGe:H film with Eg of 1.65 eV prepared by Photo-CVD was applied to solar cells. The spectral response and I-V characteristics of these solar cells are studied.

Fig.6 shows the structure of the p-i-n solar cell. The i layer is a-SiGe:H deposited by Photo-CVD and p and n layers are a-Si:H by Photo-CVD. Fig.7 shows the spectral responses of collection efficiency of the a-SiGe:H solar cells by Photo-CVD and by GD, and also that of an a-Si:H solar cell by GD. The spectral response of a-SiGe:H cell by Photo-CVD shifts to longer wavelength compared with a-Si:H cell by GD. As a-Si:H cell has wide band gap p-type a-SiC:H layer (Eg = 2.0 eV) and a-SiGe:H cell has p-type a-Si:H layer (Eg = 1.8 eV), the a-SiGe:H cell by Photo-CVD has smaller response than a-Si:H cell in the short wavelength region (350 - 500 nm). The a-SiGe:H cell by Photo-CVD has larger collection efficiency around the wavelength of 500 to 700 nm than the cell by GD. This means that more photo-generated carriers are extracted to the electrodes without recombination in a-SiGe:H cells by Photo-CVD than in a-SiGe:H cells by GD. This indicates that the quality of a-SiGe:H by Photo-CVD is better than that by GD.

The cell performances of this solar cell were measured under AM 1.5 illumination of 100 mW/cm^2. I-V curve of this cell is shown as #1 in Fig.9. This is the first report of an a-SiGe:H solar cell by Photo-CVD. Shortcircuit current (Jsc) is 13.7 mA/cm^2, open circuit voltage (Voc) is 0.74 V, and conversion efficiency is 5.3%. While, Voc of a-SiGe:H cell by GD is 0.5 V. The a-SiGe:H cell by Photo-CVD has much higher Voc than the a-SiGe:H cell by GD, though the Eg of both cells are the same. This shows that a-SiGe:H by Photo-CVD has smaller state density at band tail than a-SiGe:H by GD, because the band tail states make Voc decrease [8].

Furthermore, a-SiGe:H cell by Photo-CVD was applied to an a-Si:H/a-SiGe:H stacked solar cell. A structure of this cell is shown in Fig.8. At first an a-Si:H p-i-n cell was deposited by GD on a TCO coated glass substrate, and then the a-SiGe:H p-i-n cell was deposited by Photo-CVD on the first a-Si:H cell. The I-V curve of this cell is shown as #2 in Fig.9. Jsc, Voc and efficiency of this cell was 6.7 mA/cm^2, 1.49 V and 5.1% respectively, though the optimization of the cell structure has not been done yet.

Hence, we expect that a high efficiency solar cell can be made by optimization of this Photo-CVD technique.

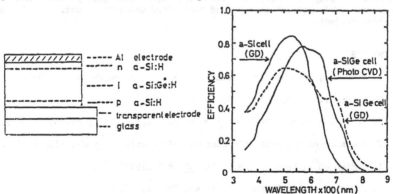

Fig. 6. Solar cell structure of a-SiGe:H.

Fig. 7. Spectral responce of collection efficiency of solar cells.

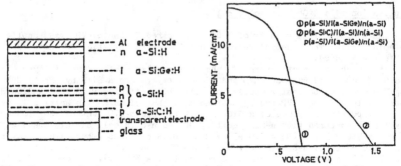

Fig. 8. Structure of a-Si:H/a-SiGe:H stacked solar cell.

Fig. 9. I-V characteristics of solar cells.

SUMMARY

(1) High quality a-SiGe:H films were obtained by mercury sensitized Photo-CVD of SiH4 and GeH4, and energy gap was controlled from 0.9 to 1.9 eV.

(2) The structure of a-SiGe:H by Photo-CVD was investigated. IR absorptions shows that a-SiGe:H by Photo-CVD has Si-H, Ge-H and also Si-H2 bonds. TEM observation shows a-SiGe:H by Photo-CVD is uniform and amorphous.

(3) High collection efficiency at longer wavelength has been obtained by a-SiGe:H solar cells by Photo-CVD. The a-SiGe:H solar cell and the a-Si:H/a-SiGe:H stacked solar cell have efficiency 5.3% and 5.1%, respectively.

ACKNOWLEDGEMENT

We wish to thank Associate Professor M. Konagai (Tokyo Institute of Technology) for the useful discussions. We acknowledge Professor Y. Hamakawa (Osaka University) for helpful advice. This work was partly supported by the Agency of Industrial Science and Technology under the contract of the Sunshine Project.

REFERENCES

1. Y. Hamakawa, Materials Research Society 1985 Spring Meeting (1985).

2. C. H. Henry, J. Appl. Phys. 51, 4494 (1980).

3. J. C. C. Fan et al. 16th IEEE Photovoltaic Specialist Conf. 692 (1982).

4. G. Nakamura et al. 16th IEEE Photovoltaic Specialist Conf. 1331 (1982).

5. T. Saitoh et al. Appl. Phys. Lett. 42, 678 (1983).

6. T. Inoue et al. Appl. Phys. Lett. 43, 774 (1983).

7. T. Tanaka et al. Appl. Phys. Lett. 45, 865 (1984).

8. T. Tiedje, Appl. Phys. Lett. 40, 627 (1982).

AN ASSESSMENT OF a-SiGe:H ALLOYS WITH A BAND GAP OF 1.5eV AS TO THEIR SUITABILITY FOR SOLAR CELL APPLICATIONS

B. VON ROEDERN[*], A.H. MAHAN, T.J. MCMAHON AND A. MADAN[*]
Solar Energy Research Institute, Golden, CO 80401
[*]Glasstech Solar, Inc., 12441 W. 49th Ave., Wheatridge, CO 80033

INTRODUCTION

Hydrogenated amorphous silicon germanium alloys a-$Si_{1-x}Ge_x$:H are being actively investigated for their application as a low band gap material in cascade solar cells [1,2]. To date, such alloys produce material of reasonable electronic quality only if the Ge-content is kept low (<40 at.%) such that the band gap is not decreased much below 1.5 eV. Conversion efficiencies of ~5% have been obtained with alloys having such a band gap, and tandem cells have shown conversion efficiencies which are lower than those of good quality single layer a-Si:H devices. Thus, the performance of alloys is well below that necessary to achieve conversion efficiencies of >16%, which are ultimately hoped to be obtained using the cascade approach [2]. Other low band gap alloys such as a-$Si_{1-x}Sn_x$:H have been shown to be even less suitable with regard to their electronic properties [3]. The cause of the degradation in electronic properties with increased alloying is not yet understood. Factors such as preferential attachment of H to Si rather than Ge [4] or microstructure observed in alloys have been suggested as a cause for the electronic degradation, [5,6] but no unique correlations have been established between such findings and the electronic properties.

The purpose of this paper is to try to understand the reasons for the amount of electronic degradation observed upon Ge incorporation. In order to do so, we first report on solar cell devices produced from device quality a-$Si_{1-x}Ge_x$:H material, noting certain deposition parameters which critically affect cell performance. We comment that even though our devices demonstrate state of the art device parameters similar to those reported elsewhere, such devices are still limited by material performance. To study in detail the reason for these limitations, we have gone back and produced a series of intrinsic a-$Si_{1-x}Ge_x$:H samples where the gap (E_g) was reduced systematically from 1.7eV to 1.4eV, and report quantitative studies of the defects that are introduced upon alloying.

From the changes of the density of states at the Fermi level and the subband gap absorption (in excess to the exponential tail ~0.5 eV above the valence band edge) and from modeling studies of device performance as a function of E_g we derive speculative information about the nature of states which are introduced in the alloying process.

EXPERIMENTAL

Samples for this study were produced on the cathode of a capacitively coupled radio frequency (rf) glow discharge system described elsewhere [7]. Pure SiH_4 and GeH_4 gases were used for the deposition of the alloys, while SiH_4 premixed with PH_3 or B_2H_6 was used to produce n^+ and p^+ layers in the devices. The deposition parameters used were similar to those reported by

Mackenzie et al. [5], where high deposition temperatures ($320^{\circ}C$) and low rf powers ($0.04W/cm^2$) resulted in enhanced electronic properties and a more homogeneous microstructure. Total flow rates in the present work were typically 10 sccm, and the deposition chamber pressure was ~250 mT. For the material studies, intrinsic layers were deposited simultaneously on glass, n^+/stainless steel, and crystalline silicon substrates, enabling measurements of dark-and photoconductivities, optical absorption, photothermal deflection spectra, space charge limited currents on identically prepared samples.

From these measurements, we have derived the mobility lifetime product of electrons $(\eta\mu\tau)_e$, the density of states at the Fermi level $g(E_F)$, the density of states ~0.5eV above the valence band edge $g(E_v)$, and the characteristic energy of the Urbach edge in a similar way as reported elsewhere [3]. By modeling the J(V) characteristics and photovoltaic performance for these alloys using a carrier generation recombination statistic which we have previously applied to a-Si:H pin junctions, [2] we derived further information about the nature of states introduced upon alloying.

RESULTS ON SCHOTTKY BARRIERS AND PIN JUNCTIONS

Schottky barriers were made with alloys of E_g=1.45 to 1.55eV on stainless steel substrates using $n^+(a-Si_{1-x}Ge_x:H:P)/i/Pd$ type structures. It turned out to be crucial to fabricate the n^+ layers from a doped alloy rather than a-Si:H; Schottky barriers fabricated using P doped a-Si:H layers as the n^+ back contact showed no rectification in the dark, and short circuit current densities (J_{sc}) and open circuit voltages (V_{oc}) of less than 1.5mA/cm^2 and 0.25V respectively under 100mW/cm^2 ELH illumination. As soon as we used P-doped a-Si$_{1-x}$Ge$_x$:H, the Schottky barriers exhibited parameters such as V_{oc} = 0.36V, J_{sc} = 4-6mA (not corrected for the ~50% transmission loss in the Pd

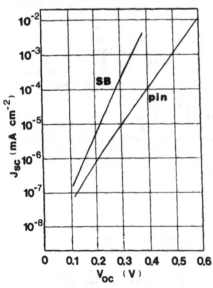

Figure 1

J_{sc} versus V_{oc} curves for a-SiGe:H (having a band gap E_g=1.5eV) in Schottky barrier (SB) and pin configuration.

contact) and rectification ratios of 10^3 to 10^4 in the dark. In Fig. 1, we show a J_{sc} (V_{oc}) curve for a Schottky barrier, which is obtained by reducing the incident illumination with neutral density filters. Diode quality factors n derived from such plots were typically 1.1 to 1.2 for the Schottky barrier structures. We note that our alloy layers tend to oxidize faster than a-Si:H; Schottky contacts could not be formed after the alloy surface had been exposed to air for several days, possibly because of the formation of a very resistive surface layer.

Solar cells with a pin structure of the type glass/SnO:F/pin/Al were fabricated where both doped layers were made with the alloy material. Again, use of the doped alloy material was crucial to device performance. We have obtained the following parameters under 100 mW/cm^2 ELH illumination: V_{oc}=0.60V, J_{sc}=13.5mA/cm^2, FF=0.46, thus $\eta \approx 3.7\%$. These numbers represent unoptimized parameters, as we have only deposited a limited number of pin solar cells to explore what kind of performance could be achieved with our alloy material with $E_g \approx$ 1.5eV. These films showed good diode behavior in the dark. We have noted 5 orders of rectification at V_{oc}, with n≈1.7, which is slightly above the n factors observed for a-Si:H pin cells. A $J_{sc}(V_{oc})$ curve is included in Figure 1. These solar cell parameters are close to the best obtained by the Mitsubishi group [1], and one has to conclude that material properties limit the performance of the alloy solar cells.

MATERIAL PROPERTIES OF a-Si$_{1-x}$Ge$_x$:H

In Table 1, we list the properties of a series of alloys with band gap varying from 1.70 to 1.41eV along with an intrinsic a-Si:H sample included for comparison purposes. In the table, we list the Ge content x, the ratio of the photoconductivity σ_L (under 100mW/cm^2 ELH illumination) to the dark conductivity, the activation energy of the dark conductivity ΔE, the $(\eta\mu\tau)_e$ product for electrons, the density of states at the Fermi level $g(E_F)$, the characteristic energy of the Urbach edge, E_o, and the density of states derived from the excess absorption in the PDS spectrum at an energy E_g-0.5eV, $g(E_v)$ [8]. The table lists samples in order of their decreasing band gap.

Table I

Properties of a-Si$_{1-x}$Ge$_x$:H. All films were typically 1.0 to 1.5 µm thick. The a-Si:H sample was deposited under slightly different conditions than those described for the alloy depositions.

x at.%Ge	σ_L/σ_D	ΔE eV	$(\eta\mu\tau)_e$ cm^2/Vsec	$g(E_F)$ cm^{-3}eV^{-1}	E_o meV	$g(E_v)$ cm^{-3}eV^{-1}	E_g eV
0	$2 \cdot 10^5$	0.69	$6 \cdot 10^{-6}$	$9 \cdot 10^{15}$	50	$1.5 \cdot 10^{17}$	1.70
7	$6 \cdot 10^4$	0.76	$4 \cdot 10^{-6}$	$4 \cdot 10^{16}$	52	$1.5 \cdot 10^{17}$	1.62
32	$5 \cdot 10^3$	0.82	$5 \cdot 10^{-7}$	$9 \cdot 10^{16}$	52	$2.1 \cdot 10^{17}$	1.52
44	$2 \cdot 10^3$	0.77	$2 \cdot 10^{-7}$	$1.3 \cdot 10^{17}$	56	$2.6 \cdot 10^{17}$	1.48
60	$4 \cdot 10^2$	0.74	$4 \cdot 10^{-8}$	$2.8 \cdot 10^{17}$	58	$4.7 \cdot 10^{17}$	1.41

In Fig. 2, we have plotted $(\eta\mu\tau)_e$, $g(E_F)$ and $g(E_v)$ as a function of E_g. As can be seen, $(\eta\mu\tau)_e$ decreases at a faster rate than $g(E_F)$ increases as the band gap is lowered. The increase in $g(E_v)$ is less rapid than that of $g(E_F)$. The activation energy ΔE indicates that the Fermi level moves to a lower position in the gap upon alloying.. We previously pointed out that this is a common trend observed in alloys [7], arguing that possibly some of the states near E_v act as acceptors.

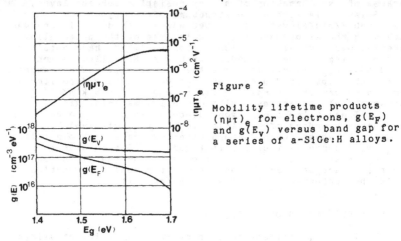

Figure 2

Mobility lifetime products $(\eta\mu\tau)_e$ for electrons, $g(E_F)$ and $g(E_v)$ versus band gap for a series of a-SiGe:H alloys.

MODELING AND DISCUSSION

In order to gain some insight on the nature of states that deteriorate the electronic properties upon alloying, we have modeled the performance of our alloy materials and have obtained an estimate of the solar cell performance that can be expected. We used a recombination statistic described earlier [2,10]. In the model, the density of states distribution is approximated by exponential tails with their respective characteristic energies E_{ch}^v and E_{ch}^c, and a flat bottom distribution g_{min}. This model has been very useful to calculate light and dark $J(V)$ characteristics (and thus diode quality factors), the photoconductivity and V_{oc} values for a-Si:H pin solar cells. In Table II, we present some calculated data, which show the influence of deep localized states (represented by g_{min}) on the diode quality factors, V_{oc} and the photoconductivity. We used the same values for both the characteristic energies of the band tails for the a-Si:H as well as the a-SiGe:H calculation ($E_{ch}^v = 0.06$eV and $E_{ch}^c = 0.03$eV). This is justified by the fact that the Urbach edges in our series of alloys show only a very minor broadening upon alloying as can be seen in Table 1. The first entry for each material represents a sample where no additional deep localized states were assumed to be present. The most important finding is that this calculation confirms the ~40% drop of V_{oc} when the band gap is narrowed, which more than offsets the increase in J_{sc} (roughly 30%) due to the increased generation rate, even when no additional deep defects are present.

Table II

V_{oc}, diode quality factors n and photoconductivity values calculated for a-SiGe:H and a-Si:H materials with varying density of deep localized states

	Eg eV	g_{min} $cm^{-3}eV^{-1}$	V_{oc} V	n	σ_{AM1} $\Omega^{-1}cm^{-1}$
Si	1.75	10^{13}	.84	1.39	$2.8 \cdot 10^{-4}$
	1.75	10^{15}	.825	1.53	$1.6 \cdot 10^{-4}$
	1.75	10^{16}	.793	1.63	$3.3 \cdot 10^{-5}$
	1.75	10^{17}	.734	1.74	$4.1 \cdot 10^{-6}$
SiGe	1.5	10^{14}	.604	1.39	$3.6 \cdot 10^{-4}$
	1.5	10^{16}	.574	1.56	$6.9 \cdot 10^{-5}$
	1.5	10^{17}	.519	1.74	$9.0 \cdot 10^{-6}$

When we analyze the data shown in Fig. 2, we find that the decrease in the $(\eta\mu\tau)_e$ products does not match closely the increases in $g(E_F)$ or $g(E_v)$, as the band gap is lowered. This could be due to a gradual increase in the capture cross section areas of the deep states, which is possibly accelerated as the band gap is lowered below 1.5eV. As can be seen in Table II from the calculated photoconductivity values, a state density of $10^{17}cm^{-3}eV^{-1}$ would actually lead to a higher photoconductivity in the 1.5eV material than in the E_g=1.75 a-Si:H sample. This is opposite of what is generally observed, thus suggesting a change in the nature of the deep states. However, the photoconductivity is also influenced by changes in the mobility and in the shallow trapping times, which makes it difficult to draw such a conclusion from the photoconductivity results alone [11].

In conclusion, alloy solar cells are limited by the decrease in V_{oc} which more than offsets the increased J_{sc} to the increased generation rate when the band gap is decreased. The fill factors and short circuit currents are limited by the fact that the number of deep defects is increased upon alloying, and there are indications that these defects might provide higher recombination rates than the same number of defects in a-Si:H [12].

REFERENCES

1. Y. Higaki, M. Aiga, S. Terazono, and Y. Yukimoto, Technical Digest of the Int. PVSEC-1, Kobe, (1984), p.209.
2. A. Madan, presented at ASES conference Minneapolis, 1983. Also in A. Madan "Silicon Processing for Photovoltaics", Ed. C.P. Khattak and K.V. Ravi; to be published by North Holland Publishing.
3. A.H. Mahan, B. von Roedern, D.L. Williamson and A. Madan, ibid, p.92.
4. W. Paul, D.K. Paul, B. von Roedern, J. Blake and S. Oguz, Phys. Rev. Lett. 46, (1981), 1016.
5. K.D. Mackenzie, J.R. Eggert, D.J. Leopold, Y.M. Li, S. Lin and W. Paul, Phys. Rev. B31, (1985), 2198.

6.   B.G. Yacobi, B. von Roedern, A.H. Mahan, and K.M. Jones, Phys. Rev. B., accepted for publication.

7.   B. von Roedern, A.H. Mahan, D.L. Williamson and A. Madan, J. Non-Cryst. Solids <u>66</u>, (1984) 13.

8   W.B. Jackson and N.M. Amer., Phys. Rev. B <u>25</u>, (1982) 5559.

9.   T.J. McMahon, B.G. Yacobi, K. Sadlon, J. Dick and A. Madan., J. Non-Cryst. Solids <u>66</u>, (1984) 375.

10.   T.J. McMahon and A.Madan, to be published.

11.   R.L. Weisfield (Thesis, Harvard U. 1983) reported for example that the photoluminescence intensity (which is limited by the competing non-radiative recombination process) of sputtered a-SiGe:H with $g(E_F)=10^{17}cm^{-3}eV^{-1}$ is a factor of ~7 less than that of a-Si:H with the same $g(E_F)$.

12.   If we assume that charged dangling bond states in addition to their neutral counterparts provide recombination centers (represented by an average capture rate in our recombination model), a decrease in the correlation energy upon alloying would increase the ratio of charged to neutral states, which could possibly account for the increased average capture rates observed in the alloys.

PART VI

Si-C Alloys

PHOTO-CONDUCTIVE AMORPHOUS SILICON CARBIDE PREPARED BY
INTERMEDIATE SPECIES SiF_2 AND CF_4 MIXTURE

HIDEKI MATSUMURA, TAKASHI UESUGI AND HISANORI IHARA

Department of Physical Electronics, Hiroshima University, Saijo,
Higashi-Hiroshima 724, Japan

Abstract

A new type of hydro-fluorinated amorphous silicon-carbide (a-SiC:F:H)
is produced by the glow discharge decomposition of gas mixture of CF_4, H_2
and intermediate species SiF_2. The electrical, optical and structural
properties of this a-SiC:F:H are studied and the results are compared with
the similar results for a-SiC:F:H produced from gas mixture of CH_4, H_2 and
SiF_2 and also for hydrogenated amorphous silicon carbide (a-SiC:H) produced
from CH_4 and SiH_4 gas mixture. It is found that the optical band gap can be
increased without degradation of photo-conductive properties only when
amorphous silicon carbide is produced from CF_4, H_2 and SiF_2 gas mixture.

1. Introduction

It is strongly required to control the optical band gap of amorphous
semiconductors without the degradation of material properties, such as
photo-conductivity, for its application on solar cells. However, inspite of
many trials when amorphous silicon-carbide (a-SiC) [1] or amorphous
silicon-germanium (a-SiGe) [2] is produced by the glow discharge
decomposition of SiH_4 and CH_4 gas mixture or SiH_4 and GeH_4 gas mixture, the
material properties are always degraded during the change of optical band
gap due to the incorporation of carbon or germanium.

Recently, it is reported that the material properties are not degraded
even when the optical band gap is narrowed by forming a-SiGe if it is
produced from gas mixture of SiF_4, GeF_4 and H_2 [3]. This report encourages
to use hydro-fluorinated amorphous semiconductors to control the optical
band gap.

On the other hand, we have already reported that hydro-fluorinated
amorphous silicon (a-Si:F:H) can be deposited stably under any plasma
conditions when it is produced from gas mixture of H_2 and intermediate
species SiF_2 instead of gas mixture of H_2 and conventional SiF_4. And we
have revealed that, a)this newly produced a-Si:F:H is heat-resistant up to
450 °C [4], b) the diffusivity of impurity in a-Si:F:H is much smaller than
that in the conventional hydrogenated amorphous a-Si (a-Si:H) produced from SiH_4 [5],
and that c) a-Si:F:H can be deposited with a rate near to 20 Å/sec without
degradation of material properties [6].

Therefore, we tried to use this SiF_2 system to produce a-SiGe and a-
SiC. Particularly, here, we tried to produce a-SiC:F:H films by using SiF_2,
CF_4 and H_2 gas mixture. The electrical, optical and structural properties
of this a-SiC:F:H are studied and the results are compared with the similar
results for a-SiC:F:H produced from CH_4, H_2 and SiF_2 gas mixture and also
for a-SiC:H produced from SiH_4 and CH_4 mixture. It is found that the
optical band gap can be widened without apparent degradation of photoconductive properties only when a-SiC is produced from SiF_2, CF_4 and H_2 gas
mixture.

2. Fundamentals for Experiments

A deposition apparatus of a-SiC:F:H is schematically illustrated in Fig. 1. Solid silicon pieces of a size of about 5 mm are packed inside an intrinsic poly-silicon tube, this silicon tube is inserted just inside a quartz tube, and all these are heated by an electric furnace. SiF_4 gas is introduced into the tube and converted to intermediate species SiF_2 by following the chemical reaction: $SiF_4 + Si \rightarrow 2SiF_2$. The reaction rate to produce SiF_2 is increased as the temperature of furnace Tf is elevated. In the present experiment, Tf is kept at 1150 °C. It is estimated that about 60% of SiF_4 is converted to SiF_2 at Tf = 1150 °C [6]. The SiF_2 is immediately mixed with H_2 and CF_4 or CH_4 gases and the gas mixture is introduced into a deposition tube and decomposed to make a-SiC:F:H by radio-frequency (RF) plasma.

The deposition conditions to prepare a-SiC:F:H from CF_4, H_2 and SiF_2 mixture (CF_4 based a-SiC:F:H) and also from CH_4, H_2 and SiF_2 mixture (CH_4 based a-SiC:F:H) are summarized in Table I. Here, the notations FR(G) such as FR(SiF_4) or FR(H_2), Ts, Pg and P.D. refer to the flow rate of gas G, the temperature of a substrate holder, the gas pressure during deposition and the power density of RF glow discharge, respectively. As shown in Table, Ts for CF_4 based a-SiC:F:H is different from Ts for CH_4 based one, because Ts is fixed at the temperature at which the photo-conductivity and the photo-sensitivity become maximum in each case. The total flow rate of CF_4 and SiF_4 at input and that of CH_4 and SiF_4 are both fixed at 50 sccm. Although a-SiC:F:H films can be grown in any conditions of mixing ratios of H_2 when SiF_2 is used instead of SiF_4, the flow rate of H_2 is also fixed at 50 sccm. The mixing ratio of CF_4 or CH_4 into SiF_2 and H_2 gas mixture is adjusted by changing the ratio X, defined as FR(CF_4)/[FR(CF_4)+FR(SiF_4)] or FR(CH_4)/[FR(CH_4)+FR(SiF_4)].

3. Structural Properties

At first, the carbon contents in both CF_4 based and CH_4 based a-SiC:F:H films were measured by the X-ray photo-electron spectroscopy (XPS). The results are shown in Fig. 2 as a function of the gas mixing ratio X. For CH_4 based a-SiC:F:H, a result of Ts = 300 °C is also plotted. In this figure, the similar results for a-SiC:H produced from SiH_4 and CH_4 gas mixture, measured by the Auger electron spectroscopy (AES) and reported by

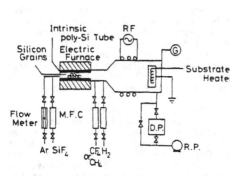

Fig.1 Schematic diagram of deposition Apparatus. M.F.C. refers to the mass flow controller.

	a-SiC:F:H	
	CF_4 based	CH_4 based
FR(CF_4)+FR(SiF_4)	50 sccm	——
FR(CH_4)+FR(SiF_4)	——	50 sccm
$\dfrac{FR(CF_4)}{FR(CF_4)+FR(SiF_4)}$	0 to 0.8	——
$\dfrac{FR(CH_4)}{FR(CH_4)+FR(SiF_4)}$	——	0 to 0.7
FR(H_2)	50 sccm	50 sccm
Ts	275 °C	350 °C
Pg	10 Pa	10 Pa
P.D.	0.32 W/cm²	0.32 W/cm²

Table I Deposition conditions of CF_4 and CH_4 based a-SiC:F:H.

Tawada et al. [1], are also shown for comparison. From this figure, it is known that carbon atoms are incorporated in CF_4 based a-SiC:F:H by adjusting the mixing ratio X as similar as in a-SiC:H, and that carbon atoms are likely to be more incorporated in CH_4 based a-SiC:F:H than in CF_4 based one.

To know the difference between CF_4 based and CH_4 based a-SiC:F:H films, next, the bond structure of these films were observed by infrared (IR) absorption measurments. The results are shown in Fig. 3. In this figure, an IR spectrum of CH_4 based a-SiC:F:H deposited at Ts = 300 °C is again shown to check the influence of the difference of Ts on the bond structure. From this figure, first, it is confirmed that the IR spectrum of CH_4 based a-SiC:F:H deposited at Ts = 300 °C is not so different from that at Ts = 350 °C. It is also known that major bond configurations of carbon atoms in CH_4 based a-SiC:F:H are Si-CH_3 bonds althouth those in CF_4 based a-SiC:F:H are Si-C bonds. That is, the carbon atoms are more incorporated in the CH_4 based film than in the CF_4 based one as shown in Fig. 2, but the carbon atoms in the CH_4 based film do not appear to be well incorporated in Si-Si atomic net-work. This implies the superiority of using CF_4 to using CH_4 as the carbon sources.

4. Electrical and Optical Properties

Figure 4 shows the plots of $\sqrt{\alpha h\nu}$ as a function of $h\nu$ for both CF_4 based and CH_4 based a-SiC:F:H, where α and $h\nu$ refer to the optical absorption coefficient of the film and the photon energy, respectively. It is clear from this figure that the optical band gap, which is evaluated from the intersection between the axis and a line extraporated from the plots, is likely to increase as the mixing ratio X increases. It is also found that the gradient of plots, which relates to the quality of films [7], is kept almost constant for the increase of X in CF_4 based a-SiC:F:H although it is likely to decrease for the increase of X in CH_4 based a-SiC:F:H. The gradient of plots for a-

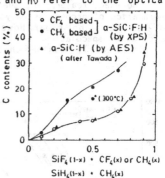

Fig.2 Carbon content in CF_4 and CH_4 based a-SiC:F:H and a-SiC:H.

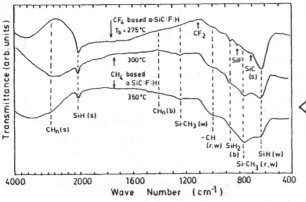

Fig.3 IR spectra of CF_4 based and CH_4 based a-SiC :F:H.

178

Si:H is about 700 to 800 (cm eV)$^{-1/2}$ [8], while that for CF$_4$ based a-SiC:F:H is about 800 (cm eV)$^{-1/2}$. The gradient for CH$_4$ based a-SiC:F:H decreases from 800 (cm eV)$^{-1/2}$ to 500 (cm eV)$^{-1/2}$ as X increases from 0 to 0.7. This figure implies again the superiority of using CF$_4$ to using CH$_4$, and the figure appears to imply that the optical band gap is controlled by mixing CF$_4$ with SiF$_2$ and H$_2$ without degradation of film quality.

The optical band gap E_{gopt}, the photo-conductivity $\Delta\sigma_p$ and the photo-sensitivity $\Delta\sigma_p/\sigma_d$ for our a-SiC:F:H films are shown in Fig. 5 as a function of the mixing ratio X. Here, the notation σ_d refers to the conductivity in dark, and the photo-conductivity $\Delta\sigma_p$ is defined as the differece between the conductivity in the AM-1 light of 100 mW/cm^2 and σ_d. In this figure, E_{gopt} and $\Delta\sigma_p$ for a-SiC:H, reported by Tawada et al. [1], are also drawn with broken curves. It is found in the figure that the optical band gap of CF$_4$ or CH$_4$ based a-SiC:F:H is controlled by the change of FR(CF$_4$) or FR(CH$_4$) just as that of a-SiC:H is controlled by FR(CH$_4$). However, the photo-conductivity and the photo-sensitivity of only CF$_4$ based a-SiC:F:H are almost kept constant. For instance, the photo-sensitivity of CF$_4$ based a-SiC:F:H is fixed at about 10^4, while that of CH$_4$ based a-SiC:F:H tends to decrease from 10^4 to 10^2 or 10^3 as X increases. Additionally, the photo-conductivity of CF$_4$ based a-SiC:F:H is higher than that of CH$_4$ based one or a-SiC:H by about 2 orders of magnitude at X = 0.5.

5. Discussions

The exact reason, why the optical band gap can be controlled without degradation of properties only when a-SiC is produced by mixing CF$_4$ with SiF$_2$ and H$_2$, is not clear. However, the results shown in Fig. 3 lead us to a following speculation.

It is easily imagined that the surface of the fluorinated amorphous

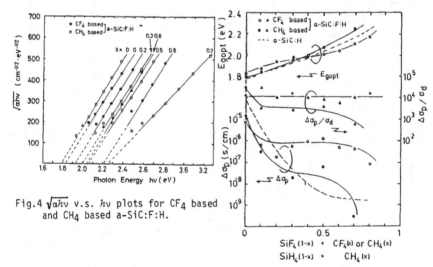

Fig.4 $\sqrt{\alpha h\nu}$ v.s. $h\nu$ plots for CF$_4$ based and CH$_4$ based a-SiC:F:H.

Fig.5 Optical band gap E_{gopt}, photo-conductivity $\Delta\sigma_p$ and photo-sensitivity $\Delta\sigma_p/\sigma_d$ as a function of gas mixing ratio X.

silicon films is covered with Si-F bonds during the growth of films since the bonding energy of Si-F bonds is so large. And it is also imagined that a key point of growth of fluorinated silicon films is how to replace these fluorine atoms to other net-work atoms such as silicon or carbon. When hydrogen atoms attack at the growing surface, fluorine atoms bonded with silicon or carbon are easily removed by forming volatile H-F bonds, since the bonding energy of H-F bonds are larger than that of Si-F or C-F bonds. And similarly, when activated CH_4 molecules or decomposed species CH_n (n = 1,2,3) attack at the growing surface, the reaction of (Si-F + CH_4 → Si-CH_3 + H-F) or the reaction of (Si-F + CH_n → Si-CH_{n-1} + H-F) could occur. On the other hand, in SiF_2 and CF_4 system, it is difficult to consider the reaction of (Si-F + CF_4 → Si-CF_3 + F-F), since the bonding energy of F-F bonds are so small and the electron-negativity of fluorine atoms at Si-CF_3 bonds is so large. Probably, in this case, decomposed species of CF_4 such as CF_2 are likely to bond with the silicon atoms at surface, whose fluorine atoms have been already removed by hydrogen attack. That is, we can speculate that as long as hydrogen atoms exist in the gas mixture, fluorinated gases such as CF_4 are more desirable to form Si-C net-work than hydrogenated gases such as CH_4, which are likely to form other undesirable bonds such as Si-CH_3.

One may ask about other properties such as the transport properties and the impurity doping properties for our CF_4 based a-SiC:F:H. We have not measured the transport properties. However , they could be expected to be almost similar to those of the conventional a-Si:H, since the doping efficiency of impurities such as boron or phosphorus in our films is comparable to that of a-Si:H and thus since the profile of localized state density in band-gap might be expected to be comparable to that in a-Si:H.

The conductivity of our CF_4 based a-SiC:F:H is well controlled by mixing B_2H_6 or PH_3 gas with CF_4, SiF_2 and H_2 mixture. For instance, the dark-conductivity can be increased from 10^{-11} S/cm to 10^{-4} S/cm as the ratio of the flow rate of B_2H_6 gas to the total flow rates of SiF_4 and CF_4 gases increases from 0 to 10^{-2}, for the CF_4 based a-SiC:F:H deposited at X = 0.1. The optical band gap of this boron doped film is about 1.8 eV. The conductivity of this boron doped film appears to be higher than that of the boron doped conventional a-SiC:H by one order of magnitude at least [1], if it is compared at the same optical band gap. Detailed study on the impurity doping properties will be reported soon.

6. Conclusions

From the above studies, the following are concluded.
1) CF_4 gas is superior to CH_4 gas as the carbon source to produced high quality a-SiC:F:H films.
2) The optical band gap of only CF_4 based a-SiC:F:H can be increased up to 2.2 eV without degradation of the material properties by incorporating carbon atoms, while the properties of CH_4 based a-SiC:F:H or a-SiC:H are degraded by the increase of the optical band gap.

Acknowledgement

The authors are grateful to Prof. M. Hirose at Hiroshima University for his support in the experiments. This work is partially supported by the Mitsubishi Foundation, by Iwatani Naoji Foundation Research Grant and also by NHK Hoso-Bunka Foundation.

References

1) Y. Tawada, Amorphous Semiconductor Technologies and Devices, ed. by Y. Hanakawa, (Ohm-North-Holland, Tokyo, 1983), Chap. 4.
2) G. Nakamura, K. Ishihara, M. Usui, K. Okaniwa and Y. Yukimoto, J. Non-Cryst. Solids, 59&60, 1111, (1983)
3) K. Nozawa, Y. Yamaguchi, J. Hanna and I. Shimizu, J. Non-Cryst. Solids, 59&60, 533 (1983)
4) H. Matsumura and S. Furukawa, Jpn. J. Appl. Phys. Suppl. 22-1, 523 (1983)
5) H. Matsumura, M. Maeda and S. Furukawa, Jpn. J. Appl. Phys., 22, 771 (1983)
6) H. Matsumura and S. Furukawa, J. Non-Cryst. Solids, 59&60, 739 (1983)
7) N.F. Mott and E.A. Davies, Electronic Processes in Non-Crystalline Materials, 2nd ed. (Clarredon Press, Oxford, 1979), Chap. 7.
8) A. Matsuda, M. Matsumura, S. Yamasaki, H. Yamamoto, T. Imura and H. Okushi, Jpn. J. Appl. Phys, 20, L183 (1981)

THE INFLUENCE OF PLASMA EXCITATION FREQUENCY ON THE PROPERTIES OF a-SiC:H PRODUCED IN A GLOW DISCHARGE PLASMA[*]

W. D. PARTLOW AND H. HERZIG
Westinghouse R&D Center, Pittsburgh, PA 15235
NASA Goddard Space Flight Center, Greenbelt, MD 20771

We have deposited a-SiC:H thin films from a silane-methane glow discharge plasma in a planar reactor using high (13Mhz and low (12kHz) frequency plasma excitation. Films produced at these two frequencies have significantly different properties as measured by vacuum UV reflectivity measurements, XPS, and infrared spectroscopy. Samples produced at the low excitation frequencies, where higher ion fluxes and energies are expected, have higher VUV reflectivities, higher bulk plasmon energies, and fewer IR bands due to organic fragments. In order to relate the VUV reflectivity data to the material properties, it was fitted to a simple Lorentz model using parameters taken from XPS data. We will discuss our interpretation of these data and also the dependence of film properties on the reactant gas composition.

INTRODUCTION

Besides the many electronic applications that are emerging for a-SiC:H thinfilms, several potential optical applications exist. These hard, high refractive index films may be used in multilayer coatings for visible and near infrared applications. If the band structure of a-SiC:H can be made to resemble that of crystalline SiC, an exceptionally useful optical application as a normal incidence vacuum UV reflecting film would be possible.

We have made a-SiC:H films in a plasma reactor under a wide range of conditions and have measured the influence of deposition parameters on the properties of the films. The variations in film properties that we were able to achieve fall roughly into two categories, chemical changes which were obtained by varying the mix of the reactant gases, and changes in the physical deposition processes obtained by varying the energy and flux of the positive ions bombarding the growing film. The latter was accomplished by using low (12kHz) and high (13.6 MHz) plasma excitation frequencies. We concentrated in this work on characterizing the differences in the high and low frequency-produced films and modeling the reflectivity and the XPS data with a simple Lorentz model which was found to give a reasonably good fit to the data.

DEPOSITION PROCESS

The films were deposited in a transverse flow, parallel electrode reactor as shown in Figure 1. Two RF generators were used to excite the plasmas: a 13.6MHz source with a "T" impedance matching network and a 12kHz source with a transformer matching network. Operating powers were selected to be the lowest possible that would maintain a uniform, stable plasma: 25W for the 13.6Mhz source and 10W for the 12kHz source. Silane and methane reactant gases with no buffer gases were used as reactants, and electrode temperatures were maintained at 300C. Silicon wafers, 1-0-0

---

[*]Work supported by NASA Goddard Space Flight Center, Greenbelt, MD 20771

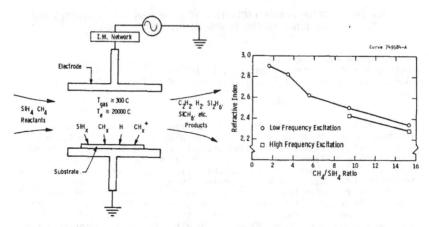

Figure 1. Schematic representation of the plasma deposition reactor illustrating the important physical processes.

Figure 2. Dependence of refractive index on reactant gas ratio.

orientation, .25mm thickness and fused silica discs 3.2mm thickness were used as substrates. The silane flow rate was the same for all experiments, and the methane was varied to obtain different reactant ratios, so the total flow rate increased with methane/silane ratio.

Because of higher electron losses at the electrodes for plasmas excited at low frequencies, higher positive ion fluxes are expected, and the energies of the ions bombarding the growing surface of the films are higher. Estimates of the ion bombardment energies at the substrates for the two excitation frequencies were made by measuring the voltage waveforms on a 10mm diameter floating probe located 7 mm above the grounded substrate electrode. For the low frequency excitation, the probe potential had an average value of 48V with peak voltages of 150V within the cycle, whereas under high frequency excitation average potential was 0V within experimental error, and the peak was no greater than 5V. Clearly higher sheath voltages, and thus ion bombardment energies, were obtained in the low frequency case.

EFFECTS OF REACTANT CHEMISTRY

The ratio of methane to silane was varied from 1.60 to 15.3 in these experiments. The refractive index, measured by spectrophotometry over the visible and near IR decreases with this ratio as has been reported elsewhere[1], and is shown in Figure 2. It is noteworthy that the refractive index for the high frequency samples is lower than that for the low frequency samples for the same reactant gas flows. We also observe a shift of the optical absorption edge toward the blue with methane/silane ratio[1]. The dependence of the vacuum UV reflectivity curves upon the reactant ratio is shown in Figure 3, showing a shift of the high energy cutoff toward high energies with increasing methane/silane ratio and a concommittant decrease in the low energy reflectivity.

Figure 4 shows the deposition rates measured for the different reactant flows. The data show a decrease in the deposition rate with

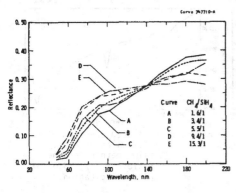

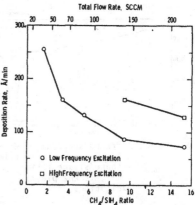

Figure 3. Vacuum UV reflectance spectra for silicon carbide thin films produced with different reactant gas ratios using 12 kHz plasma excitation.

Figure 4. Dependence of deposition rate on reactant gas flow.

increasing methane/silane ratio, but as we pointed out earlier, the total flow rate increases with this ratio, and this deposition rate dependence is expected to decrease with flow rates because of flow kinetics: higher flows result in a lower density of reactive species in the plasma and thus lower diffusion rates to the growing film surface.

INFLUENCE OF EXCITATION FREQUENCY

Significant differences in the properties of films produced with high and low frequency plasma excitation were observed when the reactant gas flows were the same. These differences reflect the higher fluxes and energies of the positive ions incident on the films produced at low frequencies compared to those produced at high frequencies.

Besides the higher refractive indices obtained for low frequency samples than for the high frequency ones, differences are also seen in the infrared absorption spectra, as shown in Figures 5 and 6. The low frequency sample in Figure 5 has a broad peak with little structure centered at 750nm consisting of Si-C vibrational modes, and few peaks from undissociated reactant species. The spectrum has an "inorganic" character, as compared to the data shown in Figure 6 which has sharper structure and many peaks attributable to reactant fragments. Similar effects have been reported for samples produced at varying pressures[2] and by varying the polarity of DC discharges[3].

Reflectivity data in the 50 to 200nm range for high and low frequency samples having the same reactant flows are shown in Figure 7, along with data for polished CVD silicon carbide[4] for comparison (The UV reflectivity of polished CVD SiC is very similar to that of single crystal SiC). It can be seen that the low frequency sample has a higher reflectivity than the high frequency sample, but still lower than the polished CVD SiC. This will be discussed further.

We also performed XPS measurements on high and low frequency samples to help understand the differences. This technique measures the energy

184

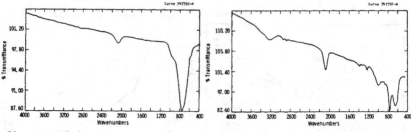

Figure 5. Infrared spectrum of a silicon carbide thin film, methane/silane ratio: 9.4/1, 12 kHz plasma excitation.

Figure 6. Infrared spectrum of a silicon carbide thin film, methane/silane ratio: 9.3/1, 13.6 MHz plasma excitation.

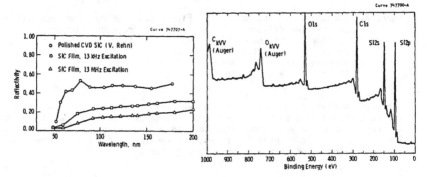

Figure 7. Vacuum UV reflectivity of plasma deposited silicon carbide thin films produced at low and high frequencies compared to data for polished CVD silicon carbide. Methane/silane ratio for plasma films: 9.4/1.

Figure 8. Wide range XPS spectrum of an air-exposed plasma deposited silicon carbide thin film. Plasma excitation frequency: 12 kHz; Methane/silane ratio: 9.4/1.

distribution of electrons that have been ejected from the sample by x-ray irradiation. It is used to identify atomic species and their chemical bonding states via the positions of electronic transitions from the valence band. These transitions also have side-bands due to volume plasmons in the conduction band. The energies of these plasmons correspond to the plasma frequency of the material, which influences the high energy reflectivity of the material. Figure 8 shows a wide range XPS scan of one of the low frequency films. In addition to silicon and carbon transitions, large oxygen peaks are seen. The oxygen peaks are due to adsorbed gases from exposure to the atmosphere, and are nearly removed by sputtering a few hundred angstroms of material from the surface, as shown in Figure 9. The sputtering in the apparatus that we used is done at normal incidence, which implants some of the surface oxygen into the sample where it does not escape[5]. The oxygen peak seen in Figure 9 is about the same magnitude as seen in the XPS of single crystal silicon carbide sputtered in the same way, so we believe that it is in fact implanted oxygen and not a bulk

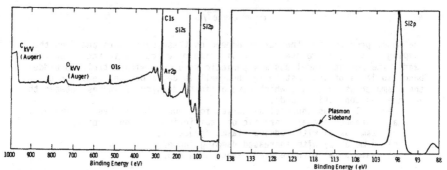

Figure 9. Wide range XPS spectrum of the sample measured in Figure 8 after 45 minutes of normal incidence argon sputtering in high vacuum.

Figure 10. High resolution XPS spectrum of the sample from Figure 9 showing the plasmon sideband of the silicon 2p line.

impurity. Implanted argon can also be seen in Figure 9. A high resolution scan of the silicon 2p peak of the low frequency sample is shown in Figure 10. The plasmon sideband is seen to be 20.5eV to the high energy side of the major peak. The plasmon sideband energy is 18.5eV in the high frequency sample prepared with the same reactant proportions. These values compare with the plasmon energy of 22.5eV measured for single crystal silicon carbide.

To summarize, samples prepared with low plasma excitation frequencies have higher refractive index, have fewer organic fragments in the IR spectrum, and have higher plasmon energy, and higher VUV reflectivities than comparable samples prepared with high excitation frequencies. The properties of the low frequency samples are more like those of equilibrium silicon carbide, and since they are bombarded by more energetic ions, this is consistent with the measurements and models for sputtered films[6,7] that relate positive ion bombardment to an increased effective surface temperature during film growth. This is thought to lead to higher surface mobilities and more complete surface reaction of the ad-species during formation of the film, producing more dense materials that behave more like their bulk, equilibrium counterparts.

THEORETICAL REFLECTIVITY CURVES

We will compare the reflectivity data for silicon carbide to the quantum mechanical modification of the classical Lorentz dispersion theory of solids[8]. This theory gives very realistic predictions of the gross features of the optical constants of metals, semiconductors, and insulators in terms of the interaction of electromagnetic radiation with a continuum of harmonic oscillators. The one-dimensional theory can not take into consideration the variation of the valence and conduction bands within the Brillouin zones, and therefore predicts simple shapes for the reflectivity curves with little structure except the rise and the drop in reflectivity near the direct band gap (ω_0) and the plasma frequency (ω_p) respectively (not to be confused with the excitation frequency of the plasma used to deposit the films). The plasma frequency is the characteristic oscillation frequency of a gas having mass and charge. This oscillation is seen as the plasmon sideband in the XPS spectra. Its value, in terms of electron charge, mass, and density, is given by:

$$\omega_p = \sqrt{4\pi n e^2/m}$$

The model predicts how the curves depend on these parameters and upon the damping factor (Γ), related to the lifetime of electrons in the conduction band. The Lorentz model assumes periodic structure with a truly forbidden band, so it is of interest to see how well it describes the behavior of these amorphous materials which have a finite density of states between the valence and conduction bands.

Figure 11 shows theoretical curves for four sets of these parameters:
a) parameters characteristic of equilibrium silicon carbide,
b) same as a) with reduced plasma frequency,
c) same as a) with increased damping factor,
d) same as a) with reduced direct band gap.

The three parameters effect the curves in qualitatively different ways. The band gap controls the low energy limit of high reflectivity, the plasma frequency controls the high energy limit, and the damping factor controls the maximum value. In addition, the maximum value is reduced when the band gap is close to the plasma frequency, irrespective of the damping factor.

The reflectivity data are compared to curves from the theory in Figure 12. Values for the direct band gaps used were: 7.5eV for equilibrium SiC [9] and 2.0eV for a-SiC:H [2]. Precise values of these parameters are not necessary because they do not strongly influence the reflectivity curves in the high energy region of the spectrum where we have data. We used values or the plasma frequency that were experimentally determined from the XPS data. The damping factor was varied to give the best fit to the reflectivity data. The best fitting values for Γ were 6.0eV for equilibrium SiC, 13.0eV for low frequency a-SiC:H, and 16.0eV for the high frequency material. These two fundamental bulk parameters we have obtained give us a further comparison of our films to equilibrium silicon carbide. The lower plasma frequency of the films may be a dilution effect resulting from vacancies and inclusions of undissociated reactant species or regions of high homonuclear bonding(eg. Si-Si). All defects do not, however, result in changes in the plasma frequency. Irradiating a SiC single crystal with high energy ions to make it amorphous, for example, does not change the energy of the plasmon sidebands[9]. The fitted Lorentz theory predicts a larger refractive index in the visible for the low frequency sample than for the high frequency one, but differing by a larger amount (0.5) than we observe (0.07). These values are highly sensitive to the values used for the band gaps in the theory. Close agreement to our measurements can be obtained by increasing the gap parameter for the low frequency case by 0.5eV or by decreasing the gap in the high frequency case by the same amount. These changes do not significantly affect the predicted reflectivity curves below 200nm.

An examination of the parameter table in Figure 12 reveals the same trend that we observed earlier: the more highly ion bombarded sample has properties most like the equilibrium material.

CONCLUSIONS

The Lorentz model serves as a reasonably good approximation by pulling into a consistent picture all of the information we were able to obtain on the band structure of these materials: reflectivity curves, XPS data, and band gaps. In the two cases that we studied in detail, samples produced at high and low excitation frequencies, our data and conclusions point to the picture that the low frequency, highly ion bombarded sample was able to form somewhat closer to the equilibrium state that is obtained when silicon carbide is produced slowly at high temperatures.

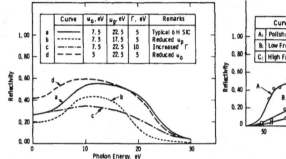

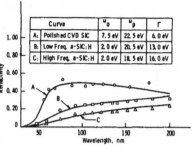

Figure 11. Dependence of the theoretical reflectivity curves
upon the three parameters
of the Lorentz dispersion
theory.

Figure 12. Comparison of plasma
deposited and polished
CVD silicon carbide
reflectivity data to
theoretical curves with
best fitting Γ.

ACKNOWLEDGEMENTS

The authors wish to acknowledge R. A. Madia for technical assistance
in the thin film deposition. The XPS measurements made by Singh Manocha
and Gary Giddick and the infrared absorption measurements made by Don
Lemmon are gratefully acknowledged. These people, along with W. J. Choyke,
J. V. R. Heberlein, and John Osantowski also provided valuable technical
input to the program.

REFERENCES

1. Y. Catherine and G. Turban, Thin Solid Films 60, 193 (1979).

2. B. von Roedern, A. H. Mahan, D. L. Williamson, and A. Madan,
    Proceedings of the 5th Symposium on Materials and New Processint
    Technologies for Photovoltaics, New Orleans, LA October, 1984.

3. Y. Catherine, A. Zamouche, J. Bullot, and M. Gauthier, Thin Solid Films
    109 14 (1983).

4. Victor Rehn and W. J. Choyke, Nuclear Instruments and Methods 177, 173
    (1980).

5. F. Bozso, L. Muelhoff, M. Trenary, W. J. Choyke, and J. T. Yates,
    J. Vac. Sci. Technol. (in press).

6. J. A. Thornton, Ann. Rev. Mat. Sci. 7, 239 (1977).

7. R. C. Ross and R. Messier, J. Appl. Phys. 52, 6329 (1981).

8. Frederick Wooton, "Optical Properties of Solids", Academic Press, New
    York (1972).

9. W. J. Choyke, Private communication.

BONDING AND RELEASE OF HYDROGEN IN a-Si:C:H ALLOYS

W. BEYER*, H. WAGNER* AND H. MELL**
* Institut für Grenzflächenforschung und Vakuumphysik, Kernforschungsanlage Jülich GmbH, D-5170 Jülich, Federal Republic of Germany
** Fachbereich Physik, Univ. Marburg, Renthof 5, D-3550 Marburg, Federal Republic of Germany

ABSTRACT

Hydrogen evolution and infrared absorption studies of a-Si:C:H films prepared from mixtures of hydrocarbon (CH_4, C_2H_6, C_2H_4) and silane gases (SiH_4, Si_2H_6) by the glow-discharge process show that both the absolute hydrogen content and the hydrogen bonding is uniquely determined by the carbon content. Up to a carbon concentration of ~ 45 % the widening of the optical gap is mainly due to an increase of the hydrogen content. For a hydrogen density $N_H > 1-2 \times 10^{22} cm^{-3}$ the incorporated hydrogen atoms lead to the formation of voids, resulting in thermal instability of the bonded hydrogen and in a deterioration of the electronic quality of the amorphous material.

INTRODUCTION

The development of a high quality amorphous alloy with an optical bandgap of 2 eV or more is of considerable interest for several applications, like photosensors, electrophotography, image tubes or multijunction thin film solar cells. Promising materials for this purpose are hydrogenated silicon-nitrogen (a-Si:N:H), silicon-oxygen (a-Si:O:H) and silicon-carbon (a-Si:C:H) alloys, the latter attracting considerable attention since its application by Hamakawa [1] and coworkers for wide gap window layers in a-Si:H-based solar cells. High-quality glow-discharge a-Si:H films having an optical bandgap near 1.7 eV, the admixture of hydrocarbon gases in the glow-discharge process results in an increase of the optical gap to above 2 eV upon carbon incorporation of 10-20% only [2]. Yet, concomitant with the widening of the optical gap, a decay of the electronic quality of the amorphous material is generally observed, evidenced by a decrease of the photoconductivity [3] and an increase of the dangling bond electron spin resonance (ESR) signal of the undoped material [4] as well as by a decreased sensitivity to doping. Possible reasons for this decay involve a defect production due to differences in the bond lengths between Si-Si and C-C bonds or due to a tendency of carbon atoms for threefold coordination [4,5,6]. Yet, another possible origin for a decreasing electronic quality is a deleterious influence of hydrogen incorporation. For glow-discharge a-Si:H films such an effect has been observed for a hydrogen concentration of 20 at.% and above when the amorphous silicon network loses its connectiveness and voids are formed [7]. It is the aim of the present work to study the role of hydrogen in a-Si:C:H films employing hydrogen evolution and infrared absorption measurements. While the latter experiment gives information on the bonding configurations of hydrogen, the former is particularly sensitive to the film structure. Moreover, hydrogen evolution experiments allow conclusions about the thermal stability of hydrogen incorporation which may limit the stability of an a-Si:C:H-based device.

EXPERIMENTAL

The a-Si:C:H films were prepared in capacitively coupled glow-discharge systems from undiluted silane and hydrocarbon gases. Typical deposition conditions were: 0.1-0.2 mbar gas pressure, 2-5 sccm gas flow, 25 W rf power at 13.5 MHz, yielding deposition rates of 2-5 Å/s. One series of films was also

prepared at an rf power of 10 W. As a substrate, sapphire, quartz and crystalline silicon platelets were used. The infrared (IR) transmission measurements
were performed with a Digilab FTS-20 Fourier-transform spectrometer. For the
hydrogen evolution experiments the films were inserted into a quartz tube
evacuated by a turbomolecular pump and heated to 1000°C at a constant heating
rate of typically 20°C/min. The partial pressure of H, H_2 and other species
was monitored by a quadrupole mass analyzer. As the pumping speed of the turbomolecular pump is constant over a wide range of pressure, the partial pressure is directly proportional to the evolution rate. The silicon to carbon
concentration ratio in the films was determined by Auger electron spectroscopy
(AES) using the silicon and carbon (KLL) transitions for Si and C detection
and a crystalline SiC sample as a reference. The electronic properties of the
a-Si:C:H films were characterized by measurements of the optical absorption,
the photo- and dark conductivity and the thermopower.

RESULTS

The chemical composition of a-Si:C:H films deposited from mixtures of
silane (SiH_4), disilane (Si_2H_6) with methane (CH_4), ethane (C_2H_6) and ethylene
(C_2H_4) at a substrate temperature of T_s = 250°C is shown in Fig. 1a,b. In
agreement with other authors [2,8] we find that the use of higher hydrocarbon
(and silane) gases leads to an increased carbon incorporation into the films
(Fig. 1a). It has been argued by Tawada et al. [3] that the structure of
methane- and ethylene-based films may be different, the former being a rather
ideal amorphous SiC alloy and the latter an organosilane-like material. In
this case, one would expect different hydrogen contents for films of equal
carbon concentration prepared from the different hydrocarbon gases. Yet, Fig.
1b demonstrates that the hydrogen density N_H is mainly determined by the carbon content, largely independent of the type of hydrocarbon and silane gases
used. There is a strong increase of the hydrogen density with rising carbon
content y up to y ~ 0.45 where the curve saturates. The maximum hydrogen
density equals a hydrogen concentration of about 50 at.%.

Not only the absolute hydrogen concentration but also the hydrogen bonding in a-Si:C:H appears to be largely independent of the hydrocarbon gases
used. Fig. 2 shows the infrared transmission spectra of two films of approximately equal carbon content prepared from methane-silane and from ethylene-
silane mixtures. Essentially the same IR absorption peaks are observed in both
cases: the C-H stretching modes near 2900 cm^{-1}, the Si-H stretch mode near

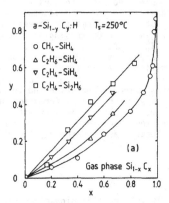
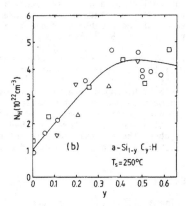

Fig. 1 (a) Composition parameter y as a function of the gas phase mixture.
(b) Hydrogen density versus composition parameter y.

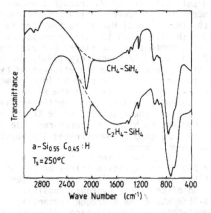

Fig. 2 IR transmission spectra for methane- and ethylene-based a-Si:C:H films.

2100 cm^{-1}, C-H bending modes in the 1200-1500 cm^{-1} range and Si-H and Si-CH$_3$ rocking and wagging modes as well as the Si-C stretch mode in the 600-800 cm^{-1} range [9]. The two spectra differ only slightly in the relative magnitude of the absorption bands. Our data, therefore, do not corroborate Tawada's conjecture [3] of the different structure of methane and ethylene-based films.

The hydrogen evolution spectra for a series of methane-based alloys (T$_s$ = 250°C) are shown in Fig. 3a. Plotted is the hydrogen evolution rate as a function of temperature for a heating rate of β = 20°C/min. With rising carbon content a transition from a one-peak to a two-peak spectrum is observed. The additional peak arises in the low temperature flank of the peak present for a-Si:H. Both evolution peaks shift with rising carbon content to higher temperatures. Fig. 3b shows the temperature T$_M$ of the evolution peaks for the methane-based films of Fig. 3a as well as for mixtures of other gases as a function of the carbon content. All data can be approximated by two curves demonstrating that just two evolution processes are active in the a-Si:C:H alloy system: a low-temperature (LT) and a high-temperature (HT) process. For y = 0, the LT curve extrapolates to T$_M$ ~ 370°C, the temperature of the LT hydrogen evolution peak of glow-discharge a-Si:H films deposited at T$_s$ ≤ 150°C. Here, the LT peak has been attributed to the desorption of hydrogen bound to

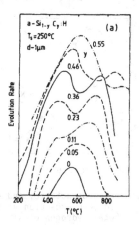 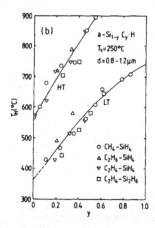

Fig. 3 (a) Hydrogen evolution rate versus temperature and (b) temperature of maximum evolution rate versus composition parameter y.

internal surfaces, followed by rapid out-diffusion of H_2 through a network of
voids [7]. The actual desorption process is considered to involve the rupture
of two Si-H bonds concomitant with the formation of H_2. It is tempting to associate the LT evolution peak in a-Si:C:H films with the same process, attri-
buting the peak shift with rising carbon content to the higher C-H binding
energy compared to Si-H ($E_B \sim 3.5$ eV and 3.1 eV, respectively [10]). The HT
peak, on the other hand, has been assigned in a-Si:H to the effusion of hydrogen from bulk material limited by diffusion of atomic hydrogen. In agreement,
we find for a-Si:C:H films similar to a-Si:H an increase of the HT peak temperature with rising film thickness. The hydrogen diffusion coefficient apparently decreases upon carbon (and hydrogen) incorporation since the HT effusion peak moves with rising carbon content to higher temperatures. The appearance of both LT and HT peaks on the same sample demonstrates the ability
of the a-Si:C:H network to reconstruct broken bonds.

Having established the presence of a continuous void structure in
a-Si$_{1-y}$C$_y$:H films with $y > 0.2$ ($T_s = 250°C$) the question arises of the origin
of this structure. In Fig. 4 we plot the hydrogen density released under the
LT and HT evolution peaks as a function of the total hydrogen density. The results are very similar to those of a-Si:H where the hydrogen content is
changed by varying the substrate temperature of glow-discharge films or by adding different amounts of hydrogen to the sputtering gas of sputtered a-Si:H
films [7]. In all cases, only a HT peak is observed up to a hydrogen density
of $1-2\times10^{22}$ cm^{-3}; above this density the low-temperature evolution, i.e. the
void structure appears. The effect has been explained by the predominant incorporation of hydrogen at positions breaking bonds of the host material until
the latter loses its connectiveness so that voids are formed [7].

The void structure, no doubt, leads to the observed decay of the electronic properties. Fig. 5 demonstrates this effect for the photoconductivity
$\Delta\sigma_{ph}$ (illumination 50 mW/cm^2 of white light) of undoped material (rf power
10 PW for a-Si:C:H films and 5 W for a-Si:H). For both a-Si:C:H and a-Si:H
films $\Delta\sigma_{ph}$ starts to decay when the LT evolution appears.

An improvement of the electronic quality of the a-Si:C:H films, therefore, requires the minimization of the void structure. For a-Si:H films this
is achieved by raising the substrate temperature, reducing the hydrogen concentration. Fig. 6 demonstrates that this is possible for a-Si:C:H films, too.
In Fig. 6a the hydrogen density is plotted as a function of substrate temperature for a-Si:C:H films prepared from 2:1 mixtures of SiH$_4$ with C$_2$H$_6$ and C$_2$H$_4$
as well as for a-Si:H films. The dashed line indicates the critical hydrogen
density N_H^c for void formation: for a higher hydrogen density, a low-tempera-

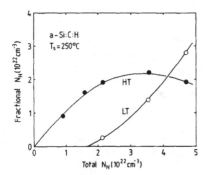

Fig. 4 Fractional release of hydrogen
under LT and HT peaks versus
total hydrogen density for
methane-based a-Si:C:H films.

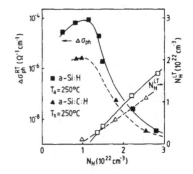

Fig. 5 Photoconductivity and LT
hydrogen evolution versus
total hydrogen density.

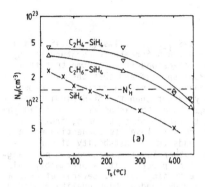 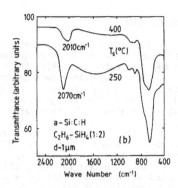

Fig. 6 (a) Hydrogen density of a-Si:C:H and a-Si:H films versus substrate temperature; (b) IR transmission spectra of a-Si:C:H films for two substrate temperatures.

ture evolution i.e. a void-structure is present; for a lower hydrogen density, the material is rather compact and evolution proceeds by the HT process only. The result of a substrate temperature-dependent film structure is also supported by IR absorption measurements. Previously we have attributed the Si-H stretching mode at 2100 cm^{-1} to the presence of voids whereas hydrogen embedded in bulk a-Si:H is considered to lead to the 2000 cm^{-1} absorption [11]. The IR transmission spectra in Fig. 6b demonstrate for ethane-based films that an increase of the substrate temperature indeed results in a transition from a 2100 to a 2000 cm^{-1} Si-H stretch mode. Thus, from the structural point of view a higher substrate temperature, i.e. a lower hydrogen content leads to an improvement of the a-Si:C:H films.

Yet, a decrease of the hydrogen concentration has also the effect of decreasing the optical gap E_G. Fig. 7 shows E_G (as determined by a Tauc plot [12]) as a function of the hydrogen density for the a-Si:C:H films of Fig. 6a (open symbols). The agreement with data of a-Si$_{1-y}$C$_y$:H films prepared at a fixed substrate temperature of 250°C, with y < 0.45 as a free parameter (closed symbols) and of films from SiH$_4$ and Si$_2$H$_6$ gases (with T_s as a free parameter) demonstrates that it is practically the hydrogen content which determines the optical gap of our a-Si:C:H films up to about 45 % of carbon. Only for y > 0.45 the carbon content presumably affects the optical gap: in this alloy range the hydrogen content decreases slightly with rising carbon content (Fig. 1b) while the optical gap increases further [2].

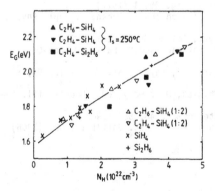

Fig. 7 Optical gap E_G versus hydrogen density.

DISCUSSION

The data presented suggest that for a-Si$_{1-y}$C$_y$:H films (y < 0.45) it is the carbon content which determines at a given substrate temperature the hydrogen concentration. This effect can be understood if we assume that the films form from rather hydrogen-rich plasma species and that the hydrogen content is determined by the degree of hydrogen elimination during film growth. As the C-H binding energy is considerably larger than the Si-H binding energy, at an elevated temperature mainly hydrogen bound to carbon will remain in the film. Since incorporated hydrogen atoms terminate Si or C bonds, from a certain hydrogen density the amorphous network will lose its connectiveness and voids are formed. The presence of voids leads to an instability of surface-bound hydrogen as it can desorb as H$_2$ at rather low temperatures. Since dangling bonds are left behind by this desorption process, it is presumably this effect which leads to the decay of the electronic quality. Several efforts appear promising for reaching the goal of a wide bandgap high quality a-Si:C:H material: (1) Improvement of the gap widening effect of carbon incorporation by changed plasma conditions. (2) Preparation of films of low hydrogen (N$_H$ < 10^{22} cm^{-3}) but high carbon content, by e.g. reactive sputtering. (3) Addition of foreign atoms in the glow-discharge process to inhibit a void formation. Of particular interest are here nitrogen and oxygen atoms. Preliminary evolution results on a-Si:N:H and a-Si:O:H films suggest that such alloys can incorporate much higher hydrogen concentrations than a-Si:H or a-Si:C:H films without void formation.

ACKNOWLEDGEMENTS

The authors wish to thank G. Wichner for carrying out the infrared absorption measurements, F. Birmans for performing the AES measurements and W. Hilgers and W. Knörchen for technical assistance. The disilane used was kindly provided by Prof. F. Fehér, University of Cologne. The work was partly supported by the Bundesministerium für Forschung und Technologie.

REFERENCES

1. Y. Hamakawa, J. de Physique Colloq. 42, C 4-1131 (1981)
2. Y. Tawada, in: Japan Annual Reviews in Electronics, Computers and Telecommunications, Vol. 6, Y. Hamakawa, ed. (Omusha, Japan, 1983), p. 148
3. Y. Tawada, K. Tsuge, M. Kondo, H. Okamoto and Y. Hamakawa, J. Appl. Phys. 53, 5273 (1982).
4. A. Morimoto, T. Miura, M. Kumeda and T. Shimizu, J. Appl. Phys. 53, 7299 (1982)
5. D.A. Anderson and W.E. Spear, Philos. Mag. 35, 1 (1977)
6. A.H. Mahan, B. v. Roedern, D.L. Williamson and A. Madan, J. Appl. Phys., to be published
7. W. Beyer and H. Wagner, J. Non-Cryst. Solids 59-60, 161 (1983)
8. S. Nitta, A. Hatano, M. Yamada, M. Watanabe and M. Kawai, J. Non-Cryst. Solids 59-60, 553 (1983)
9. H. Wieder, M. Cardona and C.R. Guarnieri, Phys. Stat. Solidi (b) 92, 99 (1979)
10. K.P. Huber, in: AIP Handbook of Physics, D.E. Gray, ed. (McGraw-Hill, New York, 1972), p. 7-168
11. H. Wagner and W. Beyer, Solid State Commun. 48, 585 (1983)
12. J. Tauc, R. Grigorovici and A. Vancu, Phys. Stat. Solidi 15, 627 (1966)

GAP STATES IN HYDROGENATED AMORPHOUS SILICON-CARBON ALLOYS

P.FIORINI, F.EVANGELISTI and A.FROVA

Dipartimento di Fisica, Universita' "La Sapienza", 00185 Roma, Italy

ABSTRACT

Tail and defect states in the gap of $a-Si_x C_{1-x}$:H alloys have been studied by measurements of spectral photoconductivity. The variation of defect-state density versus x is found to be negligible. By comparison with PDS results the $\eta\mu\tau$ product has been determined and found to be almost independent of photon energy and to strongly decrease with inclusion of carbon. This effect is attributed to changes in the transport mechanism combined with an increased recombination rate associated with the widening of the gap.

INTRODUCTION

Considerable interest has developed recently in wide-gap alloys based on a-Si:H, like a-SiC:H and a-SiN:H. This is mainly because they can be used as window layers in the construction of solar cells [1] or as part of stacked cells [2]. Moreover, a-SiN:H is of great interest in the production of Field Effect Transistors [3] and a-SiC:H is promising for large-area white luminescent devices [4].

In this work we have focused our attention onto $a-Si_x C_{1-x}$:H, for the purpose of investigating the change in defect and tail states with respect to pure a-Si:H. As these are conceivably responsible for the poorer phototransport properties of this material, their study is of great importance towards the improvement of film quality through changes in growth conditions and, possibly, methods.

EXPERIMENTAL TECHNIQUE AND RESULTS

Samples of $a-Si_x C_{1-x}$:H were prepared by glow discharge in a mixture of SiH_4 and CH_4 diluted in 80% H_2. The reactor was capacitively coupled to a R.F. generator by means of 10 cm-diameter electrodes separated by 3.5 cm. Substrate temperature of 250 °C, R.F. power of 4 watts, pressure and flow rate of the gas, respectively 0.6 Torr and 10 sccm/min, were kept constant for all samples. The gas ratio r = SiH_4 /(SiH_4 + CH_4) was varied between 0.9 and 0.2. The growth rate correspondingly decreased from 2 Å/s to 0.16 Å/s, until for r=0 no deposition took place.

The films were investigated by measurements of optical transmission and photoconductivity in the region from ~ 0.6 eV above to ~ 0.8 eV below the optical gap.

Transmission measurements, performed with a Varian 2300 spectrophotometer, allowed us to deduce the value of absorption coefficient α in the range $\alpha > 10^4$ cm^{-1}. To obtain the optical gap E_g the data were analyzed using the so-called Tauc plot - i.e. constant $|p|^2$ approximation - for historical reasons [5]. The values of E_g are listed in Table I. An estimate of carbon content has been obtained using the expression

$$E_g(x) - E_g(1) = 1.7 \ (1-x) \tag{1}$$

deduced from the data of Sussman and Ogden [6]. Eq. (1) seems to apply very generally [6-10], at least for large values of x.

Photoconductivity measurements were performed in a planar configuration, by evaporating onto the film two ohmic chromium electrodes 20 mm long and 1 mm apart. The applied electric field was varied from 10 to 10^4 V/cm depending on sample conductivity. The photoconductivity depends on the incident photon flux F via the power law $\Delta\sigma \propto F^\beta$, with β varying from 0.8 to 0.9 in different samples. The knowledge of this dependence allows the absorption coefficient α to be deduced from the relationship:

$$\Delta\sigma = e\eta\mu\tau(F) \frac{F(1-R)}{d} \left[1 - \exp(-\alpha d) \right] \tag{2}$$

under the assumption that both the quantum efficiency η and the mobility-lifetime product $\mu\tau$ are energy independent (e is the electron charge, R the reflectivity, and d the sample thickness). The resulting spectral dependence of α , as obtained by matching photoconductivity and transmission data above the optical gap, is shown in Fig. 1 for a few samples. It is seen that the intermediate absorption range is well described by the exponential law $\alpha \propto \exp(-h\nu /E_o)$, where E_o , usually referred to as the Urbach energy, increases remarkably with the inclusion of carbon (see also Table I). This behavior has been explained in terms of band-tail widening due to compositional disorder and found to be consistent with the potential variance of an uncorrelated two-component alloy [11]. The absorption at lower energy is due to the defect states lying deep into the gap. Its integral, after subtraction of the extrapolated Urbach tail, gives the defect density [12] (see Table I).

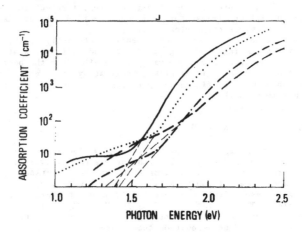

Fig. 1 Behavior of absorption coefficient for samples 214(——), 216(...), 217(·---), 218(- - -).

DISCUSSION

It is of interest to compare the Urbach energy E_o and the defect density N_g obtained here and those determined, using a set of simultaneously deposited samples, by photothermal deflection spectroscopy PDS [11], also shown in Table I. The agreement is quite satisfactory, confirming that photoconductivity is a good technique for the absolute measurement of weak absorption coefficients [13-15].

By introducing the PDS values for α into Eq. (2), we can make a spectral analysis of the $\eta\mu\tau$ product for our a-SiC alloys. The results

Table I

Estimated alloy composition x, "Tauc" optical gap E_g , Urbach energy E_0 and defect DOS, N_g . The values obtained by spectral photoconductivity are compared with PDS results [11]. Photoconductivity for sample 219 is too weak to allow spectral investigation.

SAMPLE NUMBER	x	E_g (eV)	E_0(meV)		$(N_g/10^{16})(cm^{-3})$	
			PC	PDS	PC	PDS
213	0.98	1.80	56	58	7	4
214	0.96	1.83	69	61	6	2.5
215	0.93	1.88	74	78	3	7.3
216	0.92	1.90	65	72	6	4.5
217	0.83	2.06	81	83	3	5.9
218	0.79	2.13	119	115	9	8.6
219	0.75	2.20		108		4.5

are shown in Fig. 2, where we have plotted $\eta\mu\tau$ normalized to its value for $\alpha = 10^3$. The value is very close to unity in the whole explored range, with variations showing no definite trend and no particular sample dependence. The differences are more likely to fall within the experimental uncertainty in both photoconductivity and PDS measurements. It is worth stressing here that the constancy of $\eta\mu\tau$ below gap (already known for a-Si:H [15]) is found to hold for alloys differring in their $\eta\mu\tau$ product by six orders of magnitude, as discussed below in connection with Fig. 3.

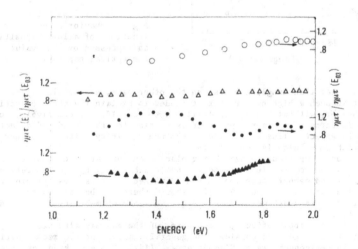

Fig. 2 Behavior of $\eta\mu\tau$ normalized to its value for $\alpha = 10^3$ for samples 213(▲), 215(●), 216(△), 218(○).

Table I shows that N_g is the same within a factor three in all samples, and comparable with the one of a-Si:H grown in the same conditions. There seems to be no trend in N_g with carbon content, different values of

N_g beeing probably due to uncontrolled changes in the deposition process. The same conclusion can be drawn from another set of samples – grown without H_2 in the gas mixture – where the compositional index x reaches up to ∿ 0.4. This is somewhat in disagreement with the ESR results by Morimoto et al. [17] who find – for the same x-interval – a factor of 100 increase. It is instead in agreement with data by Mahan et al. [10] for a-SiC:H not involving graphitic C-coordination. It is also consistent with the observation of a virtually composition-independent luminescence efficiency at room temperature [6,18], and with the notion that carbon – as opposed to germanium – does not originate large numbers of dangling bonds in the alloy [17]. In spite of these features, the photoconductivity is found to decrease dramatically with the inclusion of carbon (see Fig. 3, where $\eta\mu\tau$ is reported as a function of E_g).

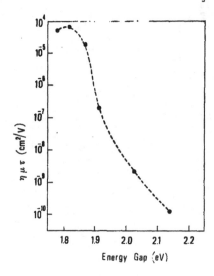

Fig. 3 Behavior of $\eta\mu\tau$ as a function of alloy composition as expressed by the value of the optical gap E_g .

Defects other than dangling bonds, on the other hand, are not likely to be present in such a high number as that needed to explain the photoconductivity behavior, without being optically active. If, alternatively, one thinks in terms of a giant capture cross section for the C-induced defects, one would expect a linear decrease of photoconductivity with C-content – i.e. with E_g – which is not the case.

The only important change taking place when the carbon content becomes larger, is the remarkable increase in the Urbach energy E_o , which reflects the deeper and deeper extension of the localized tail states within the gap. The drop in photoconductivity should therefore be related to variations in the transport mechanisms – e.g. decrease of mobility due to disorder [17].

Beside the above effects, the opening of the gap may also play an important role. Let us consider the general analysis of the recombination process in semiconductors by Simmons et al. [19]. For each kind of trap, they define two quasi-Fermi levels – for trapped electrons and trapped holes – such that the occupancy of states lying between them is constant and dependent only on the free-carrier densities n,p and on the capture cross sections for electrons σ_n and holes σ_p . They also show that electrons recombine via states between the quasi-Fermi levels of trapped electrons E_{tn} and the dark Fermi level E_{F_o} .

If we suppose that n is larger than p in a-Si_xC_{1-x}:H - as happens in a-Si:H - the distance of E_{tn} from the conduction band depends only on n. This means that, when the gap opens, for equal n (i.e. keeping the photoconductivity level constant) the number of states between E_{F_O} and E_{tn} increases, thus reducing the recombination lifetime. The actual decrease of τ depends on the particular shape of the density of states. To evaluate the decrease of τ we have used a model proposed by Hack et al. [20] for the density of states and for the recombination process in intrinsic and n-doped a-Si:H. They considered two kinds of traps, acceptor-like in the upper half of the gap and donor-like in the lower half. Both kinds of traps are supposed to have the same ratio between the probability of capturing a free carrier when they are charged and when they are neutral. Scaling this model up for a gap of 2.2 eV of the highest-x sample, a decrease of almost two orders of magnitude in photoconductivity is obtained.

To definitely establish the weight of this process in the decrease of the electron lifetime in a-Si_xC_{1-x}, a more detailed knowledge of the spectral distribution of gap states is necessary, together with an accurate determination of the parameters involved in the calculation.

CONCLUSIONS

Spectral photoconductivity has allowed the determination of band-tail and defect states in a-Si_xC_{1-x}:H. It is found that the inclusion of carbon remarkably increases the localized-state density in the band tail, but does not significantly affect the density of defect states. The results are in excellent agreement with PDS experiments on identical samples.

The value of $\eta\mu\tau$, found to be independent of photon energy, decreases by six orders of magnitude with the inclusion of $\sim 25\%$ carbon. The data prove that the decrease in $\eta\mu\tau$ is not related to the density of defect states in the gap. Mechanisms such as decrease in mobility and increase in recombination, are proposed. However, the analysis suggests the need for more quantitative understanding of the photoconductive behavior of a-Si_x:C_{1-x}:H.

REFERENCES

1) Y.Tawada, H.Okamoto, and Y.Hamakawa, Appl. Phys. Lett. **39**, 237 (1981).
2) Y.Kuwano, M.Ohuishi, H.Nishiwaki, S.Tsuda, T.Fukatsu, K.Enomoto, Y.Nakashima, and H.Tarui, Proc. of the 16th IEEE Photovoltaic Specialists Conference, 1982, p1338.
3) A.J.Snell, W.E.Spear, P.G.LeComber, and K.Mackenzie, Appl. Phys. A**26**, 83 (1981).
4) H.Munekata and H.Kukimoto, Appl. Phys. Lett. **42**, 432 (1983).
5) W.B.Jackson, S.M.Kelso, C.C.Tsai, J.W.Allen and S.-J.Oh, to be published.
6) R.S.Sussman and R.Ogden, Philos. Mag. B**44**, 137 (1981).
7) D.A.Anderson and W.E.Spear, Phil. Mag. **35**, 4 (1977).
8) Y.Catherine and G.Turban, Thin Sol. Films **60**, 193 (1979).
9) T.Shimada, Y.Katayama, and F.Komatsubara, J. Appl. Phys. **50**, 5530 (1981).
10) A.H.Mahan, B. von Roedern, D.L.Williamson, and A.Madan, to be published in J. Appl. Phys.
11) A.Skumanich, A.Frova and N.M.Amer, Solid State Commun., 1985.
12) W.B.Jackson and N.M.Amer, Phys. Rev. B**25**, 5559 (1982).
13) F.Evangelisti, P.Fiorini, G.Fortunato, A.Frova, C.Giovannella, and R.Peruzzi, J. Non-Cryst. Solids **55**, 191 (1983).

14) G.Moddel, D.A.Anderson and W.Paul, Phys. Rev. B22, 1918 (1980).
15) W.B.Jackson, R.J.Nemanich, and N.M.Amer, Phys. Rev. B27, 4861 (1983).
16) J.Bullot, M.Gauthier, M.Schmidt, Y.Catherine, and A.Zamonche, Phil. Mag. B49, 489 (1984).
17) A.Morimoto, T.Miura, M.Kumeda, and T.Shimizu, J. Appl. Phys. 53, 7299 (1982).
18) R.Carius, K.Jahn, W.Siebert, and W.Fuhs, private information.
19) J.G.Simmons, and G.W.Taylor, Phys. Rev. B4, 502 (1971).
20) M.Hack, S.Guha and M.Shur, Phys. Rev. B30, 6991 (1984).

PART VII

Si-N Alloys

PROPERTIES OF a-SiN$_x$:H FILMS PREPARED BY GLOW DISCHARGE

OF Si$_2$H$_6$-NH$_3$ GAS MIXTURE

Y. KOJIMA, S. NARIKAWA, T. MATSUYAMA, E. IMADA, H. NOJIMA, T. HAYAKAWA AND
S. EHARA
Central Research Laboratories, Sharp Corporation, Tenri-shi, Nara, 632, JAPAN

ABSTRACT

The electrical and optical properties of a-SiN$_x$:H films prepared by the
glow discharge of Si$_2$H$_6$-NH$_3$ gas mixture have been investigated in a wide
range of the molar ratio of NH$_3$ to Si$_2$H$_6$. The maxima of the dark- and the
photo-conductivity are observed at the molar ratio of about 10^{-1}. At this
molar ratio the deposition rate is about 33 Å/sec. The optical gap and the
compositional ratio of nitrogen to silicon change smoothly from those of
a-Si:H to those of a-Si$_3$N$_4$. It is found that the valency control is possible in a-SiN$_x$:H films prepared at a high deposition rate.

1. INTRODUCTION

A considerable number of investigations have been directed towards the
study of the properties of hydrogenated amorphous silicon nitride, a-SiN$_x$:H,
films, because a-SiN$_x$:H films have high photoconductivity and thermal stability. So far a-SiN$_x$:H films have been mostly prepared by the glow discharge
of SiH$_4$-NH$_3$ [1,2] or SiH$_4$-N$_2$ [1,3] gas mixture. However these films have
been prepared at a low deposition rate. Therefore, in order to prepare
a-SiN$_x$:H films at a higher deposition rate, Si$_2$H$_6$ and NH$_3$ gases are used in
this experiment. In this paper the electrical and optical properties of
a-SiN$_x$:H films prepared by the glow discharge of Si$_2$H$_6$-NH$_3$ gas mixture and
the effect of boron and phosphorus doping on the electrical properties of
a-SiN$_x$:H films are reported.

2. EXPERIMENTAL

a-SiN$_x$:H films were prepared by the glow discharge of Si$_2$H$_6$-NH$_3$ gas
mixture in a capacitively coupled reactor with parallel electrodes of 21 cm
diameter separated by 2 cm. The preparation conditions are summarized in
Table I. The film thickness was measured by Tallysurf gauge. The conductivity was measured in a gap cell configuration with evaporated Al electrodes 0.2 mm apart and in a vacuum chamber evacuated at about 10^{-2} Torr.
The photoconductivity was measured at room temperature under the illumination of 0.3 mW of He-Ne laser with the wave length of 6328 Å. The optical
gap was obtained by extrapolating a plot of $(\alpha h\nu)^{1/2}$ vs. $h\nu$. The compositional ratio of nitrogen to silicon, N/Si, was determined by the peak to peak
heights of N(KLL) and Si(LMM) Auger, AES, signals by comparison with the

Table I Preparation conditions

rf frequency	13.56 MHz
rf power	200,500 W
flow rate	
Si_2H_6	5 – 30 SCCM
NH_3	0 – 200 SCCM
PH_3(0.3 % in H_2)	0 – 100 SCCM
B_2H_6(0.3 % in H_2)	0 – 100 SCCM
total pressure	1 Torr
substrate temperature	250 °C

corresponding AES signal of pyrolytic   a-Si_3N_4. The IR spectroscopy measurements were carried out by a FT-IR spectrometer, FT/IR-3 by JASCO. The contents of hydrogen bonded to Si atom and N atom were obtained from Si-H stretching mode and wagging mode, and N-H stretching mode, respectively.

3. RESULTS AND DISCUSSION

Figure 1 shows the deposition rate of a-SiN_x:H films as a function of the molar ratio of NH_3 to Si_2H_6, $N_{NH_3}/N_{Si_2H_6}$. A circle mark indicates the deposition rate where Si_2H_6 gas flow rate is fixed at 30 sccm and at RF power of 200W. A triangle mark indicates the deposition rate where NH_3 gas flow rate is fixed at 200 sccm and at RF power of 200 W. A square mark indicates the deposition rate at RF power of 500 W. The deposition rate of a-SiN_x:H depends on the molar ratio and changes from 37 Å/sec to 8 Å/sec with increasing the molar ratio of NH_3 to Si_2H_6. It is considered to be attributed to the decrease of the partial pressure of Si_2H_6 which dominates the film formation, since the total pressure in the reactor is fixed to be 1 Torr in the experiments. A similar tendency has been observed in the case of the chemically vapor deposited a-SiN_x:H films [4]. With increasing RF power up to 500 W the deposition rate increases because of further dissociation of Si_2H_6 and NH_3 gases.

The dark conductivity, σ_d, the photoconductivity, σ_p, and the activation energy, E_a, for a-SiN_x:H are plotted against the molar ratio in Fig. 2. The E_a was estimated from the temperature dependence of the σ_d by the following equation:

$$\sigma_d = \sigma_0 \exp(-E_a/kT) \qquad (1)$$

in the temperature range from 25°C to 200°C. As shown in Fig. 2 the σ_d and

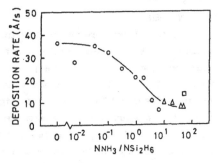

Fig.1 Deposition rate of a-SiN_x:H films plotted as a function of the molar ratio of NH_3 to Si_2H_6.

the σ_p are improved by nitrogen incorporation and attain the maxima at $N_{NH_3}/N_{Si_2H_6} \simeq 10^{-1}$, where the compositional ratio N/Si is found to be about 0.03 by AES measurement. The E_a reaches to the minimum nearly at the same molar ratio. At this molar ratio the deposition rate of 33 Å/sec is attained. The variation of the σ_d corresponds to that of the E_a, implying that nitrogen acts as donor. A similar behavior has been observed in the case of a-SiN$_x$:H films prepared from SiH$_4$-NH$_3$ gas mixture [2] but there is a large difference in the deposition rates between the two systems.

Figure 3 shows the optical gap, E_g^{opt}, and the compositional ratio of nitrogen to silicon as a function of the molar ratio. The marks in the figure indicate the same as those in Fig. 1. Up to $N_{NH_3}/N_{Si_2H_6} \simeq 3 \times 10^{-1}$ the E_g^{opt} remains unchanged at 1.75 eV which is equal to that of a-Si:H. Beyond this ratio the E_g^{opt} increases linearly to 4.7 eV with increasing the molar ratio and with increasing RF power the E_g^{opt} further increases to be 5.1 eV, which corresponds to that of a-Si$_3$N$_4$. The atomic ratio N/Si is approximately proportional to $N_{NH_3}/N_{Si_2H_6}$ at the molar ratio less than 10 and then tends to reach nearly the value of a-Si$_3$N$_4$. These results suggest that the variation of E_g^{opt} cannot be explained only by nitrogen content in the film.

Figure 4 shows the hydrogen content as a function of the molar ratio. In the IR spectrum of a-SiN$_x$:H, Si-N asymmetric stretching mode can be observed even at the smallest molar ratio. At $N_{NH_3}/N_{Si_2H_6} \lesssim 1$ Si-N asymmetric

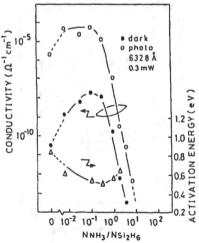

Fig.2 Dark conductivity, photo-conductivity and activation energy plotted as a function of the molar ratio of NH$_3$ to Si$_2$H$_6$.

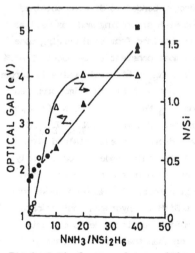

Fig.3 Optical gap and compositional ratio plotted as a function of the molar ratio of NH$_3$ to Si$_2$H$_6$.

stretching bands shows two absorption peaks, i.e. 840 cm^{-1} and 790 cm^{-1}. The former absorption is assigned to the stretching mode for Si-N site in which there is at least one H atom as a second neighbor and the latter is for the isolated Si-N site with no hydrogen second neighbor [5]. At $N_{NH_3}/N_{Si_2H_6} \gtrsim 1$ the absorption band centered at 790 cm^{-1} disappears and also the frequency of Si-N bond stretching vibration is shifted from 840 cm^{-1} toward 890 cm^{-1}. Si-H stretching mode is observed in a whole range of the molar ratio and the frequency of Si-H stretching vibrations is shifted from 2000 cm^{-1} ($N_{NH_3}/N_{Si_2H_6} \lesssim 10^{-1}$) to 2160 cm^{-1} ($N_{NH_3}/N_{Si_2H_6} \gtrsim 20$). This frequency shift is considered to be due to the increase of the effective force constant, that is, the increase of the electronegativity sum of Si-H bond which is induced by the nitrogen neighbor [6].

The content of hydrogen bonded to Si atom was estimated from the integrated absorption by using the following equation:

$$N_H = A \int \frac{\alpha(\omega)}{\omega} d\omega \qquad (2)$$

where $\alpha(\omega)$ is the absorption coefficient at wave number ω, A=1.4x10^{20} cm^{-2} for Si-H stretching mode [7] and A=1.6x10^{19} cm^{-2} for Si-H wagging mode [8] assuming that A is independent of the composition of the film. Also the content of hydrogen bonded to N atom was estimated from the integrated absorption of N-H stretching mode at ~3350 cm^{-1} using eq.(2) where A=2.6x10^{20} cm^{-2} derived from the data by W.A. Lanford et al. [9].

As shown in Fig. 4 the content of hydrogen bonded to Si atom obtained from Si-H stretching mode as well as that obtained from Si-H wagging mode is almost constant to be about 25 at.% at $N_{NH_3}/N_{Si_2H_6} \lesssim 1$. This value is larger than that of a-SiN$_x$:H films prepared from SiH$_4$-NH$_3$ [2] or SiH$_4$-N$_2$ [10] gas mixture. However, at $N_{NH_3}/N_{Si_2H_6} \gtrsim 1$ the hydrogen content obtained from Si-H stretching mode is considerably larger than that obtained from Si-H wagging mode. It has been shown that in a-Si:H the hydrogen content obtained from Si-H wagging mode is correct rather than that obtained from Si-H stretching mode, because the oscillator strength of Si-H wagging mode is independent of preparation conditions [11].

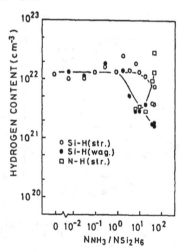

Fig.4 Hydrogen content plotted as a function of the molar ratio of NH$_3$ to Si$_2$H$_6$.

In a-SiN$_x$:H the oscillator strength of both Si-H stretching mode and wagging mode may be changed by nitrogen incorporation, so that the determination of the hydrogen content by using the same proportionality constant as in a-Si:H in a whole range of the molar ratio is considered not to be appropriate. However, since the oscillator strength is considered to be hardly affected by a small amount of nitrogen incorporated into the film, it is probable that at $N_{NH_3}/N_{Si_2H_6} \lesssim 1$ the hydrogen content provides a true value.

N-H stretching mode can be observed only at $N_{NH_3}/N_{Si_2H_6} \gtrsim 6.7$. The content of hydrogen bonded to N atom is estimated in this region and increases with increasing the molar ratio.

Figure 5 shows the σ_d, the σ_p and the E_a plotted as a function of doping ratio, $N_{B_2H_6}/(N_{Si_2H_6}+N_{NH_3})$ and $N_{PH_3}/(N_{Si_2H_6}+N_{NH_3})$. The molar ratio of NH_3 to Si_2H_6 is maintained at 10^{-1}, where the σ_d and the σ_p attain the maxma as shown in Fig. 2. The σ_d decreases with increasing doping ratio of B_2H_6 to attain the lowest conductivity; $\sigma_d=10^{-12}$ $(\Omega cm)^{-1}$ and the highest activation energy; $E_a=0.75$ eV at doping ratio of 10^{-3}. Beyond this ratio the σ_d increases and the E_a decreases with increasing doping ratio of B_2H_6. Therefore the incorporation of boron effectively shifts the Fermi level of a-SiN$_x$:H, so that the film changes from n-type to p-type a-SiN$_x$:H. With increasing doping ratio of PH_3 the σ_d increases monotonically and changes by four orders of magnitude at doping ratio of 10^{-2}. The E_a approaches the value of about 0.3 eV. Comparing these results with those in a-Si:H reported by Spear and Le Comber [12], the doping efficiency of boron and phosphorus in a-SiN$_x$:H is still poor. The lowest conductivity of the film prepared from SiH_4-NH_3 or SiH_4-N_2 gas mixture is about 10^{-14} $(\Omega cm)^{-1}$ [3,13], but that obtained in this experiment is about 10^{-12} $(\Omega cm)^{-1}$. However, by adding the same amount of PH_3, the change of the conductivity is larger in a-SiN$_x$:H films prepared from Si_2H_6 -NH_3 gas mixture than that in a-SiN$_x$:H films prepared from SiH_4-NH_3 or SiH_4- N_2 gas mixture. Also the E_g^{opt} is investigated and is constant to be about 1.75 eV in a whole range of doping ratio. Narrowing of the E_g^{opt}

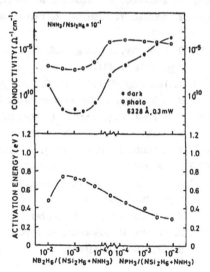

Fig.5 Dark conductivity, photo-conductivity and activation energy plotted as a function of doping ratio.

by doping B_2H_6 is not observed in this experiment. The deposition rate is about 30 Å/sec in a whole range of doping ratio.

4. CONCLUSIONS

a-SiN$_x$:H films have been prepared at a high deposition rate by the glow discharge of Si_2H_6-NH$_3$ gas mixture. The dark- and the photo-conductivity are improved by incorporating a small amount of nitrogen. The optical gap can be changed controllably from 1.75 eV to 5.1 eV by varying the gaseous molar ratio and RF power. It is found that the valency control of a-SiN$_x$:H films is possible and p-type or n-type a-SiN$_x$:H films can be produced at a high deposition rate. This can lead to more extensive applications of this material.

ACKNOWLEDGEMENTS

The authors would like to express their gratitude to Dr. T. Sasaki, the vice president, and Mr. K. Hayashi, the general manager of the engineering center of Sharp Corp. for stimulating this study and for continuous encouragement.

REFERENCES

[1] T. Noguchi, S. Usui, A. Sawada, Y. Kanoh and M. Kikuchi, Jpn. J. Appl. Phys. 21, L485 (1982).

[2] H. Kurata, M. Hirose and Y. Osaka, Jpn. J. Appl. Phys. 20, L811 (1981).

[3] H. Watanabe, K. Katoh and M. Yasui, Thin Solid Films 106, 263 (1983).

[4] Y. Ashida, Y. Mishima, M. Hirose and Y. Osaka, J. Appl. Phys. 55, 1425 (1984).

[5] G. Lucovsky, J. Yang, S.S. Chao, J.E. Tyler and W. Czubatyj, Phys. Rev. B28, 3234 (1983).

[6] G. Lucovsky, Solid State Commun. 29, 571 (1979).

[7] M.H. Brodsky, M. Cardona and J.J. Cuomo, Phys. Rev. B16, 3556 (1977).

[8] C.J. Fang, K.J. Gruntz, L. Ley, M. Cardona, F.J. Demond, G. Müller and S. Kalbitzer, J. Non-Cryst. Solids 35/36, 255 (1980).

[9] W.A. Lanford and M.J. Rand, J. Appl. Phys. 49, 2473 (1978).

[10] H. Watanabe, K. Katoh and M. Yasui, Jpn. J. Appl. Phys. 21, L341 (1982).

[11] H. Shanks, C.J. Fang, L. Ley, M. Cardona, F.J. Demond and S. Kalbitzer, Phys. Stat. Sol. (b) 100, 43 (1980).

[12] W.E. Spear and P.G. Le Comber, Solid State Commun. 17, 1193 (1975).

[13] H. Kurata, H. Miyamoto, M. Hirose and Y. Osaka, Jpn. J. Appl. Phys. 21 Suppl. 21-2 205 (1982).

PHOTOCONDUCTIVITY OF BORON DOPED a-SiN$_x$:H

H. Nojima, T. Hayakawa, E. Imada, Y. Kojima, S. Narikawa, T. Matsuyama and
S. Ehara
Central Research Laboratories, Sharp Corporation, Tenri-shi, Nara, JAPAN

ABSTRACT

The photoconductivity of boron doped a-SiN$_x$:H(x=0.07)has been studied
by steady state secondary photocurrent, xerographic photodischarge, the time
of flight measurement, and ESR. The dominant carrier of the photocurrent in
a-SiN$_x$:H is found to be electron. The $\mu\tau$ product of the electron decreases
with increasing doping amount of boron, while that of the hole increases and
becomes larger than 10^{-8}cm^2/V. From these results it is found that the life
time of the hole becomes longer with increasing the amount of doping boron.
The intensity of the ESR signal due to the neutral dangling bonds decreases
monotonically with doping boron, and this decrease of the neutral dangling
bond density is considered to be due to not only change of the occupation
statistics, but mainly to the chemical or electrical interaction between
boron and nitrogen.

1. INTRODUCTION

Hydrogenated amorphous silicon nitride, a-SiN$_x$:H is a material with
potential for application to electronics, because of the easiness of controlling the optical band gap as well as conductivity by changing the compositional ratio[1] and the possibility of controlling the electric properties by substitutional doping[2]. It also has excellent thermal stability[2].
Therefore it is important to clarify the photoconductivity of a-SiN$_x$:H alloy
system. In this experimental study the photoconductivity of boron doped
a-SiN$_x$:H has been investigated comprehensively by the steady state secondary
photocurrent measurement, xerographic photodischarge experiment, the time of
flight experiment, and ESR.

Flow Rate		
SiH$_4$		60 sccm
NH$_3$		4 sccm
B$_2$H$_6$(0.3%in H$_2$)		0 - 50 sccm
RF Power		100 W
Substrate Temperature		235°C
Total Pressure		0.25 Torr

Table 1. Deposition Conditions.

2. EXPERIMENTAL

2. 1. deposition conditions

a-SiN$_x$:H films used in this
study were prepared by an RF glow
discharge CVD system with a SiH$_4$-
NH$_3$-B$_2$H$_6$(0.3% in H$_2$) gas mixture.
The film deposition conditions are
listed in table 1.

The compositional ratio of nitrogen
to silicon, N/Si, was obtained to be

	T.O.F.	Second. Photocon.	Xero. Photodis.	E.S.R.
Substrate	Al Foil	Corning 7059	Al Plate	Quartz Plate
Thickness	~5 μm	1 ~ 2 μm	~ 7.5 μm	1 ~ 2 μm
Electrode	Au semitrans-parent	Coplanar Al interdigital	none	none
Structure	Parylene block-ing layer		Sandwiched between SiN_x blocking layers	

Table 2. Sample Structures.

0.07 by AES, at which conductivity is maximum.

2. 2. Sample structures

The sample structures used in this study are listed in Table 2.

3. RESULTS AND DISCUSSION

3. 1. Steady state secondary photoconductivities

Steady state secondary photocurrents have been measured. In this measurement the linearity of current-voltage characteristics is ensured. The observed photoconductivities were ascertained to be proportional to the incident light intensity. Figure 1 shows the normalized photoconductivities, $\eta\mu\tau$, as a function of the doping ratio of B_2H_6, where η is the photocarrier generation efficiency, μ is the drift mobility and τ is the carrier life time. As shown in Fig. 1 $\eta\mu\tau$ decreases with increasing the doping boron up to the doping ratio of 2.3×10^{-4}, and then increases with increasing the doping ratio beyond 2.3×10^{-4}. Generally the secondary photocurrent consists of two contributions from the electron and the hole, which can not be measured separately in secondary photocurrent measurements.

3. 2. Xerographic photodischarge

In order to elucidate the carrier transport properties of the electron and the hole in boron doped a-SiN_x:H (x=0.07) separately, xerographic photodischarge has been measured. The surface potential decay by light exposure for negative or positive charging represents the carrier

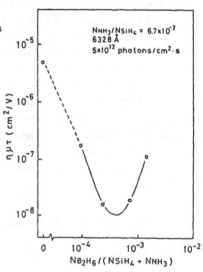

Fig. 1. Normalyzed photoconductivity, $\eta\mu\tau$, estimated by the steady state secondary photocurrents plotted as a function of the doping ratio of B_2H_6.

transport of the electron or the hole respectively. Figure 2 shows the surface potential decays of boron doped a-SiN$_x$:H (x=0.07) by light exposure. It is found that the transport properties of the hole in undoped a-SiN$_x$:H (x=0.07) are very small in comparison with that of the electron and with doping boron the transport properties of the hole are increased, and finally exceed that of the electron to become dominant.

The xerographic photodischarge experiments are analyzed quantitatively[3]. The photocurrent through the sample film is obtained from the rate of the surface potential decay as $J_p = C\frac{dV}{dt}$, where C is the sample capacitance. The surface potential decay is analyzed by Hecht rule as follows:

$$J_p = \eta Fe \frac{\mu\tau E}{L}\left[1-\exp(-\frac{L}{\mu\tau E})\right]$$

where F is the incident light flux, L is the sample thickness, and E is the electric field in the sample. From the result of this analysis, the $\mu\tau$ product can be estimated. Figure 3 shows the electrophotographic gain, J_p/eF vs. the electric field E for the intrinsic a-SiN$_x$:H (x=0.07) compensated by doping boron. The magnitude of the $\mu\tau$ product of the hole is estimated to be about 10^{-8}cm^2/V. In this estimation it is assumed that η is equal to unity. If the effect of geminate recombination in a-SiN$_x$:H (x= 0.07) is considerable and η depends on electric field[4], the $\mu\tau$ product of the hole obtained by xerographic photodischarge should be higher than 10^{-8}cm^2/V. This experiment was performed under the space charge free condition. In order to avoid the perturbation of space charge, the incident light flux was kept sufficiently small to be 1.9×10^{12}photons/cm^2·sec. This ensures the emission limited condition.

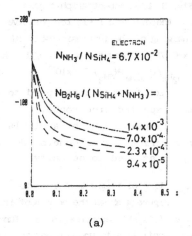

(a)

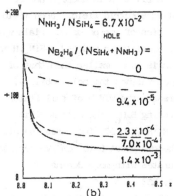

(b)

Fig. 2. Xerographic photodischarge characteristics for (a) negative and (b) positive surface charging with various doping amount of boron. Dominant transport carrier is electron and hole respectively.

Fig. 3. Electrophotographic gain
vs. the electric field for positive
charging of the intrinsic a-SiN$_x$:H
(x=0.07) compensated by doping boron
$(N_{B_2H_6}/(N_{SiH_4}+N_{NH_3})=7.0\times10^{-4})$.

The solid lines correspond to
the expected electrophotographic gain by Hecht analysis
for various $\mu\tau$ products, where
η is assumed to be unity.

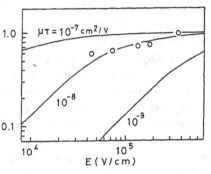

3. 3. Time of flight measurements

Figure 4 shows the hole mobilities obtained in this experiment by the
time of flight measurements as a function of the doping ratio of B$_2$H$_6$. The
hole mobilities in undoped and boron doped a-SiN$_x$:H (x=0.07) increase with
applied electric field. Such field dependent mobility is common characteristics of amorphous materials which exhibit dispersive transport. The
hole mobility in undoped a-SiN$_x$:H is observed to be about 5×10^{-6} cm^2/V·sec ,
which is much smaller than that of a-Si:H. It is found that nitrogen alloying diminishes the hole mobility. It is considered that nitrogen alloying
increases the width of tail states of the valence band. With small amount
of doping B$_2$H$_6$, the hole mobility in
a-SiN$_x$:H (x=0.07) increases sharply
to become about 5×10^{-5} cm^2/V·sec and
saturates with more doping.

3. 4. ESR

From the extensive work on defects in a-Si:H, it is concluded that
the dominant deep trapping center is
dangling bond[5) 6)]. In order to clarify the doping effect on the transport
properties of a-SiN$_x$:H (x=0.07), the ESR
has been measured.

The ESR measurements were performed at room temperature with a spectrometer, JES FE2XG by JEOL at X band.
The signal with the g value of about
2.0055 and the peak to peak line-
width,ΔH_{pp}, of about 7 gauss was
observed and was almost unchanged
in the range of the doping ratio

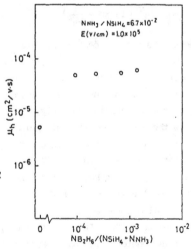

Fig. 4. The hole mobility obtained
by the time of flight measurements as
a function of the doping ratio of
B$_2$H$_6$.

Fig. 5. The spin density
plotted vs. the energy
difference of the Fermi level
from the mobility edges.
The broken line corresponds
to the results on a-Si:H by
Dersh et al[7].

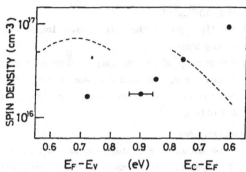

of B_2H_6 up to 1.4×10^{-3}. This signal is corresponding to the neutral silicon
dangling bond. The spin density is plotted vs. the energy difference of the
Fermi level from the mobility edges as shown in Fig. 5. The activation
energy was obtained by the temperature dependence of the electrical conductivity. It is found that the spin density of boron doped a-SiN$_x$:H (x=0.07)
decreases as the Fermi level shifts toward the mid gap, and this result is
different from that of a-Si:H which is shown in the same figure[7]. The
observed decrease of the spin density indicates either (1) the decrease of
dangling bond density by chemical or electrical compensation of boron and
nitrogen, or (2) change of the occupation statistics of dangling bond centers
by the shift of the Fermi level, resulting in the decrease of neutral dangling bond density, where neutral dangling bond centers are changed to positively charged dangling bond centers. From the experimental results of carrier
transport properties in a-SiN$_x$:H (x=0.07) discussed above, it is found that
by doping boron the $\mu\tau$ product of the hole is greatly increased and that of
the electron is slightly decreased. This result is considered to indicate
that the negatively charged dangling bonds are changed to the neutral dangling bonds and it is in contrast with the latter mechanism mentioned above.
Therefore the decrease of the ESR intensity in a-SiN$_x$:H (x=0.07) by doping
boron is mainly due to the decrease of dangling bond density rather than
change of the occupation statistics.

4. CONCLUSION

The photoconductivity of boron doped a-SiN$_x$:H (x=0.07) has been systematically investigated by steady state secondary photocurrent, xerographic
photodischarge, the time of flight measurement, and ESR. From these measurements, it is concluded the following;
1) The dominant carrier of the photocurrent in a-SiN$_x$:H (x=0.07) is electron.
2) The $\mu\tau$ product of the electron decreases with increasing the amount of

doping boron, while that of the hole increases greatly and becomes larger than $10^{-8} cm^2/V$.

3) The life time of the hole becomes longer with increasing the amount of doping boron.

4) The density of the dangling bond centers decreases monotonically with doping boron, and this decrease of the dangling bond centers is considered to be mainly due to the chemical or electrical compensation between boron and nitrogen.

ACKNOWLEDGEMENTS

The authors are grateful to Dr. T. Sasaki, the vice president, and Mr. K. Hayashi, general manager of the engineering center of Sharp Corp. for their kind encouragements throughout this study.

REFERENCES

1) H. Kurata, M. Hirose and Y. Osaka: Jpn. J. Appl. Phys. 20(1981) L811.

2) H. Kurata, H. Miyamoto, M. Hirose and Y. Osaka: Proc. 3rd Photovoltaic Science and Engineering Conf. in Japan, 1982, Jpn. J. Appl. Phys. 21 (1982) Suppl. 21-2 205.

3) S. Oda, Y. Saito, I. Shimizu and E. Inoue: Philos. Mag. B43 (1981) 1079.

4) J. Mort, A. Troup, M. Morgan, S. Grammatica, J. C. Knights, and R. Lujan: Appl. Phys. Lett. 38(1981) 277.

5) R. A. Street: Appl. Phys. Lett. 41(1982) 1060.

6) R. A. Street, J. Zesch and M. J. Thompson: Appl. Phys. Lett. 43(1983) 672.

7) H. Dersch, J. Stuke and J. Beichler: Phys. Stat. Sol. B105 (1981) 265.

DEFECT STATES IN SILICON NITRIDE

JOHN ROBERTSON* AND MARTIN J. POWELL**
* Central Electricity Research Labs., Leatherhead, Surrey, UK
** Philips Research Labs., Redhill, Surrey, UK

ABSTRACT

The energy levels of defect centers in amorphous silicon nitride have been calculated. The results are related to recent photoemission and light-induced electron spin resonance data. The Si dangling bond is argued to be the memory trap in MNOS devices and to be responsible for the electron accumulation at interfaces with amorphous silicon and for the n-type charge-transfer doping of amorphous silicon-silicon nitride superlattices.

Silicon nitride is used as the memory medium in non-volatile metal-nitride-oxygen-semiconductor (MNOS) devices [1], as the gate dielectric in thin-film transistors (TFTs) based on hydrogenated amorphous silicon (a-Si:H) [2-4], and as the wider-gap layer in a-Si:H/a-SiN$_x$:H superlattices [5]. Each of these applications depends on the nature and behaviour of the defect gap states in the nitride layers. Silicon nitride is an amorphous 4:3-coordinated random network which also contains hydrogen as an important impurity. We recently calculated the energy levels associated with the expected defects in silicon nitride [6], namely the Si and N dangling bonds ($\equiv$Si and $=$N), the $\equiv$SiH and $=$NH bonds and the $\equiv$Si-Si$\equiv$ bonds. We found that the Si dangling bond gives a state near midgap, and the Si-Si bond and the negative $=$N$^-$ center introduce gap states just above the valence band edge, E_v, while the other centers do not produce gap states. We now use these results to review recent work on bulk silicon nitride and to interpret aspects of a-Si:H/a-SiN$_x$:H interfaces in TFTs and superlattices.

The concentration of the various defects and indeed the nitride's precise composition depend on the deposition process: chemical vapour deposition (CVD), plasma deposition (PD) or sputtered (SP). CVD films are used for MNOS devices and tend to be closest to a Si$_3$N$_4$ stoichiometry [7]. PD films tend to be Si-rich although both Si-rich and N-rich films are possible [8]. For TFT applications, near-stoichiometric material is used to maximise the band gap, whereas superlattices are usually made with Si-rich alloys. Both CVD and particularly PD films contain chemically bonded hydrogen, as seen from their infra-red vibrational spectrum [9]. As in a-Si, the hydrogen saturates both Si and N dangling bonds and according to our calculations the resulting Si-H and N-H states lie outside the gap.

The chemical ordering in a-SiN$_x$ and a-SiN$_x$:H alloys is high but not perfect. As the composition x approaches stoichiometry, Si-Si bonds are replaced progressively by Si-N bonds, leaving only a small concentration at stoichiometry. Beyond x = 1.33 excess N would normally enter as N-N bonds, but this probably does not occur as single N-N bonds are chemically unstable with respect to molecular N$_2$. Indeed, CVD nitride cannot be prepared with x > 1.33 (see Fig. 2 in ref. [7]), while hydrogen-rich PD alloys can be prepared with x > 1.33 [8], presumably because N enters as $=$NH and $-$NH$_2$ groups. Interestingly though, unhydrogenated sputtered a-Si$_3$N$_4$ nitride is believed to contain appreciable numbers of Si-Si bonds, according to its photoemission core level spectra [10], which is only possible if compensating $=$N or N-N groups are also present.

216

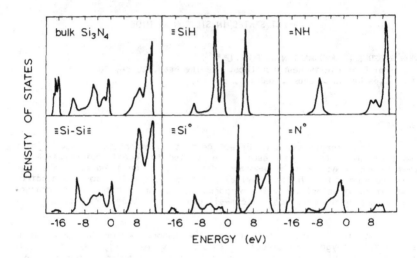

Fig. 1. The calculated local density of states for various defect configurations in Si₃N₄

The local density of states (DOS) of each center was calculated by the tight-binding recursion method [6]. Si uses its four sp³ hybrids for bonding in both Si₃N₄ and Si, but a hybrid must be assigned a higher energy in a Si-N bond than in a Si-Si bond to give the observed 1.9 eV conduction band discontinuity at the Si₃N₄ interface with crystalline (c-)Si [11]. This same self-energy shift is then used for Si hybrids in Si-Si and Si-H bonds and Si dangling bonds in bulk silicon nitride. The Si-Si bonding (σ) state is found to lie just in the gap, at about 0.1 eV above E_v, and its antibonding (σ*) state lies in the conduction band (Fig. 1).

Recent photoemission spectra [10] confirm the DOS shown in Fig. 1. Firstly, they find that the highest valence band of Si₃N₄ is indeed due to N pπ states which were missed in early photoemission spectra [12]. The pπ lone pair states arise from the planarity of the N site and are a distinctive feature of Si₃N₄ [13]. Secondly, photoemission [10] finds the Si-Si σ states at 0.4 eV above the valence band edge, E_v, in good agreement with the calculated value of 0.1 eV in Fig. 1, but much below our earlier erroneous value of 3 eV [13]. Finally, Si-H σ states are located around -6.3 eV below E_v, close to their calculated position, with no evidence of occupied hydrogen-related gap states. This confirms that hydrogen does indeed passivate ≡Si sites, without leaving states in the gap.

The evolution of the band edge energies with alloy composition is of great interest for work on superlattices. Photoemission has been used to trace the evolution of the valence band edge [10]. The Si-Si σ states were found to form the upper valence band in a-SiN$_x$ for x < 1.2, at which point these states split off as isolated, localized gap states, leaving the N pπ states as the top of the main valence band for x > 1.2. The variation in the conduction band edge was found by adding the gap to the valence band energy [10], and a similar dependence was deduced by resonant photoemission [14]. These results were interpreted as showing that near stoichiometry the σ* states of isolated Si-Si bonds lay in the gap, but our calculations find them to lie in the conduction band.

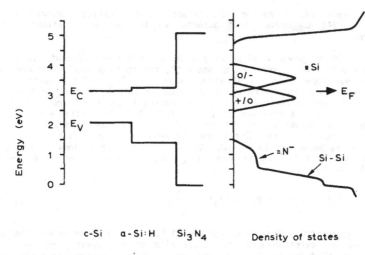

c-Si   a-Si:H Si_3N_4 Density of states

Fig. 2.   A schematic density of states due to Si dangling bonds in plasma
deposited a-Si_3N_4:H

We proposed [6] that ≡Si centers were the long term memory traps
in MNOS devices.   This identification is favoured because they are present
in the correct concentrations (~10^{19} cm^{-3} in CVD material [15,16]), and
because they are also amphoteric, deep and energetically aligned with the
gap of c-Si.   Thus, they trap either electrons or holes during the write
process and release them during reading.   The Si-H bond had also been
suspected as a possible memory trap after experimental correlations between
hydrogen and trap densities had been inferred [17].   However, if memory
action is attributed to only a single center, it must be amphoteric in
nature but Si-H bonds are not.   Furthermore, both the present calculations
and the photoemission data confirm that Si-H σ states lie below E_v.   Thus,
we must conclude that the Si-H bond is not the memory trap although at
present we cannot account for the experimental correlation with H
concentration.

The Fermi level lies within the Si dangling bond levels for all
nitride compositions.   Experimentally, the density of states at E_F,
$N(E_F)$, is lower in PD nitride than in CVD nitride.   We found that $N(E_F)$
~ 10^{18} cm^{-3} eV^{-1} from an analysis of hopping conduction [18] and a similar
density was inferred from measurements on superlattices [19].   Interestingly
this density is much larger than that of intrinsic a-Si:H, which is about
10^{17} cm^{-3} eV^{-1} or less [20].

Our calculations place the +/o ≡Si level· in Si_3N_4 at about 3.1 eV.
The o/- level lies at an energy U higher, where U is the correlation energy.
We have argued that U is positive and U = 0.6-1.0 eV, by comparison with
U = 0.5 eV for ≡Si P_b centers at Si:SiO_2 interfaces [21] and U = 0.35 eV
for the ≡Si center in a-Si:H [22].   The two levels are each expected to have
some disorder broadening but nevertheless to be relatively sharp (Fig. 2).
The major source of broadening is from variations in the bond angle, but
electron spin resonance (ESR) shows that the ≡Si center in a-SiO_2 (the E'
center) retains its tetrahedral bond angles with very little disorder [23],
so that a large broadening is not expected.   On the other hand, hopping
conduction suggests that the width of these levels exceeds 0.3 eV [18].
These are both consistent with relatively sharp levels of width 0.3-0.6 eV.

However recently, a rather featureless sub-gap optical absorption was measured [24] suggesting that defects may be less well defined in CVD nitride.

Other group V elements act as substitutional dopants in a-Si:H when present in concentrations of under 5%, but it is generally assumed that N does not dope in a-SiN$_x$:H alloys with x > 0.1. The possibility of doping would seriously effect the behaviour of interfaces because there may be a layer where a low N content allows doping to occur. Experimentally, N doping is sometimes observed in bulk a-Si:H, but its inefficiency and its dependence of deposition conditions [25] could be used to argue against a substitutional mechanism. The absence of N-doping in c-Si is a insufficient guide to its behaviour in a-Si:H. The prototype group V elements such as P are now believed to dope by forming P_4^+-Si_3^- site-pairs, according to the equilibrium

$$P_3^\circ + Si_4^\circ \rightleftharpoons P_4^+ + Si_3^- \qquad \cdots (1)$$

(where the subscript denotes coordination) [26,27]. We have argued that N_4^+ sites do not exist because the small size of N prevents four Si atoms surrounding N at its equilibrium bond length, thereby preventing doping [6]. However, we now suggest that those N_4^+ sites with three H and perhaps even two H neighbours are sterically possible and are estimated to be shallow [28], so perhaps a minimal substitutional doping is possible from those rare sites with a high hydrogen coordination.

A dark ESR signal from $\equiv Si^\circ$ sites is seen in a-Si$_3$N$_4$:H, but an $=N^\circ$ signal is not [29-31]. We argued that as Si dangling bonds outnumber N dangling bonds, and as their levels lie higher in energy, charge transfer occurs so that only $\equiv Si^\circ$, $\equiv Si^+$ and $=N^-$ sites remain [6]. However, some aspects of the ESR spin density data are less readily understood. Fig. 3 shows the composition dependence of the spin density for SP a-SiN$_x$ alloys and SP, CVD and PD a-SiN$_x$:H alloys. The spin density decreases unremarkably with increasing x for the unhydrogenated SP alloys. However, the spin density increases quite rapidly with x for the hydrogenated PD and SP alloys, as does their total defect density, which is slightly unusual in that defect densities often decrease with mean coordination number. Finally, the behaviour of CVD films is quite remarkable in that the spin density apparently drops sharply from 10^{19} cm^{-3} to below 10^{16} cm^{-3} as x approaches stoichiometry, while the total defect density as measured electrically in MNOS devices [15,16] is believed to change little. This may be due to a very close compensation of Si and N dangling bond densities at stoichiometry, as above, or perhaps to some reconstruction which occurs between the Si dangling bonds themselves. This apparent absence of a Si_3° ESR in stoichiometric CVD nitride is an inconsistency with the above arguments for the amphoteric memory trap which is not yet fully understood.

Recently, a light-induced ESR (LESR) signal from $\equiv Si^\circ$ was found in a-SiN$_x$:H, which increases with x and which is up to six times larger than the dark signal [33]. The LESR could be due to the excitation:

$$Si_3^+ + e^- \rightarrow Si_3^\circ \qquad \cdots (2)$$

Thus the dark ESR would be due to Si_3° and the LESR would be due to both Si_3^+ and Si_3° sites. However, Kumeda et al. [33] noted that the reason for the composition dependence is not obvious in this model.

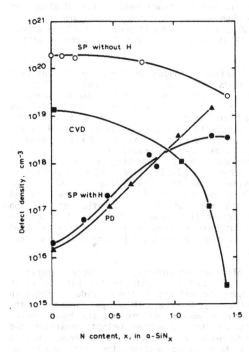

Fig. 3. Spin densities of a-SiN_x and a-SiN_x:H films, after Shimizu [29]. The value for CVD a-Si is from Hirose [32] and those for CVD alloys are adapted from Fujita et al. [31] by converting gas flow ratios to alloy compositions using the data of ref. [7]

A large LESR signal often indicates the presence of negative U centers which become excited from their stable diamagnetic configuration by light. The most likely candidate is the valence alternation pair [27,33]:

$$N_3{}^\circ + Si_4{}^\circ \rightleftharpoons N_4{}^+ + Si_3{}^- \qquad \qquad \text{... (3)}$$

with the paramagnetic center formed by electron trapping:

$$N_4{}^+ + Si_4{}^\circ + e^- \rightarrow N_3{}^\circ + Si_3{}^\circ \qquad \qquad \text{... (4)}$$

However, we believe that $N_4{}^+$ sites do not exist in sufficient quantities, so this mechanism is not responsible. We propose instead that the equilibrium

$$N_3{}^\circ + Si_4{}^\circ \rightleftharpoons N_2{}^- + Si_3{}^+ \qquad \qquad \text{... (5)}$$

occurs. Also, the higher defect density in stoichiometric PD nitride than in a-Si:H indicates that equilibrium (5) lies further to the right than does the equilibrium

$$2Si_4{}^\circ \rightleftharpoons 2Si_3{}^\circ \qquad \qquad \text{... (6)}$$

In this way, N introduces extra Si dangling bonds into a-Si:N$_x$:H alloys in such a way that they just outnumber N dangling bonds. Thus our original assignment of LESR to reaction (2) still holds as it does give a LESR signal which is larger than the dark ESR and which is expected to increase with N content, as observed [33].

The a-Si:H/a-SiN$_x$:H interface produces an electron accumulation layer in the a-Si:H, i.e. the bands bend downwards near the interface at zero bias [34,35]. Superlattices containing alternating ~100 Å layers of intrinsic a-Si:H and a-SiN$_x$:H are found to have high conductivities because E$_F$ is raised in the a-Si:H layers by the electron accumulation [19,36,37]. There has been a charge-transfer doping of the a-Si:H layers which leaves a net positive charge in the nitride layers. The band bending could be due to either a mismatch of the work functions of the two materials or to defect states near the interface. The band bending is found to depend on the order of deposition; Street et al. [34,35] find that accumulation is largest if the nitride is deposited after the a-Si:H (top nitride) rather than before (bottom nitride), while Roxlo et al. [36] find the reverse and in our own work we find that the magnitude of accumulation at the top nitride varies with the nitride deposition conditions [3]. A mechanism based on direct doping by substitutional N sites at the interface is ruled out because the conductivity of the superlattices [19] is much greater than that of bulk N-doped a-Si:H [25].

We propose that the electron accumulation at these interfaces arises from the higher energy of Si dangling bonds in the nitride which causes a local charge transfer from the defects in the nitride to those in the a-Si:H. In our model (Fig. 4), the band line-ups depend on the difference of the silicon and nitride work-functions and the band-bending and the position of E$_F$ depends on the respective defect densities and their energy levels. Taking the band-edge discontinuities to depend only on work-function differences is a fair approximation for lattice-matched crystalline semiconductor heterojunctions [38-40]. It should be equally valid for amorphous semiconductor heterojunctions where topology does not require the presence of mismatch dislocations. The discontinuities can be calculated from bulk band structure parameters, following Harrison [41]. (In fact, we have already assumed the validity of this approximation and used the reversed process to find the self-energy shifts for Fig. 1.) Thus, in constructing the band line ups in Fig. 3, we take E$_g$ for Si$_3$N$_4$, c-Si and a-Si:H to be 5.1, 1.1 and 1.8 eV respectively and the conduction band discontinuity of Si$_3$N$_4$/c-Si as 1.9 eV [11]. The gap of a-Si:H varies strongly with composition and deposition conditions, but the c-Si/a-Si:H conduction band discontinuity always remains small and is taken to be -0.1 eV. Turning now to the band bending, E$_F$ lies in the Si dangling bond levels in both bulk a-Si:H and a-SiN$_x$:H. This level in a-SiN$_x$:H lies higher [42] than in a-Si:H and is adjacent to the conduction band edge E$_c$ of a-Si:H. Additionally, the defect density is greater in a-SiN$_x$:H, as discussed earlier, (even allowing for the possibly higher defect density at the surface of a-Si:H [43,44]). Thus, E$_F$ is forced towards E$_c$ of the a-Si:H at the interface. Taking E$_F$ = E$_v$ + 0.9 eV in bulk a-Si:H and E$_F$ = E$_v$ + 3.0 eV in bulk a-Si$_3$N$_4$:H [10] gives the electron accumulation shown in Fig. 3. Charge transfer leaves Si$_3^+$ states in the nitride and Si$_3^-$ states in the silicon. Hence, we propose that the higher density and higher energy of Si dangling bonds in a-SiN$_x$:H causes the electron-accumulation at the interface and the charge transfer doping of the superlattices.

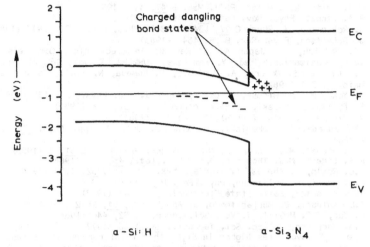

Fig. 4. Band line ups and electron accumulation at a-Si:H/a-SiN$_x$:H interfaces. Charge is transferred from Si$_3$ sites in the nitride to Si$_3$ sites in the silicon

References

1. P.C.Y. Chen, IEEE Trans. ED-24, 584 (1977)
2. M.J. Powell, B.C. Easton, O.F. Hill, App. Phys. Let. 38, 794 (1981)
3. M.J. Powell, in Materials Research Society Symposium Proceedings, Vol. 33, "Comparison of Thin Film Transistor and SOI Technologies", eds. H.Y. Lam, M.J. Thompson (North Holland, New York) (1984)
4. M.J. Powell, in Proc. 4th Int. Display Research Conf., Paris (1984)
5. B. Abeles, T. Tiedje, Phys. Rev. Let. 51, 2003 (1983)
6. J. Robertson, M.J. Powell, App. Phys. Let. 44, 415 (1984)
7. R.S. Bailey, V.J. Kapoor, J. Vac Sci. Technol. 20, 484 (1982)
8. M. Maeda, Y. Arita, J. App. Phys. 53, 6852 (1982)
9. P.S. Peercy, H.J. Stein, B.L. Doyle, S.T. Picraux, J. Electron Mater, 8, 11 (1979)
10. R. Karcher, L. Ley, R.L. Johnson, Phys. Rev. B 30, 1896 (1984)
11. D.J. DiMaria, P.C. Arnett, App. Phys. Let. 26, 711 (1975)
12. Z.A. Weinberg, R.A. Pollak, App. Phys. Let. 27, 254 (1975)
13. J. Robertson, J. App. Phys. 54, 4490 (1983)
14. L. Ley, R. Karcher, R.L. Johnson, Phys. Rev. Let. 53 710 (1984)
15. P.C. Arnett, B.H. Yun, App. Phys. Let. 26, 94 (1975)
16. S. Fujita, H. Tryoshima, M. Nishihara, A. Sasaki, J. Electron Mat, 11, 795 (1982)
17. V.J. Kapoor, R.S. Bailey, H.V. Stein, J. Vac. Sci. Technol A1, 600 (1983)
18. M.J. Powell, App. Phys. Let., 43, 597 (1983)
19. T. Tiedje, B. Abeles, App. Phys. Let. 45, 179 (1984)
20. D.V. Lang, J.D. Cohen, J.P. Harbison, Phys. Rev. B 25, 5285 (1982)
21. E.H. Poindexter, G.J. Gerardi, M.E. Rueckel, P.J. Caplan, N.M. Johnson, D.K. Biegelsen, J. App. Phys. 56, 2844 (1984)
22. H. Dersch, J. Stuke, J. Beichler, Phys. Stat. Solidi, b105, 265 (1981)
23. D.L. Griscom, E.J. Friebele, G.H. Sigel, Solid State Commun., 15, 479 (1974)
24. C.H. Seager, J.A. Knapp, App. Phys. Let. 45, 1060 (1984)

25. G.J. Smith, W.I. Milne, Phil. Mag. B $\underline{47}$, 419 (1983)
26. R.A. Street, Phys. Rev. Let. $\underline{49}$, 1187 (1982)
27. J. Robertson, J. Phys. C $\underline{17}$, L349 (1984), Phys. Rev. B $\underline{31}$, 3817 (1985)
28. J. Robertson, Phys. Rev. B $\underline{28}$, 4647 (1983)
29. T. Shimizu, in "Japan Annual Reviews in Electronics, Computers and Telecommunications", ed. Y. Hamakawa, (OHMSHA, Tokyo), $\underline{16}$, 21 (1984); T. Shimizu, S. Oozora, A. Morimoto, M. Kumeda, N. Ishii, Solar Energy Mats. $\underline{8}$, 311 (1982)
30. S. Yokoyama, M. Hirose, Y. Osaka, Japan J. App. Phys., $\underline{20}$, L35 (1981)
31. S. Fujita, A. Sasaki, J. Electrochem Soc., $\underline{132}$, 398 (1985)
32. M. Hirose, J. Phys. (Paris), $\underline{42}$, C4-705 (1981)
33. M. Kumeda, H. Yokomichi, T. Shimizu, Jpn. J. App. Phys. $\underline{23}$, L502 (1984)
34. R.A. Street, M.J. Thompson, N.M. Johnson, Phil Mag. B $\underline{51}$, 1 (1985)
35. R.A. Street, M.J. Thompson, App. Phys. Let., $\underline{45}$, 769 (1984)
36. C.B. Roxlo, B. Abeles, T. Tiedje, Phys. Rev. Let. $\underline{52}$, 1994 (1984)
37. N. Ibaraki, H. Fritzsche, Phys. Rev. B $\underline{30}$, 5791 (1984)
38. R.L. Anderson, Solid State Electronics, $\underline{5}$, 341 (1962)
39. A.D. Katnani, G. Margaritondo, J. App. Phys., $\underline{54}$, 2522 (1962)
40. A. Zur, T.C. McGill, J. Vac. Sci. Technol., B2, 440 (1984)
41. W.A. Harrison, J. Vac. Sci. Technol., $\underline{14}$, 1016 (1977)
42. The $Si_3{}^\circ$ level lies higher in Si_3N_4 than in Si because it is repelled upwards by the valence p states on its adjacent N sites, while in Si itself it is repelled equally by both valence and conduction states.
43. M.J. Powell, J.W. Orton, App. Phys. Let., $\underline{45}$, 171 (1984)
44. W.B. Jackson, D.K. Biegelsen, R.J. Nemanich, J.C. Knights, App. Phys. Lets. $\underline{42}$, 105 (1983)

PART VIII

Miscellaneous Alloys

ON THE ELECTRONIC PROPERTIES OF SILICON-TELLURIUM FILMS

F.A. FARIS, A. AL-JASSAR, F.G. WAKIM AND K.Z. BOTROS
Physics Department, Faculty of Science
Kuwait University, P.O. Box 5969, Kuwait

ABSTRACT

Silicon and Silicon-Tellurium films were prepared in an ultrahigh
vacuum system. The d.c. electrical conductivity of these films were
measured under vacuum before exposing them to the air. The thermal activation energies of Silicon-Tellurium films were always found to be greater
than that of Silicon films. Furthermore the optical measurements on
Silicon-Tellurium films indicated that the optical gap of these films
decreases as the concentration of tellurium in the films increases. Electron microscope examinations revealed that the films were amorphous in
structure and homogeneous in composition.

INTRODUCTION

Hydrogenated amorphous silicon (a-Si:H) films [1] are promising for
producing p-n junctions and solar cells. This is due to the fact that the
mobility gap of such films contains low density of defect states [2], and
therefore possible to be doped to produce p or n-type semiconductors.
Recent work of Wakim et al [3,4] has indicated that amorphous silicon with
selenium (a-Si:Se) films have some of their electrical and optical properties (d.c. conductivity and band gap) similar to those of (a-Si:H) films.
The presence of hydrogen or selenium in amorphous silicon (a-Si) seems to
passify and saturates the dangling bonds, therefore reduces the density
of undesired states in the gap [3,5].

In the present work the effect of adding tellurium to amorphous
silicon is investigated in order to explore the presence of similar
effects to that of hydrogen or selenium on amorphous silicon.

EXPERIMENTAL METHOD

The starting materials employed in the present work (Si and Te) were
both of highest purity (99.999%). Si-Te films were deposited by evaporation on silica and glass substrates for electrical and optical measurements respectively. For subsequent electrical measurement the substrates
were apriori provided with suitable aluminium contacts with a contact
separation of about 1 mm. The actual evaporation was carried out in an
ultrahigh vacuum system providing a base pressure of 10^{-9} Torr. During
evaporation Si was heated by an electron gun and Te separately heated
inductively in a silica crucible. The geometry of the two sources with
respect to the substrates provided a method of changing the percentage
of Te/Si.

The film structure was determined by using electron diffraction in a
transmission electron microscope [6] and composition was determined by
x-ray fluorescence in a scanning electron microscope. All measurements
indicated that the prepared films were amorphous in structure. Furthermore, the composition of each film was found always to be homogeneous.
The concentration of Te/Si was varied in the range from 0 to 26 at %.

In order to avoid effects of film surfaces on the electrical and optical properties, only films of thickness not less than about 0.4 μm were considered.

The electrical measurements were carried out before exposing the films to the atmosphere in order to minimize effects of contamination and film oxidation. The optical measurements were carried out under atmospheric pressure by using HP 8450 UV/VIS spectrophotometer permitting wavelength range between 200-800 nm.

RESULTS AND DISCUSSION

In order to assess the effect of Te in Si-Te films, it was important first to carry out measurements on silicon film deposited under identical conditions.

The a-Si films were deposited on a substrate held at 538°K. This was done to ensure the production of films with lower defect concentrations[7] In early attempts to perform the evaporation using a hot substrate in the case of Si-Te system, a problem was encountered. This problem resulted from the instability of the composition of the prepared films due to tellurium effusion. In order to minimize such undesired effects, the Si-Te films were deposited on a substrate held at room temperature and subsequent measurements were carried out on the films only after annealing at high temperature.

D.C. Electrical Conductivity

Electrical conductivity (σ) as a function of temperature ($T^{o}K$) is shown in fig. (1). As can be seen from the figure, a linear relationship results between log (σ) and 1000/T in the temperature range 300-583°K. This indicates a thermally activated process in the form:

$$\sigma = \sigma_o \, e^{-E_T/KT}$$

where E_T is the thermal activation energy of the process, K is the Boltzman constant and σ_o is a parameter related to the mobility and the density of states. In the present work it was found that the activation energy for this process for a-Si films was 0.15 $\mp$ 0.02 eV (see fig. 1.a).

Fig. (1.b and c) are typical results of the d.c. electrical conductivity (σ) as a function of 1000/T for annealed Si-Te films. It was found that the activation energy in these films increases with the increase of annealing temperature. For example, a

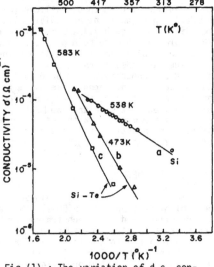

Fig.(1) : The variation of d.c. conductivity (σ) with 1000/T (K^{-1}).
(a) For Si film deposited at 538°K.
(b) and (c) For Si-Te film deposited at room temperature and annealed at 473° K and reannealed at 583° K respectively.

film annealed at 473°K gave activation energy of 0.44 ∓ 0.04 eV (see fig.1.b) and when reannealed at 583°K gave activation energy of 0.53 ∓ 0.05 eV (see fig. 1.c). The increase in the activation energy with the increase of annealing temperature is taken to indicate that annealing has the effect of reducing the density of defect states in the film by readjustment of the random network. These deductions are further substantiated by the reduction of the d.c. conductivity with the increase of the annealing temperature (fig. 1b and c). The present results are in many ways similar to the results of Thomas and Flachet [7] in their work on a-Si:H films.

Optical Measurements

Fig. (2) shows the dependence of the absorption coefficient $\alpha(cm^{-1})$ on the photon energy $h\nu(eV)$ in both a-Si and a-Si:Te films. As can be seen from the figure, the absorption coefficient of a-Si:Te has shifted towards lower photon energies, which may indicate a reduction in the optical gap when tellurium is introduced in the film. As can be seen in fig. (2) the absorption coefficient for both a-Si and a-Si:Te films is greater than 10^4 cm^{-1}, therefore Tauc's relation of the form:

$$(\alpha h\nu)^{\frac{1}{2}} \sim (h\nu - E_g)$$

can be applied [8]. E_g is the optical gap of the prepared film.

Fig. (3) shows the variation of $(\alpha h\nu)^{\frac{1}{2}}$ versus $(h\nu)$ from which E_g can be obtained as an intersect of the extrapolation of the linear part with $h\nu$ axis.

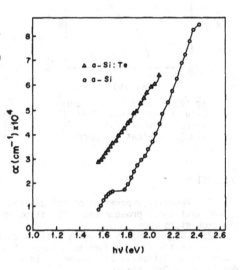

Fig. (2) : Variation of the optical absorption coefficient (α) for different films as a function of photon energy ($h\nu$), (0) a-Si film and (Δ) a-Si:Te film. (Te content is approx. 24 at %).

The value of E_g obtained for a-Si was 1.5 eV and that obtained for a-Si:Te films was always found to be smaller than that of a-Si but dependent upon the concentration of Te in the film. This is seen clearly from Fig. (4) which presents E_g for a-Si:Te versus Te concentration. It is also seen from the figure that no further reduction in E_g results for concentrations beyond about 20 at % of Te. It is interesting to note that similar behaviour was reported by Shimizu et al [9] in their work on evaporated Ge-Te films. They explained the reduction in E_g due to the incorporation of Te to be due to the conversion of localized into more delocalized states in the pseudo gap. The present results might be explained in a similar manner.

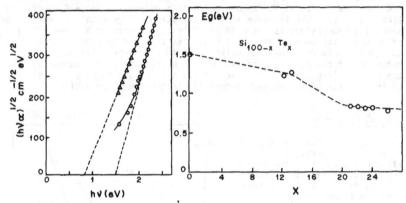

Fig.(3) : Variation of $(\alpha h\nu)^{\frac{1}{2}}$ versus photon energy $(h\nu)$, (0) silicon film and (Δ) a-Si:Te film (Te content is approx. 24 at %).

Fig.(4) : Variation of the optical band gap for a-Si:Te films with the tellurium atomic percentage (X).

CONCLUSIONS

1) Thermal evaporation of silicon and tellurium in an ultrahigh vacuum system can produce amorphous films as revealed by transmission electron microscopy.

2) The activation energies assisting electrical conduction in Si-Te films are found higher than those of a-Si prepared under identical conditions. This suggests that a-Si:Te films have lower density of defect states.

3) The results of the optical measurements revealed that the optical gap of a-Si:Te system is always lower than that of a-Si films (1.5 eV) and lies in the range 0.8 - 1.5 eV depending upon the tellurium concentration in the range up to 26 Te at %.

4) The reduction of the density of defect states point to the possibility of doping the Si:Te films.

ACKNOWLEDGEMENT

The authors would like to acknowledge the financial support of Kuwait University through Project SP008.

REFERENCES

[1] D.E. Carlson, J. Non-Cryst. Solids, 35 & 36, 719 (1980).
[2] Kaplan D., Inst. Phys. Conf. Proc. Ser., 43, 1129 (1978).
[3] F.G. Wakim, S.A. Abo-Namous, A. Al-Jassar and M.A. Hassan, App. Phys. Lett. 42 (6), 15 (1983).
[4] F.G. Wakim, A. Al-Jassar, S.A. Abo-Namous, J. Non-Cryst. Solids, 53, 11-17 (1982).
[5] M.H. Brodskey, M. Cardona and J.J. Cuomo, Phys. Rev. 13 - 16 (1977).
[6] F.A. Faris, K.Z. Botros and A. Al-Jassar (to be published).
[7] P. Thomas and J.C. Flachet, J. de Physique, Colloque C_4, Supplement au n^0 10, Tome 42, Oct. (1981).
[8] J. Tauc, Optical Properties of Solids, Ed. F. Abeles (North Holland, Amsterdam, 1970).
[9] T. Shimizu, M. Kumeda and M. Ishikawa, J. Non-Crys. Solids 33 (1979) 1 - 11.

THE FIRST TETRAHEDRALLY BONDED DIAMOND-LIKE COMPOUNDS
WITH METALLIC CONDUCTIVITY

Y. Sawan.,Kuwait University, Chemistry Department, P.O. Box 5969,Kuwait.

ABSTRACT

In this paper we report for the first time the discovery of metallic conductivity in some tetrahedrally bonded diamond-like ternary compounds with the general formula $A_3^l B^v C_4^{vl}$. Such compounds are characterized by a Valence Electron Concentration (VEC) of four valence electrons per atom. The preparation and characterization of one of the investigated compounds Cu_3AsSe_4 will be presented.

The transition metal atoms in such a compound are connected by non-transition metal atoms (Se) via coordinate locallized bonds in a tetra-hedral environment. Such an environment should lead to a splitting in the 3d filled orbitals of the transition metal because of lattice field. A delocalization of the 3d electrons is expected to take place via back donation, through an overlap with the empty antibonding orbitals on the non-metal atoms to minimize the negative formal charge on the transition metal atoms. If such an overlap leads to continuous electron path through the whole material, metallic conductivity could be visualized. The detailed bonding model will be discussed.

INTRODUCTION

In metals the conduction electrons (bonding electrons) are delo-callized and are presented by plane waves progressing in the same direc-tion as electrons. The resistivity of metals is attributed to electron scattering with imperfection in lattice periodicity. Such imperfec-tions are due to ionic vibrations of equilibrium lattice positions because of thermal effect. The higher the metal temperature the greater will be the scattering cross-section of the vibrating ions. Lattice periodicity could also be disturbed by the presence of impurity atoms as well as other lattice defects. These two are relatively temperature independent and lead to residual metal resistivity at very low temperatures.

In this paper an attempt was made to build a model of a solid so as to minimize the probability of electron scattering by lattice thermal vibrations and, at the same time, to retain an outermost partially filled band characteristic in metals.

Bonding Model and Results

To minimize the scattering cross section due to thermal effect, we should first examine the potential energy diagram for atoms (ions) bonded either metallicaly by delocallized electrons or covalently via localllized pairs of electrons. In Fig.(1) the potential energy diagram of two nearest neighbours atoms (ions) assigned to lattice sites A and B as a function of distance X is given. $V_m(x)$ and $V_{cov}(x)$ in Fig. (1)

stands for the changes in potential energy of metallicaly and

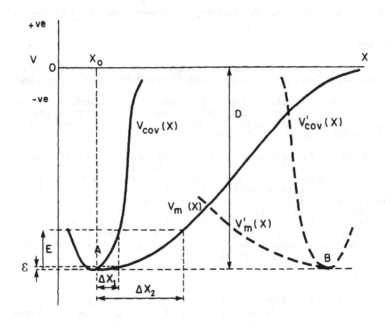

Fig. 1.   The potential energy diagram of covalently and metallicaly
          bonded atoms (ions)

covalently bonded atoms (ions) respectively and has equal value of
dissociation energy D. From the figure we notice that different amounts
of energies E and ε (E>>ε) are needed to displace an atom (ion) on site
A by a distance ΔX_1 depending on whether it is covalently bond or
metallicaly bonded, respectively. Also, it is clear from Fig. (1) that
an energy E will lead to a greater change in $V_m(x)$ (ΔX_2) than that in
the case of $V_{cov}(x)$. It is expected that the amplitude of thermal
vibrations in case of solids bonded by locallized covalent (or coordinate) electron pairs is relatively small as compared to that in
solids bonded by metallic delocallized electrons.

     This fact should lead to a smaller electron scattering cross sec-
tions in covalently bonded solids than in metallicaly. A similar presentation to that of Fig.(1) was used earlier by Myuller [1] for the
interpretation of the nature of the glassy state.

     To overcome the second problem and retain an outermost partially
filled band, the building structural units of the desired solid should
be chosen in such a way to push some of the non-bonding electrons on
certain sites to interact (overlap) with empty anti-bonding orbitals on
other sites. Under such a situation all the possible bonding orbitals
should be kept filled by localized covalent or coordinate bonds. The

system which was chosen for this study is a 2-cations, 1-anion three component $A_{(1-x-y)} B_x C_y$ system with a diamond-like structure. A, B and C stands for the group number (i.e. the total number of electrons in the outermost s and p orbitals and $C > B > A$). According to Goryuonova [2] the condition for normal valency is given by

$$A(1-x-y) + BX = (8-C)Y \dots\dots\dots \quad 1$$

For diamond-like conditions, the Valence Electron Concentration (VEC) must be equal to four electrons per atom and is given by:

$$A(1-x-y) + BX + CY = 4 \dots\dots\dots \quad 2$$

Solving 1 and 2 the formula of the system reduces to:

$$A_{(\frac{1}{2}-x)} B_x C_{\frac{1}{2}}, \quad \text{where}$$

$$0 < x = \frac{8-A-C}{2(B-A)} < \frac{1}{2}$$

According to [2] only 5 systems satisfy these conditions and we have chosen the system $I - V - VI$ with the formula $A_3 B C_4$. One of the investigated compounds was Cu_3AsSe_4. In this compound, the Cu atoms as well as As atoms will act as monovalent cations and the Se atoms will act as monovalent anions as given by the equations:

$$3\,Cu \quad \rightarrow \quad 3Cu^{+1} + 3e$$

$$1\,As \quad \rightarrow \quad 1As^{+1} + 1e$$

$$4Se + 4e \quad \rightarrow \quad 4Se^-$$

adding these equations yields

$$3Cu + 1As + 4Se \rightarrow 3Cu^{+1} + As^{+1} + 4Se^-$$

Accordingly it is expected that the nearest neighbouring ions for both Cu^{+1} and As^+ are Se^-.

The most probable configuration for Cu^{+1} in such a compound is the one which will satisfy the 18-electron rule [3a]. We expect Cu^{+1} to be bonded tetrahedrally via coordinate bonds with 4LP (Lone pair) electrons from four nearest neighbours Se^- donor centers. Such a four-fold coordinated Cu^{+1} centers were predicted by Liang et al [4] and Hunter et al [5]. The presence of Cu^{+1} in such a tetrahedral environment should lead to a build up of negative formal charge of −3 as was stated by Mott and Street [6] and is also expected from 18-electron rule [3a]. According to "Electroneutrality Principle" proposed by Pauling [7] and verified recently by Saito [8], metals could not exist with such unfavorable negative formal charge [3b]. This is because of unequal sharing of the LP's electrons with the donor atoms (ions) since the last are usually more electronegative than the metal itself.

Pauling [7] also stated that electron density should be shifted from the metal ion via delocallization of some metal non-bonding electrons, depending on the environments, (e.g. the $3d^{10}$ electrons in case of Cu^{+1}) to empty energy states lying at higher energy on the donor centers.

It is expected that these states to be the unfilled anti-bonding states on the Se^{-1} centers. Such a delocallization process (back donation) will reduce the unfavorable high negative formal charge on the Cu^{+1}, which is in agreement with the ESCA results given by Liang et al in [4] and Sawan et al in [9]. Enhancement of such a delocallization will take place because of the splitting of the 3d orbitals of Cu^{+1} ions when present in a tetrahedral environment according to crystal field theory [3b] and as shown in Fig. (2) where the separation is given by 10Dq. This

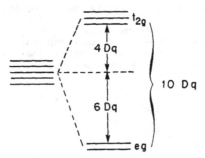

Splitting of d orbitals in a tetrahedral field.

Figure 2

splitting is a result of repulsion between the LP's from the negative donors (Se^- centers) and the 3d orbitals on Cu^{+1}. The more the σ coordinate bond the Se^{-1} ions can donate, the greater the affinity of the t_{2g} orbitals of the Cu^{+1} to be delocallized via back donation. If such a delocallization leads to the formation of a continuous electron path through the whole material, metallic conductivity could be expected.

In Cu_3AsSe_4 structural units, it is assumed that Se^{-1} anions will make use of the S-like LP's to form sp^3 hybrid orbitals with 7 electrons (3LP's and 1 unpaired electron). The 3LP's electrons will form three coordinate σ bonds with 3-nearest neighbours Cu^{+1} ions and the unpaired electron will interact with one of the unpaired electrons on the As^{+1} (P_4^+ according to Kastner et al [10].

All the four electrons on the sp^3 hybrid of As^{+1} ($\overset{+}{P_4}$) will be bonded tetrahedrally with four nearest Se^{-1} anions. In Fig.(3) a two dimensional representation of the building structural units in the $Cu_3As\,Se_4$ is presented.

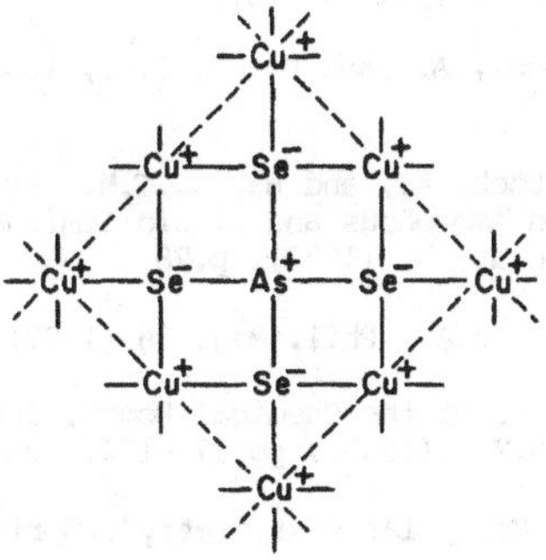

Figure 3. The structural units of Cu_3AsSe_4

Samples were prepared from highly pure elemental row materials. Eight gram total, were sealed in silica ampoules (I.D. 10 mm) at 10^{-5} torr. Ampoules were heated at 950°C for 6 hrs. and subsequently quenched. The melting temperature for this compound was found to be 460°C. Studied thermal analysis exhibited a slight endothermal effect at about 10°C higher than the melting temperature which may be due to partial dissociation of the compound. X-ray diffraction analysis indicated that the structure was sphalerite with lattice constant 5.496 ± 0.002 A. All the prepared samples were found to be metallic in nature from room temperature down to liquid nitrogen temperature. The conductivity varies slightly from one preparation to another depending on the cooling regime. The optimum metallic conductivity ($\Omega^{-1}\,cm^{-1}$) of one of the samples was found to be 1.1×10^5 at 298 K and 1.4×10^5 at $\simeq$ 80K. The measured density was found to be 5.63 gram cm^{-3} and was in good agreement with the calculated one (5.81 gram cm^{-3}). Thus it can be assumed that the requirements for tetrahedrally covalent (locallized) bonded diamond-like material with metallic conductivity are satisfied by Cu_3AsSe_3 ternary system.

ACKNOWLEDGEMENT

The author is grateful to H. Fritzsche, A.M. Kiwan and M. El-Gabaly for their fruitful discussions.

REFERENCES

[1] Myuller, R.L., Proceedings of the All-Unium Meeting on the Vitreous
     State; lzd. Akad Nauk. SSSR, Leningrad (1959) p. 63.

[2] Goryuonova, N.A.,"Chemistry of Diamond-like Semiconductors", lzd.
     LGU, (1963) p. 21.

[3] Huheey, J.E., "Inorganic Chemistry", 3rd edition, Harper and Row,
     N.Y. (1983); a (p-590); b (pp.365-378).

[4] Liang, S.H., Bienenstock, A., and Bates, C.W., Phys.Rev.B, 10
     (1974) 1528.

[5] Hunter, S.H., Bienenstock, A., and Hayas, T.M., Proceedings of the
     Seventh Inter.Conf. on Amorphous and Liquid Semiconductors,
     Edinbury, U.K., ed. Spear, W. (1977). p.78.

[6] Mott, N.F., and Street, R.A., Phil. Mag. 36 (1977) p.33.

[7] Pauling, L., "The Nature of the Chemical Bond", 3rd ed., Cornell
     Univ. Press, Ithaca, N.Y., (1960), pp.172-174.

[8] Saito, Y., "Inorganic Molecular Dissymmetry", Springer-Verlag,
     Berlin (1979).

[9] Sawan, Y., El-Gabaly, M., and Kollias, N., Physica, 117B and 118B
     (1983) p.995.

[10] Kastner, M., Adler, D., and Fritzsche, H., Phys.Rev.Lett. 37
     (1976), p.1504.

PART IX

Solar Cells

RECENT PROGRESS OF AMORPHOUS SILICON SOLAR CELL TECHNOLOGY

Y. HAMAKAWA
Faculty of Engineering Science, Osaka University
Toyonaka, Osaka 560, Japan

ABSTRACT

A review is given on recent progress in the amorphous silicon solar cells and their technologies. Firstly, some unique advantages of amorphous silicon as a low cost solar cell material are pointed out, and its significant position in the photovoltaic project are discussed. Secondly, newly developed key technologies for improving the photovoltaic performance are demonstrated from the film quality improvement to new junction structure solar cells with a wide and narrow energy gap amorphous silicon alloys. Then, current state of the art in the cell performance are summarized. In the final part, recent feature of the industrializations in both consumer and power application fields are overviewed.

INTRODUCTION

Since the first announcement on the invention of amorphous silicon (a-Si) solar cell done by Carlson and Wronski in 1976 (1), a remarkable progress has been seen in both views of device physics and technological developments as a new electronic material (2). Particularly, its significant properties such as excellent photoconductivity with high optical absorption coefficient for visible light match very timely with a strong social need for

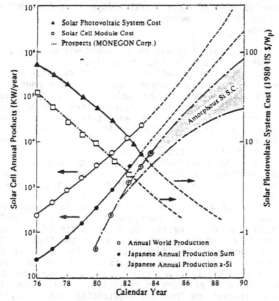

Fig. 1 *Growth of world and domestic annual production of solar cell modules and transitions of module cost and system cost.*

240

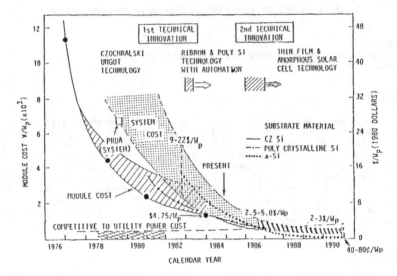

*Fig.2 Cost transition of solar cell modules and photovoltaic
systems with prospective technological innovations.*

the development of a low cost solar photovoltaic system as a new energy
resource. A remarkable advantage of a-Si alloys as a low cost solar cell
material are enumerated as following;
  a) high α and large σ_{ph},
  b) low cost,
  c) large area & non-epitaxial growth on any substrate material,
  d) large scale merit with big massproduciability.
    With aids of accelerative promotion strategy of national or seminational
level support for the photovoltaic project as a renewable energy resource, a
remarkable progress has been seen in the field from film growth technology to
cell fabrication processes on the heterojunction and/or multijunction
structure solar cells utilizing new materials such as a-SiC:H and
microcrystalline Si etc. (3). As the results, more than 12% conversion
efficiency has been attained in a small area laboratory phase cell with the
a-Si/poly c-Si stacked junction structure (4). Market size and
industrializations of a-Si solar cells are growing quite smoothly.
    A significant evidence of this remarkable progress in the photovoltaic
development is more than one order of magnitude reduction of the solar cell
price accomplished during these ten years. In fact, it was more than 80$/Wp
of the module cost in 1974, and now has come down to 5-8$/Wp as shown in
Fig.1 (5). The main reason of this cost-down is due to a big scale merit of
solar cell production with increasing about two order of magnitude of the
annual productions in recent few years as shown in the figure. In the near
future, another one order of magnitude cost down in the solar cell module
could be expected with the accomplishments of successive technological
innovations by the poly-crystalline, ribbon crystal, amorphous silicon and
other thin film solar cells as shown in Fig.2 (6). It is forecasted through
these technological innovations that the electricity cost from photovoltaics
might be competitive with present utility-generated electricity cost in the
period of 1990, and photovoltaic industry might be grown up to the same order
of electronic industry (7).

FILM QUALITY IMPROVEMENT & HIGH DEPOSITION RATES

Major efforts to understand the plasma-deposition mechanism and to improve the film quality have been made using several different approaches, such as the cross-field method (8), application of magnetic fields (9), and plasma-emission spectroscopy (10). Although a massive amount of experimental data exists (11), the plasma decomposition mechanism is still not completely understood. Recently, systematic investigation of the relation between solar cell performance and plasma decomposition configuration (12) and also plasma frequency (13) were carried out. From the analysis of these investigations, three main processes, i.e., i) plasma decomposition, ii) transport of radicals and iii) compiling of decomposed species into a-Si network for controlling the quality of the deposited film, can be identified (14).

Through these basic investigations, several new types of plasma deposition furnaces have been designed, e.g. proximity dc plasma furnaces (15), multi-chamber horizontal glow furnaces (16) and consecutive separated reaction chambers (17). Figure 3 (a) and (b) show schematic illustrations of the two typical massproduction processes of a-Si solar cell module (17-21).

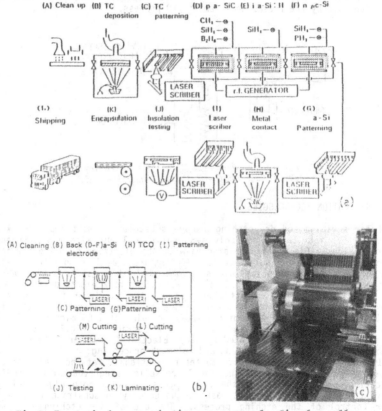

Fig.3 Two typical massproduction processes of a-Si solar cells; (a) a budge type production sequence, (b) a roll-to-roll mass production sequence and (c) an example of the roll-to-roll processing (presented by Sharp-ECD Solar Inc.).

A considerably low deposition rate (1-5 A/sec) in the plasma CVD is one of the bottle neck for the massproduceability of a-Si solar cells. To solve this problem, wide varieties of experimental trials have been made in recent few years. Named technologies are for example:
  a) Reactive sputtering & post hydrogenation (22)
  b) Sputter assisted plasma CVD
     (SAP-CVD) (23)
  c) CVD of higer silanes by
     Chronar (24)
  d) Plasma CVD of Si_2H_6 by ETL
     (25-27)
  e) Photo-CVD of Si_2H_6 (28)
  f) Plasma confinement CVD (29)
  g) ECR(Electron Cycrotron Resonance)-CVD.
Recently Hitachi/Mitsui-toatsu group have reported about 30A/sec growth rate by using 100% Si_2H_6 source gas and fabricated 7.1% efficiency solar cell with this system. Figure 4 shows a schematic diagram of the marcury sensitized photo-CVD reaction chamber by Konagai et al. (28). They got 9.64% efficiency solar cell with this chamber in a-SiC/a-Si heterojunction solar cell. Recently, Hamazaki et al. (29) has reported more than 40A/sec deposition rate by the plasma confinement CVD process.

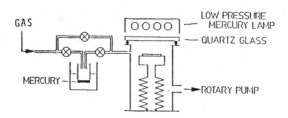

*Fig.4 A reaction chamber system of the mercury
         sensitized photo-CVD after Konagai et al.*

MICRO-FABRICATION TECHNOLOGIES

Laser scribing and dry etching processes become the most important key technology for fabricating not only a-Si solar cell but also a-Si TFT and imaging devices. Fig.5 shows enlarge view of the electrode coupling portion of an integrated a-Si solar cell. TC cutting, a-Si deposition, cell isolation and interconnection processes are made three steps scribing as shown the processes C-G-I in Fig.3 (a). While, the back electrode is formed by evaporation of aluminium in the process H. The YAG laser is now widely employed for the a-Si microfabrication due to enough absorption of its ocsillating wavelength of 0.53 micron (2nd harmonics of 1.06 micron). Fig.6 shows a schematic representation of YAG laser scriber. Fig.7 shows a-Si solar cell roofing tiles presented by Sanyo Electric Co. which is fabricated a combination of laser scribing and ordinal mask processing.

Dry etching as an alternative to the conventional, wet-chemical methods has become an essential technique for large-scale integrated-circuit (LSI) fabrication. This is because the dry process provides both high-fidelity pattern transfer from the resist mask to the underlying substrate layer and better control of the etching process. The use of dry etching is also increasing in amorphous semiconductor technology, particularly in lithographic application, which is better known as inorganic resist technology.

For the past several years, one of the major aims of resist research has been accomplished dry development of the resist. Dry development was first

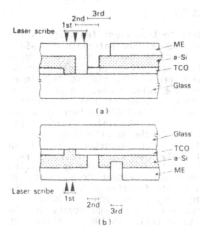

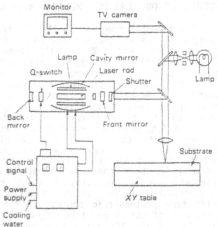

Fig.5 Cross-sectional views of the laser-scribe sequence for the integrated a-Si solar cell. TC isolation, cell isolation and side contact interconnection are made by three steps laser scribing.

Fig.6 A schematic block diagram of the laser scribing system for the integrated type a-Si solar cell.

achieved in inorganic resist systems, using the high etch selectivly between Ag-photodoped and undoped chalcogenide films in CF_4 (30,31,32) or SF_6 (32,33) gas plasma. Since previous experiments were predominantly performed under isotropic etching conditions using a barrel-type plasma reactor, under cutting was more or less unavoidable. This was because the Ag-photodoped layer was relatively thin, typically several hundred angstroms in the Ag/SeGe system, and the uderlying undoped layer was etched isotropically. Recently, laser scribed dry etching and also photo-scribed CVD deposition of a-Si have been tried elsewhere. This kind of advanced techniques might grow coming few years.

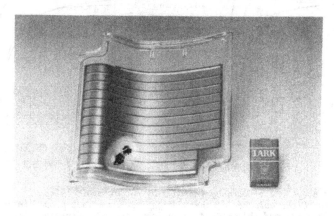

Fig.7 An outlook of a-Si solar cell roofing tile processed by MS-LS (presented by Sanyo Electric Co. Ltd.).

HETEROJUNCTION SOLAR CELL WITH NEW a-Si ALLOYS

A noticeable progress in the R&D effort in the recent few years is the opening of the amorphous mixed alloy age such as new materials of a-SiC, a-SiGe, a-SiSn etc. While, all these materials have considerably good valency controllability by the hydrogenation passivation of dangling bonds, and doping of substitutional impurity with proper gas mixture technique. Since 1978, a systematic investigations on the valency electron control of amorphous mixed alloy have been made by Hamakawa group at Osaka University. As an application of their results, they developed a-SiC/a-Si heterojunction solar cell having efficiency of more than 8% in 1980(34). Figure 8 (a) reproduces the memorial record data brokenthrough the 8% efficiency barrier. Comparing with the current density-voltage (J-V) characteristics of an ordinary a-Si solar cell shows that not only the short circuit current density but also the open circuit voltage are clearly improved by utilizing p-type a-SiC:H as the window junction material. Recent developments in material synthesis technology in which the carbon fraction in $a\text{-}Si_{1-x}C_x$ (35) is controlled and in optimum design theory (36) based upon the concept of drift type photovoltaic process (37) have made rapid improvemnts in the efficiency of this type of heterojunction solar cell. In 1982, Catalano et al. (38) of the RCA group have obtained an efficiency of 10.1% with an a-SiC/a-Si heterojunction solar cell having an active area of 1.01 cm^2. The best record of single p-i-n junction for the a-SiC/a-Si heterojunction solar cells having 11.5% by Sanyo (39) and 11.1% by Fuji (40). Figure 8(b) shows an area dependence of efficiency in the heterojunction cells.

Figure 9 shows the transitions of cell efficiency for various types of a-Si solar cells since 1976. As can be seen from this figure, a step-like increase of the cell efficiencies is seen in the region -1981, while the slope A before 1981 corresponds to the improvement of the film quality and

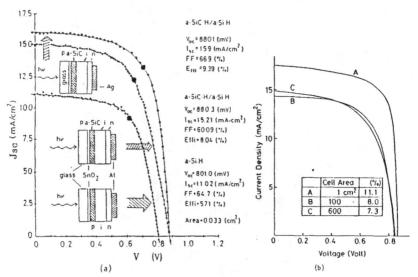

Fig.8 A step-like efficiency increase by p-type a-SiC window.
(a) Clear evidence of the wide gap window effect proved in 1980
(ref.34) and (b) in 1984 technology, the efficiency has been
improved more than 11% (ref.40).

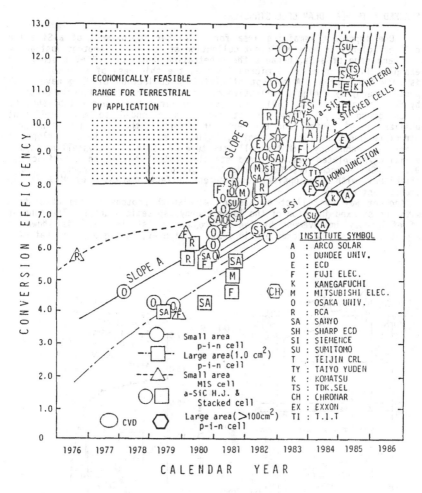

Fig.9 *Progress of cell efficiency improvement in a-Si solar cells.*

routine cell fabrication progresses. The key technologies that lead to the steep slope change from A to B at 1981, were due to development of heterojunction solar cells with a-SiC:H(35) and a-SiGe:H(41). With a full use of the wide gap window effect, an increase of the built-in potential and the minority carrier mirror effect in a-SiC/a-Si heterojunction have been intensively studied by Tawada et al.(35), Nonomura et al.(42) and Okamoto et al.(43). Due to a recent advance of material synthesis technology by controlling the carbon fraction in a-Si$_{(1-x)}$C$_x$ (44) and of the optimum design theory(36) based upon the concept of the drift type photovoltaic process, this type of heterojunction solar cell efficiency has continued to improve its efficiency with the slope B. The recent world record more than 10% efficiency by RCA (38) and following Sanyo (39), Fuji (45), Komatsu (46), TDK-SEL (47), ARCO (48) and ECD (49) records are also obtained in this type of a-SiC/a-Si heterojunction solar cell.

STACKED & TRIPLE SOLAR CELL STRUCTURE

One important remaining area for a further improvement of a-Si solar cell efficiency is more efficient collection of low energy photons just above the band edge of a-Si, because the penetration depth of 1.8 eV photon, for example, is the order of 5 micron. While, the thickness of a-Si solar cell is only 0.6 micron. A concept of efficient collection of long wavelength photon energy by a highly reflective random surface has been firstly proposed by Boer et al. in 1981 (50), and its theoretical basis was established by Exxon group (51). This concept has been extended to more efficient utilization of optical and carrier confinement in the multilayered hetero-structure junction (52). Recently, Fujimoto et al. (53) have developed a practical technology with the cell structure of ITO/n microcrystalline Si/i-p a-Si/TiO₂/Ag plated semi-textured stainless steel having an efficiency of 9.17%. Quite recently, Taiyo-Yuden/ETL groups (54) have reported 10.26 % efficiency with the optical confinement effect employed by MTG (Milky Transparent Glass).

Another way to collect the longer wavelength photons is the absorption with the stacked junction of the lower energy gap semiconductor. The concept of efficiency improvement by the heterojunction stacked cell is shown in Fig.10. Because the energy gap in a-Si, 1.7-1.8 eV, is higher than that of

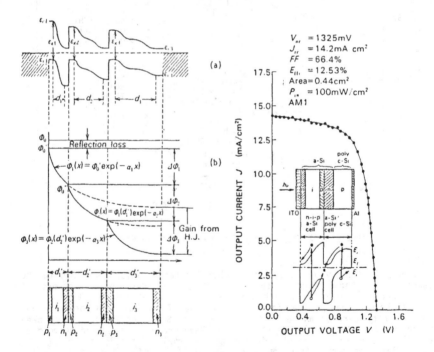

Fig.10 Band diagram explanation of the multiband gap stacked solar cell (a) and its photon energy collection (b).

Fig.11 J-V characteristic of the a-Si/poly Si stacked solar cell and its junction structure (ref.4).

Table I Summary of remarkable cell performance in various types of
a-Si solar cells.

Type	Configuration	Area (cm²)	Eff. (%)	V (mV)	J (mA/cm²)	FF	Year	Institute
SJ	ITO/nipn a-Si/p poly Si/Al	0.082	12.7	1380	14.18	0.65	1984	Sumitomo
SJ	ITO/nipn a-Si/p poly Si/Al	0.44	12.5	1325	14.2	0.66	1984	Osaka Univ.
HJ	Ag/ni a-Si/p a-SiC/text. TCO/glass	1.0	11.5	869	18.9	0.70	1984	Sanyo
HJ	Me/ni a-Si/p a-SiC/tex. SnO₂/ITO/glass	1.05	11.63	850	18.7	0.732	1985	TDK-SEI
TJ	ITO/nipnipn a-Si/i a-SiGe/ p a-Si/S.S.	1.0	11.2	-	-	-	1985	ECD
HJ	Me/ni a-Si/p a-SiC/TCO/ glass	1.0	11.1	864	17.6	0.73	1984	Fuji
HJ	Ag/ni a-Si/p a-SiC/text. SnO₂/glass	1.0	11.0	860	21.5	0.6	1985	Kanegafuchi
HJ	Me/ni a-Si/p a-SiC/text. TCO/glass	0.32	10.7	840	18.8	0.68	1984	Komatsu
HJ	Ag/ITO/ni a-Si/p a-SiC/MTG	0.045	10.26	802	22.32	0.57	1984	ETL/Taiyo
HJ	Al/ni a-Si/p a-SiC/text. SnO₂/glass	4.15	10.2	865	16.1	0.73	1984	ARCO
HJ	Ag/ni a-Si/p a-SiC/SnO₂/ glass	1.09	10.1 ·	840	17.8	0.676	1982	RCA
HJ	Ag/p a-SiC/in a-Si/SnO₂/ glass	0.084	9.64	848	17.1	0.664	1984	TIT
SJ	ITO/nipn a-Si/i a-SiGe/ p a-Si/S.S.	929	9.42	-	-	-	1985	ECD
TJ	ITO/nipnipn a-Si/i a-SiGe/ p a-Si/S.S.	0.09	8.5	2200	6.74	0.57	1982	Mitsubishi
Mod	Me/ni a-Si/p a-SiC/TCO/ glass	100	8.1	11.96V	15.6	0.61	1984	Sanyo
HJ	Me/ni a-Si/p a-SiC/TCO/ glass	100	8.0	850	14.4	0.654	1984	Fuji
SJ	Al/nipni a-Si/p a-SiC/ SnO₂/glass	4.15	7.7	1710	6.23	0.71	1984	ARCO
HJ	Al/ni a-Si/p a-SiC/SnO₂/ glass	1.0	7.7	880	14.1	0.62	1982	Osaka Univ.
Mod	Al/ni a-Si/p a-SiC/SnO₂/ glass	400	7.5	47.8V	114mA	0.55	1985	Kanegafuchi
Mod	Me/ni a-Si/p a-SiC/SnO₂/ glass	600	7.0	16.0V	12.5	0.63	1984	Sumitomo
Mod	Me/ni a-Si/p a-SiC/SnO₂/	3200	6.7	48.7V	890mA	0.50	1985	Kanegafuchi

SJ: Two stacked Junction, HJ: Heterojunction, TJ: Triple stacked junction, Mod; Module

crystalline solar cell semiconductor, e.g. 0.66 eV for Ge, 1.1 eV for Si and
1.43 eV for GaAs, the carrier collection efficiency for solar radiation
spectrum in a-Si solar cell is considerably low as compared with that of the
crystalline basis solar cells. On the other hand, the fabrication of large
area polycrystalline thin films have been already established on wide variety
of classical semiconductors with CVD, MOCVD, MBE, sputtering and ion plating
etc.. By combining these well developed technology with the low temperature
a-Si solar cell deposition technology, it is possible to make higher efficiency a-Si basis solar cell with low cost. Quite recently, Osaka University
group have developed a new type of the stacked a-Si solar cell deposited on
p-type polycrystalline silicon (55). The cell structure and J-V characteristics are shown in Fig.11. As can be seen in this figure, more than 12.5%
efficiency is easily obtained. Although a series of systematic studies on
the material selection, and economical feasibility are now in progress, more
than 15% efficiency would be obtained with the stacked a-Si solar cell on
only one micrometer of polycrystalline GaAs or Ge thin film deposited on
stainless steel substrates. In the case of polycrystalline silicon film,
about 20-30 micron thickness will be required for conventional CVD or
Photo-CVD deposition method from SiH_4 or $SiHCl_3$ (56). To sum up this
section, some remarkable cell performance are listed up in Table I.

RECENT FEATURE OF APPLICATION SYSTEMS

A wide variety of application systems are developed in a recent few
years particularly, consumer electronic applications such as showing in

Fig.12 A wide variety of consumer application
(presented by Sanyo Electric Co. Ltd.).

Fig.12 are expanding very rapidly. For instance, about 5.0 Million sets/month of a-Si drived pocketable calculators are fabricated in Japan as of 1984. On the other hand, the field of power application has been still in the experimental phase. To demonstrate the present feature in the field, some photographs of a-Si power modules developed recently are shown in Fig.13, and a-Si solar cell power application experimental plants installed in Japan are listed up in Table II.

(a) ACRO-Solar $1ft^2$ glass
    module.

(b) Kyocera Ceramic subst-
    rate module.

(c) ECD ultralight a-Si
    solar cell compared with
    stainless substrate $1ft^2$
    module.

(d) Kanaka flexible solar
    cell.

(e) Fuji NEDO sise glass
    module.

Fig.13 Photographs of a-Si solar cell module for power application system.

Table II a-Si solar cell utility power application demonstration/experimental plants in Japan (as of March 1985).

Year Installed	Application	Location	Power Size (kW$_p$)	Site Owner	Prime Contractor
1981	Residential	Moriguchi Osaka	2.0	Sanyo Electric Co. Ltd.	Sanyo Electric Co. Ltd.
1981	Test Plant	Chofu Tokyo	0.4	Tokyo Electric Power Co. Ltd.	Fuji Electric Co. Ltd.
1981	Test Plant	Hirakata Osaka	0.4	Sanyo Electic Co. Ltd.	Sanyo Electric Co. Ltd.
1982	Experimental Research	Meguro Tokyo	2.5	TIT	Sanyo Electric Fuji Electric
1984	Sea Water Distillation	Koza Okinawa	0.15	Hitachi Zosen Corp.	Kanegafuchi Chemical Ind.
1984	Demonstration Power Plant	Akihabara Tokyo	3.5	Dai-ichi Kaden Bldg.	Sanyo Electric Co. Ltd.
1984	Utility Power Experiment	Akagi Gunma	3.0	DRIEPI	Sanyo Electric Fuji Electric
1985	Utility Power Experiment	Saijo Ehime	3.6	NEDO Shikoku Electric Co. Ltd.	Sanyo Electric Fuji Electric
1985	Utility Power Experiment	Sakurajima Osaka	0.58	Hitachi Zosen Corp.	Kanegafuchi Chemical Ind.

ACKNOWLEDGEMENT

The author wishes to express his sincere thanks to all his research friends who kindly supplied the newest informations from all over the world in the field. However, due to the limitted space, he could not cite on details, even many other advanced technologies are being developed. The author also gratefully acknowledges to Drs. H. Takakura and H. Okamoto for their advices and assistance in preparation of this reviews.

REFERENCES

(1) D.E. Carlson and C.R Wonski, Appl. Phys. Lett., 28 (1976) 671.
(2) for example, Review is given in JARECT Vo.2, edited by Y. Hamakawa "Amorphous Semiconductor - Technologies & Devices", North Holland (1982)
(3) Y. Hamakawa, H. Okamoto and Y. Nitta, Appl. Phys. Lett., 35 (1979) 187.
(4) K. Okuda, H. Okamoto and Y. Hamakawa, Jpn. J. Appl. Phys., 22 (1983) L605.
(5) for example, Review is given in JARECT Vo.6, edited by Y. Hamakawa "Amorphous Semiconductor - Technologies & Devices", North Holland (1983).
(6) Paul Maycock, 4th New Energy Industrial Symposium (Tokyo, 1984) 81.
(7) Y. Hamakawa, Business Japan, 29 (1984) 36.
(8) H. Okamoto, T. Yamaguchi and Y. Hamakawa, J. Non-Crystalline Solids 35-36 (1980) 313.
(9) M. Taniguchi, M. Hirose, T. Hamasaki and Y. Osaka, Appl. Phys. Lett. 37 (1980) 787.
(10) A. Matsuda K. Nakagawa, K. Tanaka M. Matsumura, S. Yamasaki, H. Ohkushi and S. Iijima, J. Non-Crystalline Solids 35-36 (1980) 183.
(11) for example, 9th Intern. Conf. on Amorphous and Liquid Semiconductors, Grenoble (1981) part II, preparation session.
(12) S. Hotta, Y. Tawada, H. Okamoto and Y. Hamakawa, J. de Phys. 42 suppl. 10, C-4 (1981) 631.

(13) R.R. Gay, D.L. Morel and D.P. Tanner, Proc. 4th EC Photovoltaic Solar Energy Conf. Stresa (1982).
(14) S. Hotta, N. Nishimoto, Y. Tawada, H. Okamoto and Y. Hamakawa, J. Appl. Phys. 21 (1982) 289.
(15) D.E. Carlson and C.R. Wronski, in "Amorphous Semiconductors" ed. M.H. Brodsky (Springer-Verlag, New York, 1979) chap.10.
(16) Y. Kashima, H. Kida, H. Okamoto and Y. Hamakawa, J. Non-cryst. Solids 59/60 (1983) 755.
(17) Y. Kuwano, M. Ohnishi and H. Shibuya, Jpn. J. Appl. Phys. 20 (1981) 157.
(18) Y. Hamakawa, J. Non-cryst. Solids 59/60 (1983) 1265.
(19) Y. Kashima, S. Nonomura, H. Kida, H. Okamoto and Y. Hamakawa, 5th EC Solar Energy Conf., Athens (1983) 793.
(20) M. Ohnishi, H. Nishiwaki, K. Enomoto, Y. Nakashima, S. Tsuda and Y. Kuwano, J. Non-cryst. Solids 59/60 (1983) 1107.
(21) H. Okaniwa, K. Nakatani and I. Ohuchi, Annual Amorphous Silicon Contractor Meeting, Sunshine Project, April 20th, 1984.
(22) T.D. Moustakas, Proc. 5th EC PVSEC, Athens (1983) 698.
(23) H. Hitotsuyanagi, 1983 Annual Amorphous Silicon Contractor Meeting, Sunshine Project, April 20th, 1984.
(24) V.L. Dalal, M. Akhpar and A. Dalahoy, Proc.16th IEEE PV Specialists Conf., San Diego (1982) 1385.
(25) A.Matsuda, T.Kaga, H.Tanaka, L.Malhotra and K.Tanaka, Jpn. J. Appl. Phys. 22 (1980) L115.
(26) N. Fukuda, 1983 Annual Amorphous Silicon Contractor Meeting, Sunshine Project, April 20th, 1984.
(27) J. Shimada, 1983 Annual Amorphous Silicon Contractor Meeting, Sunshine Project, April 20th, 1984.
(28) M. Konagai and K. Takahashi: 1985 SERI Workshop, SERI (1985).
(29) T.Hamasaki, M.Ueda, A.Chayahara, M.Hirose and Y.Osaka, Appl. Phys. Lett. 44 (1984) 600.
(34) Y. Tawada, M. Kondo, H. Okamoto and Y. Hamakawa, Sol. Energy Mat., 6 (1982) 237.
(35) Y. Tawada, K. Tsuge, M. Kondo, H. Okamoto and Y. Hamakawa, J. Appl. Phys., 53 (1982) 5273 and Y. Hamakawa, 17th IEEE Photovol. Spec. Conf., Orlando (1984) Late News.
(36) H. Okamoto, H. Kida and Y. Hamakawa, Solar Cells, 8 (1983) 97.
(37) H. Okamoto, T. Yamaguchi and Y. Hamakawa, J. Phys. Soc. Japan 49 (1980) 1213.
(38) T. Catalano, A. Friester and B. Fanghman, Proc. 16th IEEE Photovol. Special. Conf., San Diego, 1982, IEEE, New York, p.1421.
(39) S. Nakano, H. Kawada, T. Matsuoka, S. Kiyama, S. Sakai, K. Murata, H. Shibuya, Y. Kishi, I. Nagaoka and Y. Kuwano, Tech. Digest of the Intern. PVSEC, Kobe, Japan, (1984) 583.
(40) H. Sakai, K. Maruyama, T. Yoshida, Y. Ichikawa, T. Hama, M. Ueno, M. Kamiyama and Y. Uchida, Tech. Digest Intern. PVSEC-1 (Kobe, 1984) 591.
(41) G. Nakamura, M. Kato, H. Kondo, Y. Yukimoto and K. Shirahata: J. Phys. (Orsay, Fr.) 42 C4-483 (1981)
(42) S.Nonomura, H.Okamoto, K.Fukumoto and Y.Hamakawa, J.Non-cryst. Solids 59/60 (1983) 1099.
(43) H. Okamoto, H. Kida, K. Fukumoto and Y. Hamakawa, J. Non-cryst. Solids 59/60 (1983) 1103.
(44) Y. Tawada, K. Tsuge, K. Nishimura, H. Okamoto and Y. Hamakawa, Jpn. J. Appl. Phys. 21 (1982) suppl.21-2, 291.
(45) Y. Uchida, H. Sakai, M. Nishiura, M. Miyagi and K. Maruyama, Proc. 8th Int. Vacuum Congr., Cannes, 1980, Vol.1, North-Holland, Amsterdom, 1981, p.669.
(46) Reported by G. Kagaya, Photovoltaic Insider News, Vo.III, No.1, January (1984) 4.
(47) S. Yamazaki, K. Itoh, S. Watanabe et al., Proc. 17th IEEE PV Special. Conf., Florida, May (1984) P1-1B-1.
(48) K. W. Mitchell, Annual Review of Energy #10 (1984)
(49) private communication from S. Ovshinsky.
(50) W. den Boer and R. M. Van Strijp, Proc. 4th Photovoltaic Solar Energy Conf., Stresa (1982) 764.
(51) E. Yablonovitchi and G. D. Cody, IEEE Trans. Electron Devices ED-29 (1982) 300.
(52) Y. Hamakawa, Y. Tawada, K. Nishimura, K. Tsuge, M. Kondo, K. Fujimoto, S. Nonomura and H. Okamoto: 16th IEEE Photovol. Special. Conf., San-Diego (1982) 679.
(53) K. Fujimoto, H. Kawai, H. Okamoto and Y. Hamakawa: Solar Cells 11 (1984) 357.
(54) H. Iida, T. Miyado and Y. Hayashi: Proc. 44th Fall Mtg. Soc. Japan Appl. Phys., Sendai, Sept. (1983) 25p-L-2 & 3.
(55) Y. Hamakawa, K. Okuda, H. takakura and H. Okamoto, Proc. 17th IEEE PV Spec. Conf. (1984, Florida) 1386.
(56) Y. Hamakawa and H. Okamoto, JARECT vol.16 "Amorpghous Semiconductor Technologies & Devices" (OHMSHA LTD and North-Holland Pub. Co., Tokyo, 1984) 200.

CRITICAL MATERIALS PARAMETERS FOR THE
DEVELOPMENT OF AMORPHOUS SILICON ALLOYS

STANFORD R. OVSHINSKY* AND DAVID ADLER**
*Energy Conversion Devices, Inc., 1675 West Maple Road, Troy, MI 48084
**Department of Electrical Engineering and Computer Science, Massachusetts
Institute of Technology, Cambridge, MA 02139

ABSTRACT

The desired properties of solar cells are discussed, and a relative
figure of merit for the comparison of cells fabricated using different
technologies is described. The advantages of utilizing amorphous silicon
alloys as the active material in solar cells are enumerated. Selected
materials properties of these alloys are described and the physics of their
electronic structure is discussed in detail. The necessary steps for
achieving commercially viable cells based on amorphous silicon alloys are
listed, and it is demonstrated how each of them has been achieved using a
technology that incorporates fluorine throughout the entire process. The
chemical and physical basis for the superiority of fluorinated material is
presented in detail. Continuous web large-area high-efficiency
multijunction solar cells are in production. Dual band gap multijunction
cells have been tested under continuous air mass 1 exposure for over 2000
hours and show essentially no degradation. Some recent results are
presented.

INTRODUCTION

Solar cells were invented more than 30 years ago, but have not yet
achieved the commercial success predicted many times in the past. There is
no question that they efficiently convert the eternally available sunlight
directly to electricity, they have the reliability inherent in a system with
no moving parts, they can be used as either distributed power sources or
central power stations, and they yield minimal environmental contamination
or safety hazards [1]. However, it is clear that additional criteria must
be imposed to guarantee their large-scale commercial utilization, since they
must compete successfully with conventional power sources such as coal, gas,
oil, and nuclear energy. The desired properties are straightforward to
identify:

(1) Efficiency: The solar cells must have a high initial conversion
efficiency, η_0, defined as the ratio of the electrical power generated to
the power of the sunlight intercepted by the entire module. While it has
been estimated that cells must exhibit at least 15% efficiency for use in
central power stations [2], a figure we are now approaching, there is an
equally important market, viz. distributed power, where such an efficiency
is not necessary. Distributed power has the advantage of eliminating the
cost and losses associated with transmission lines.

(2) Stability: The efficiency must remain high after at least 20
years of continuous operation. If the stability, S, is defined as the ratio
of the efficiency after 20 years of exposure to sunlight, η_f to η_0, i.e.

$$S = \eta_f / \eta_0, \qquad\qquad (1)$$

then it is necessary to optimize the product $\eta_0 S$ rather than η_0. This arises because solar-cell degradation ordinarily occurs primarily over the first few months of operation, and $\eta_f = \eta_0 S$ is thus essentially the average efficiency over the entire lifetime of the device. It is extremely misleading to quote only η_0 in presenting solar-cell data if S is significantly less than unity.

(3) Cost: It is evident that the cost of the module, C, in, e.g., dollars per peak watt, is crucial. High cost has been a major reason that high-quality crystalline Si and GaAs solar cells have not achieved market penetration to date [1], and C must be reduced sufficiently to compete with the available alternative sources of electricity. Of course, it is not only the cost of the modules that is important, but the cost of the delivered power, including the transportation and installation of the cells.

(4) Power Density: The possibility of distributed power via solar-cell use is very appealing as a means of reducing the economic and energy costs of electrical transmission over long distances. But transportation and installation of bulky solar cells are also expensive. In addition, solar cells have major value both in outer space and in remote areas. For all of these reasons, high power per unit weight, P, in peak watts per kilogram, is extremely desirable. In addition, if the cells are flexible, there are major advantages in transportation and storage.

(5) Abundance: Irrespective of cost, widespread conversion to solar-generated power requires the use of enormous quantities of material. Thus, even if, for example, GaAs solar cells could be made cost competitive using currently produced materials, the finite abundance of Ga would preclude their providing a major fraction of the world's electricity budget. In reality, partial depletion of the ores would begin to drive the cost up enormously.

(6) Toxicity: Especially in distributed residential and commercial applications, materials toxicity could have a deleterious effect on market penetration. For example, easily formed compounds of As and Cd can be lethal in relatively low concentrations, and completely reliable safety precautions in the event of fires, etc. could increase the cost of the solar power a great deal.

Although it is difficult to put the proper weight on all the above factors, especially considering the diverse potential applications, some simply calculated figure of merit would be useful to evaluate the relative merits of different solar-cell technologies. A linear coefficient has been proposed [3]:

$$F = \eta_0 \, S \, P \, / \, C \tag{2}$$

which has the virtue of ease of estimation. We shall use this figure of merit subsequently to quantify some of the current progress being made in amorphous-silicon-alloy solar-cell technology.

ADVANTAGES OF AMORPHOUS SILICON ALLOYS FOR SOLAR-CELL APPLICATIONS

There are many reasons why amorphous silicon alloys are highly desirable for solar-cell applications. These include:

(1) Silicon is essentially infinitely abundant. It is the second most common element in the earth's crust and there is a sufficient quantity

in a 1mm layer of sand over the Sahara desert to cover the entire earth's surface with amorphous solar cells.

(2)   Amorphous silicon alloys are excellent absorbers of sunlight. A layer of only 500nm is sufficient to attain high solar energy conversion efficiencies. Consequently, thin-film solar cells on inexpensive substrates are feasible. This suggests both a very low cost for the starting materials and the possibility of high power densities using low-density substrates, in sharp contrast with single-crystalline or polycrystalline silicon solar cells, which require more than 10 times the thickness to absorb the same fraction of visible light.

(3)   High-quality amorphous silicon alloys can be prepared using inexpensive starting materials, such as silicon tetrafluoride gas, another reason why low-cost solar cells are feasible.

(4)   Amorphous silicon alloys can be deposited very rapidly, further suggesting low overall solar-cell cost. Recently, photoreceptor-grade amorphous silicon alloys have been grown at the extremely rapid rate of $36 \mu m$/hour using a 2.45 GHz microwave glow discharge [4], opening up the eventual possibility of ultrarapid deposition of solar-grade material as well.

(5)   The lack of lattice mismatches in multilayered amorphous structures offers the possibility of very high solar energy conversion efficiencies via the fabrication of stacked cells of different band gaps using an array of alloys.

(6)   With appropriate alloying, the range of band gaps achievable using these amorphous materials can cover the gamut from 1.2 eV to 2.2 eV, enabling the possibility of using multilayer devices to make efficient use of the entire solar spectrum in a stacked cell configuration.

(7)   Amorphous silicon alloys do not contain any highly toxic components.

In brief, amorphous silicon alloys appear to possess all of the essential desirable characteristics for commercial sources of solar-generated electricity. The material is nontoxic, infinitely abundant, and capable of being made into stable low-cost, high-efficiency, high-power-density solar cells. In fact, they appear to be unique in this respect.

SELECTED PROPERTIES OF AMORPHOUS SILICON ALLOYS

Although amorphous silicon alloys have the impressive characteristics enumerated in the previous section, the development of solar cells based on their use is far from straightforward. In this section, we describe the origin of some of the problems that have had to be overcome to achieve commerciality.

(1)   Amorphous silicon alloys form primarily tetrahedrally bonded networks, a reflection of the chemical nature of Si, which is in Column IV of the Periodic Table and bonds optimally using four sp^3 orbitals. Because this represents the maximum possible number of bonds per atom using only s and p electrons, no low-energy defects exist. Thus, only negligible defect concentrations are required from purely thermodynamic considerations, in sharp contrast with the case of amorphous chalcogenide alloys [5]. This opens up the possibility of depositing semiconductor-grade amorphous silicon

alloys. However, strain-induced defects are present under ordinary circumstances and these degrade the materials and preclude their use in any conventional semiconductor device.

(2) The origin of the defects in amorphous silicon alloys is the intrinsic strains necessarily introduced on deposition. The fact that the optimal coordination number, z, for Si is $z = 4$ results in its forming an overconstrained network [6,7]. Most of this strain is relieved by bond-angle distortions which induce valence and conduction band tail states. Although these band tails limit somewhat the maximum open-circuit voltage, V_{oc}, in solar cells, they primarily act as shallow traps for photogenerated electrons and holes. Since the trapped carriers are almost always re-released before they recombine, their presence does not reduce the short-circuit current density, J_{sc}, or fill-factor, ff, significantly, and thus the band tails do not have a major effect on solar-cell efficiency. However, in certain regions of the film, the bond-angle distortions necessary to relieve the strain are sufficiently large that the resulting increase in total energy would exceed the energy reduction due to formation of the bond. In that case, the bond is unstable and a defect center appears. This could take the form of an isolated dangling bond, T_3, in the conventional notation [5], a two-fold-coordinated Si atom, T_2, or even a defect complex [8]. Such defect centers introduce states deep in the gap, which act as recombination centers that degrade both J_{sc} and ff, and sharply reduce solar-cell efficiency.

(3) A partial solution to the problem of strain-induced recombination centers is to alloy the amorphous silicon with hydrogen. Since the coordination number of hydrogen is $z = 1$, the average coordination, $\bar{z}$, of hydrogenated amorphous silicon, $a-Si_{1-x}H_x$, is

$$\bar{z} = 4 - 3x, \tag{3}$$

which is considerably lower than 4 for typical hydrogen concentrations. Thus, a-Si:H films are much less overconstrained than pure a-Si, and the defect concentration can be reduced from $\sim 10^{-3}$ to $\sim 10^{-8}$. In addition, the fact that the Si-H bond is approximately 40% stronger than the Si-Si bond results in the absence of fully bonded states in the gap resulting from Si-H bonds. The somewhat greater electronegativity of H relative to Si leads to a hydrogen-induced removal of states from near the top of the a-Si valence band to positions much deeper in the gap [5]. However, this results in a net increase of the gap with increasing hydrogen concentration, thus inducing a decrease in the theoretical maximum solar-cell efficiency.

(4) The last-mentioned problem could be easily solved by the development of amorphous alloys with reduced band gaps. This would have the further desirable feature of providing the opportunity for fabrication of very high efficiency stacked cells with different gaps. The most obvious answer would be the use of Si-Ge alloys, since Ge is chemically similar to Si but has a considerably smaller energy gap. However, hydrogenated amorphous Si-Ge alloys [9] and even hydrogenated amorphous Ge itself [10] have sufficiently large defect concentrations to preclude their use in efficient solar cells. The origin of this is most likely the weaker Ge-H bond relative to Si-H taken together with the greater chemical tendency of Ge towards divalency and thus T_2 defects [3,6].

In brief, although amorphous silicon alloys appear to be ideal for use in commercial solar cells, many problems have to be overcome. These primarily involve sharp reductions in the concentration of strain-induced defect centers.

CHEMISTRY AND PHYSICS OF AMORPHOUS SILICON ALLOYS

In this section, we elaborate some of the previous discussion, in order to pinpoint the chemical and physical origin of the problems with conventional amorphous silicon alloys. In particular, we discuss the basic band structure, the nature of the traps and recombination centers, the problem of material stability, and the origin of the additional defects that characterize Si-Ge alloys.

The basic band structure of amorphous silicon alloys can be derived from their known local atomic structure [5]. The vast majority of Si atoms are surrounded by four other atoms bonding primarily covalently and forming approximately a regular tetrahedron around the central atom. Disorder enters both geometrically, with the introduction of $\pm$ 10^0 distortions around the 109.5^0 tetrahedral bond angle, and chemically, with the introduction of bonded hydrogen or fluorine. A good approximation to the density of states of a-Si:H can be obtained by considering a small cluster, Si_4H (sat)$_9$, in which a central Si atom is surrounded by three other Si atoms and one H atom in a tetrahedral configuration with the outer Si atoms saturated by a monovalent atom at the appropriate Si-Si separation of 2.35Å [11]. This calculation shows that the valence and conduction bands of a-Si:H are similar in character to those of c-Si, except for the disappearance of the sharp structure (van Hove singularities) that characterize systems with long-range order. Because of the fact that the Si-H bond is about 40% stronger than the Si-Si bond and H is somewhat more electronegative than Si, the presence of bonded hydrogen lowers the energy of states near the top of the valence band, thus increasing the energy gap of a-Si:H relative to that of pure a-Si. The presence of bond-angle and dihedral-angle disorder leads to the appearance of band tails, experimentally [12] of exponential form:

$$g_c(E) = g_0 \exp [-(E_c - E)/kT_c], \qquad (4)$$

with $T_c \approx 325K$ for the conduction band, and:

$$g_v(E) = g_0 \exp [(E_v - E)/kT_v], \qquad (5)$$

with $T_v \approx 550K$ for the valence band. As mentioned previously, the band-tail states limit V_{oc} in solar cells.

Since a-Si:H forms an overconstrained network, small concentrations of defect centers are also present. These include dangling bonds, T_3 centers, which introduce two defect states near midgap, and perhaps other centers such as two-fold-coordinated Si atoms, T_2, which introduce four states in the gap, or three-center bonds, T_{3c}, with bridging hydrogen atoms [8,13]. As discussed previously, the defect states near the Fermi energy increase the recombination current and thus reduce J_{sc} and ff in solar cells.

After exposure to intense light, e.g., several hours of sunlight, the concentration of neutral dangling bonds, T_3^0, in a-Si:H films sharply rises [14], increasing the recombination current and concomitantly degrading solar-cell efficiency. This photostructural change is known as the Staebler-Wronski effect [15] and has been shown to be an intrinsic property of a-Si:H [16]. The effect is driven by a particular recombination branch [17] and can be reversed by annealing [15], so that it is self-limiting. However, the degradation in solar-cell efficiency can be severe [18], yielding up to a 40% reduction in several days of operation prior to saturation. A partial solution of this problem can be achieved by the use of thinner films. Tandem devices even with the same band-gap alloy in both junctions have demonstrated improved stability, because of a reduction in the recombination current in very thin layers (since greater concentrations of photogenerated carriers are then collected).

Finally, it is also useful to analyze the source of the problems with Si-Ge alloys. We have previously pointed out [19,3,6,5] that Ge chemically has a much stronger tendency than Si towards divalency, a state in which an s lone pair (Sedgwick pair) remains nonbonded on the Ge atom. The resulting four states in the gap sharply increase $g(E_f)$ and thus the recombination current in alloys containing Ge. Clearly, the incorporation of hydrogen cannot of itself cure this problem. In addition, the Ge-H bond is somewhat weaker than the Si-H bond, which can have two deleterious consequences. First, H preferentially attaches to Si in Si-Ge alloys, thereby leading to relatively high concentrations of Ge dangling bonds. Second, if the bond is sufficiently weak, Ge-H antibonding states may contribute to the conduction band tail or even form defect states deeper in the gap. (Note that the chemical trend of Column IV elements towards divalency increases as the Row in the Periodic Table increases, resulting in still greater defect concentrations in, e.g., alloys containing Sn and Pb [6].)

To summarize, the origin of the problems in a-Si:H based solar cells can be traced to: (1) the inherent bond-angle disorder; (2) the strain-related defects; (3) the photostructural changes; and (4) the tendency of Ge towards divalency. Since these are all problems intrinsic to the material itself, it is clear that cures must involve new materials development.

ROAD TO UTOPIA

In order to solve the scientific and technological problems necessary for the development of amorphous silicon alloy solar cells of sufficient efficiency, stability, and cost to achieve commercial success vis-a-vis conventional power sources, the following four steps must be completed:

(1) Develop a very high quality material for single-cell devices. To achieve energy conversion efficiencies beyond 10% requires band tails with both T_c and T_v below 600K and $g(E_f) < 10^{16} cm^{-3} eV^{-1}$.

(2) Develop a process for rapid, continuous production of large-area, high-efficiency modules. This requires attainment of (a) high deposition rates, (b) excellent uniformity, and (c) little or no loss in efficiency in large-area compared to small-area devices.

(3) Produce cells which retain their initial efficiency after many years of exposure to sunlight. A projected 20 years of operation in relatively sunny areas requires no significant degradation for about 20,000 hours of operation at AM1 (100 mW/cm^2) conditions.

(4) Develop very high quality stable alloys with reduced band gaps for use in high-efficiency stacked configurations. Such alloys must also have $g(E_f) < 10^{16} cm^{-3} eV^{-1}$ to provide the necessary increase in efficiency over single-junction devices.

RESULTS ACHIEVED BY ECD AND SOVONICS

In this section, we show how our theoretical understanding of amorphous silicon alloys has allowed us to accomplish all of the goals enumerated in the previous section. We describe the progress attained to date in solar-cell development, and evaluate the achievements quantitatively in terms of the figure of merit of Eq. (2).

The first step of developing a high-quality material for use in a single-cell device has been accomplished. Carefully produced a-Si:H films and especially our proprietary a-Si:F:H films with band gaps in the 1.7-1.8 eV range have been used to produce solar cells with energy conversion efficiencies in the 10% range. A typical density of states curve for such materials, as determined by a combination of optical absorption data, photothermal deflection spectroscopy, and transient and steady-state photoconductivity experiments [20] is shown in Fig. 1. Both valence and conduction band tails fall off with characteristic temperatures below 600K and the midgap density of states is less than $10^{16} cm^{-3} eV^{-1}$, as desired.

The second step, development of a high-speed continuous process for large-area production, is unique to ECD and its partners [3]. This required the bold step of designing and building a machine that had not previously been contemplated. The concept was to use rolls of an inexpensive substrate such as stainless steel to produce solar cells continuously in a completely automated manner. This eliminates both bulky glass modules and inefficient, space-consuming batch processing. Figure 2 shows the production of 16-inch wide, 1000-foot long rolls of solar cells via our proprietary continuous-web process at the Sharp-ECD plant in Shinjo, Japan where commercial cells for consumer applications have been produced for the past two years. Figure 3 demonstrates the high level of automation in this plant, in which at most four employees are necessary during the production run. Figure 4 is a photograph of the latest version of our processor, now in operation at the Sovonics plant in Troy, Michigan.

There are several noteworthy factors in our approach. The first was our choice of a one-square-foot format at a time when everyone was involved in small-area processing. This required solving the potential difficulties of adhesion of the film to the substrate, uniform large-area deposition, and maintenance of cell integrity on flexible substrates, all nontrivial problems. As an example of the uniformity achieved, Fig. 5 shows the film thickness as a function of position across a 30cm (12 in) section of substrate [21], and Fig. 6 shows the spatial distribution of solar-cell parameters over a one-square-foot tandem a-Si:F:H device with an average efficiency of 8%, in which 90% of the sub-cells exhibit efficiencies in the 7-9% range [21,22].

The second factor was to process a 1000-foot roll continuously with high yield and high efficiency. To achieve this, we had to solve the problems of uniform gas flow, homogeneity of the gas mixture, and control of the speed of the web, among many others. It is of great significance to our overall strategy that all this was achieved in a tandem configuration. Our approach has led to our being awarded the basic patents in the field, and we now have several years of production experience behind us. The impressive results attained to date include a 9.57% average efficiency over a one-square-foot single-cell device which has a peak sub-cell efficiency of 10.5% [23], and a continuous-web one-square-foot single-cell device fabricated on a production machine with an average efficiency of 8.2% [22]. In addition, a fully encapsulated one-square-foot module with an active area efficiency of 8.3% has been produced with a measured output of 7.33 Wp.

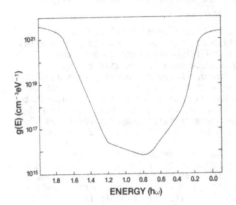

Fig. 1   Density of states of a
high-quality amorphous silicon
alloy film.

Fig. 2   Continuous one-foot-
wide rolls of mass-produced
Ovonic solar cell material.

Fig. 3   Interior of the Sharp-ECD Solar, Inc. plant.

Fig. 4   View of a portion of the Sovonics production plant, showing the Sovonics Photovoltaic
Processor.

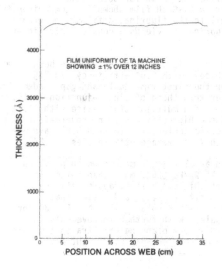

FILM UNIFORMITY OF TA MACHINE
SHOWING ±1% OVER 12 INCHES

THICKNESS (Å)

POSITION ACROSS WEB (cm)

Fig. 5   Film thicknesses as a
function of position across
a strip 40-cm wide, de-
posited by our continuous-
web process.

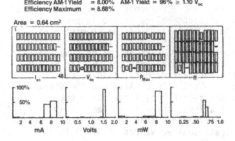

Efficiency Average = 7.69%   Power Sum = 0.24 W
Efficiency AM-1 Yield = 8.00%   AM-1 Yield = 96% ≥ 1.10 V_{oc}
Efficiency Maximum = 8.68%

Area = 0.64 cm²

I_{sc} 48 V_{oc} P_{Max} ff

100%
50%

2 4 6 8 10   0.5 1.0 1.5 2.0   2 4 6 8 10   0.25 .50 .75 1.0
mA   Volts   mW

Fig. 6   Solar-cell parameters
of the individual 48 sub-
cells on a one-square-foot
solar cell.

As we discussed previously, none of the initial efficiency data have any significance if the cells degrade during operation. However, using our knowledge of the physics of amorphous silicon alloy materials and devices, we have designed and produced multijunction solar cells with efficiencies in the 8% range which retain over 99% of their initial efficiency after 2000 hours of AM1 exposure, as is clear from Fig. 7 [24]. Extrapolation of these data indicates that such devices will retain over 95% of their initial efficiency after 20 years of actual operation in sunlight. Also shown in Fig. 7 for comparison is the recent data reported by ARCO Solar and Fuji Electric on the degradation of single-junction devices of comparable efficiencies [18].

Finally, it is of the utmost importance to develop smaller band-gap alloys of comparable quality and stability. As is evident from Fig. 8, our a-Si-Ge:F:H alloy has solved this problem. The only difference in the subgap optical absorption between the a-Si:F:H alloy and the a-Si-Ge:F:H alloy is the lower gap in the latter. A translation of the $\alpha(\omega)$ data of the a-Si-Ge:F:H film to higher photon energies by the 0.3 eV difference in band gaps results in complete superposition of the two curves. This demonstrates that $g(E_f)$ of the Si-Ge alloy is sufficiently low for solar-cell applications. Note that although the a-Si-Ge:H film shown for comparison in Fig. 8 has a similar exponential tail to the fluorinated films, the greater defect density in the absence of fluorination yields a considerably higher midgap absorption.

Using the a-Si-Ge:F:H alloy characterized in Fig. 8, we have been able to fabricate a single-junction solar cell with 9.0% efficiency [20,24]. More important, we have produced for the first time dual-band-gap tandem solar cells with efficiencies greater than those of single-junction cells. Figure 9 shows the J-V characteristic of a multijunction device with 11.2% efficiency [24]. We have also produced high-efficiency one-square-foot devices--tandem with 9.42% average efficiencies, triple with 9.83% average efficiencies [23], and quadruple with 7.1% average efficiencies [23,24].

Finally, we have also developed an ultralight module which produces power at the impressive density of 615 W_p/kg [25], more than twice the previous record. The significance of this can best be appreciated by returning to the proposed figure of merit, Eq. (2), and noting that $F \approx 0.1 - 0.2$ for conventional c-Si or a-Si:H solar cells, but $F = 34$ for our current prototype ultralight module. With further improvements in efficiency and substrate development, we are planning an ultralight module with $F = 750$. A photograph of a typical present-day ultralight device is shown in Fig. 10.

EFFECTS OF FLUORINE

It is worthwhile to emphasize that the achievement of each of the four goals discussed in the previous two sections relies on the incorporation of fluorine in the amorphous alloy. An understanding of the chemistry of amorphous silicon alloys has been clouded in the past. The high-quality "amorphous silicon" originally reported [26] turned out to be a hydrogenated alloy [27], in fact with the hydrogen responsible for the quality of the material. However, a-Si:H still has residual defects, leading to $g(E_f) \sim 10^{15}-10^{16} cm^{-3} eV^{-1}$, even in the best films. Although the dispersion in material quality of both a-Si:H and a-Si:F:H is larger than the differences between the two types of alloys, there is evidence that the best a-Si:F:H films have lower defect concentrations than the best a-Si:H films. Field-effect measurements [28] indicate smaller values of $g(E_f)$ in the

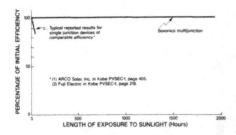

Fig. 7   Stability of a Sovonics multijunction solar cell after 2000 hours of AM1 exposure, compared to recent results on typical single-junction a-Si:H cells.

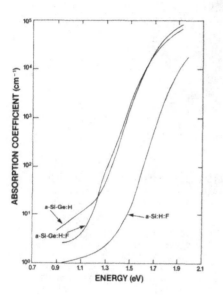

Fig. 8   Absorption coefficient as a function of photon energy for a-Si:Ge:F:H alloy films compared to that of an a-Si:F:H film with a 0.3 eV larger band gap.

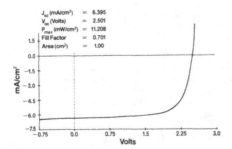

Fig. 9   J-V characteristic of a multiple band gap triple junction Ovonic solar cell with an efficiency of 11.2%.

Fig. 10   Rolls of one-foot wide
ultralight Ovonic solar cell.

fluorinated films, and independent DLTS experiments [29] suggest that it is
primarily the density of interface states that is reduced by fluorination.
Such states are particularly important in semiconductor devices such as
solar cells.

The original report [28] of the resistance of a-Si:F:H alloys to
photostructural effects has also been independently confirmed [30]. In
fact, this property has been shown to increase the relative stability of
fluorinated single-junction solar cells [31]. In addition, the use of
fluorine results in striking increases in the achievable deposition
rate [4], a very important parameter in the final cost of the completed
devices.

Finally, as we have already discussed (see Fig. 8), fluorination is
essential in the production of high-quality amorphous Si-Ge alloys. As with
all of the others, this result has been independently confirmed [32,33]. It
is of major significance that the fluorinated Si-Ge alloys are also much
more stable than comparable unfluorinated alloys. Guha [20] has recently
measured the photoconductivity of two Si-Ge alloys with the same optical
gap, and found the initial value of the fluorinated material was twice that
of the unfluorinated material. Moreover, after 16 hours of AM1 exposure in
a coplanar configuration, the photoconductivity of the a-Si-Ge:H film had
decreased by 34%, compared to only a 5% decrease in the a-Si-Ge:F:H film.
Clearly, there is a real difference between the two alloys. In addition,
through the years we have incorporated fluorine in a wide variety of
amorphous alloys, including high-quality a-Si-C:F:H alloys.

All of the above materials differences can be understood theoretically
from the following considerations:

(1)   Si-F and Ge-F bond are primarily ionic, while Si-H and Ge-H bonds are primarily covalent. Since ionic bonds do not impose any bond-angle constraints, the use of fluorine results in films with lower overall strain.

(2)   Si-F and Ge-F bonds are about half again as strong as Si-H and Ge-H bonds. There is thus no possibility that the antibonding states of the fluorine bonds are located either in the gap or in the conduction band tail.

(3)   Fluorine is the most electronegative atom in the Periodic Table. The incorporation of fluorine thus reduces the concentration of negatively charged dangling bonds (T_3^- centers) in tetrahedrally bonded materials. Since the trapping of holes by T_3^- centers induces a Staebler-Wronski degradation, the fluorinated films exhibit increased stability.

(4)   Fluorine expands the valency of both Si and Ge. The concomitant reduction in divalency, particularly in films containing Ge, lowers $g(E_f)$ sharply.

(5)   Fluorine acts in the plasma, at the surface during growth, and in the bulk to induce unique local configurations that would otherwise not be present in the final film [34].

CONCLUSIONS

We have shown how our proprietary approach which utilizes the incorporation of fluorine makes superior films of tetrahedral amorphous alloys, including amorphous silicon, silicon-germanium and silicon-carbon alloys. It is also useful in increasing the deposition rate and improving the stability of the material. These fluorinated alloys combined with our continuous-web production process have been used to fabricate high-efficiency stable multijunction solar cells. An ultralight module which delivers power at a density of 615 W_p/kg has also been produced.

ACKNOWLEDGEMENTS

We wish to acknowledge S.J. Hudgens, J. Yang, J. Hanak, S. Guha, J. Doehler, M. Izu, P. Nath, W. Czubatyj, and many others at ECD whose contributions have made possible the devices discussed in this paper, and thank them for making their data available to us prior to publication.

REFERENCES

1. See D. Adler and S.R. Ovshinsky, Chemtech, in press for a recent review.
2. M.K. Armstrong-Russell, W. Freedman, and E.E. Spittes, S.P.I.E. Proc. 407, 132 (1983).
3. S.R. Ovshinsky, Tech. Digest Intern. PVSEC-1, (Kobe, Japan, 1984) p. 577. (As presented.)
4. S.J. Hudgens and A.G. Johncock, these proceedings.
5. D. Adler, in Physical Properties of Amorphous Materials, ed. by D. Adler, B.B. Schwartz, and M.C. Steele (Plenum Press, N.Y., 1985), p.5.
6. S.R. Ovshinsky, in Physical Properties of Amorphous Materials, ed. by D. Adler, B.B. Schwartz, and M.C. Steele (Plenum Press, N.Y., 1985), p. 105.
7. S.R. Ovshinsky, A.I.P. Conf. Proc. 31, 67 (1976).
8. S.R. Ovshinsky and D. Adler, Contemp. Phys. 19, 109 (1978).

264

9. W. Paul, D.K. Paul, B. von Roedern, J. Blake, and S. Oguz, Phys. Rev. Lett. 46, 1016 (1981).
10. J.A. Reimer, B.A. Scott, D.J. Wolford, and J. Nijs, Appl. Phys. Lett. 46, 369 (1985).
11. K.H. Johnson, H.J. Kolasi, J.P. deNeufville, and D.L. Morel, Phys. Rev. B 21, 643 (1980).
12. T. Tiedje, in Semiconductors and Semimetals, ed. by R.K. Willardson and A.C. Beer (Academic Press, N.Y., 1984), vol. 21C, p. 207.
13. S.R. Ovshinsky, in Amorphous and Liquid Semiconductors, ed. by W.E. Spear (C.I.C.L., U. of Edinburgh, 1977) p. 519.
14. H. Dersch, J. Stuke, and J. Beichler, Phys. Stat. Sol. B 105, 265 (1981).
15. D.L. Staebler and C.R. Wronski, J. Appl. Phys. 51, 3262 (1980).
16. C.C. Tsai, J.C. Knights, and M.J. Thompson, J. Non-Cryst. Solids 66, 45 (1984).
17. S. Guha, J. Yang, W. Czubatyj, S.J. Hudgens, and M. Hack, Appl. Phys. Lett. 42, 5881 (1983).
18. D.P. Tanner and K.W. Mitchell, Tech. Digest. Intern. PVSEC-1, (Kobe, Japan, 1984), p. 405. Y. Uchida, M. Kamiyama, Y. Ichikawa, T. Hama, and H. Sakai, ibid, p. 217.
19. S.R. Ovshinsky and K. Sapru, in Amorphous and Liquid Semiconductors, ed. by J. Stuke and W. Brenig (Taylor and Francis, London, 1974) p. 447.
20. S. Guha, unpublished data.
21. J. Doehler, unpublished data.
22. M. Izu and K. Hoffman, unpublished data.
23. P. Nath, unpublished data.
24. J. Yang, unpublished data.
25. J. Hanak, unpublished data.
26. W.E. Spear and P.G. Le Comber, Phil. Mag. 33, 935 (1976).
27. S.J. Hudgens, Phys. Rev. B 14, 1547 (1976).
28. S.R. Ovshinsky and A. Madan, Nature 276, 482 (1978).
29. C.H. Hyun, M.S. Shur, and A. Madan, Appl. Phys. Lett. 41, 178 (1982).
30. H. Matsumura and S. Furukawa, in Amorphous Semiconductor Technologies and Devices, ed. by Y. Hamakawa (North-Holland, N.Y., 1982), p. 88.
31. Y. Kuwano, M. Ohnishi. H. Nishiwaki, S. Tsuda, H. Shibuya, and S. Nakano, in Proc. 15th IEEE Photovoltaics Specialists Conf. (1981), p. 698.
32. K. Nozawa, Y. Yamaguchi, J. Hanna, and I. Shimizu, J. Non-Cryst. Solids 59-60, 533 (1983).
33. S. Nakano, Y. Kishi, M. Ohnishi, S. Tsuda, H. Shibuya, N. Nakamura, Y. Hishikawa, H. Tarui, Y. Takahama, and Y. Kawano, these proceedings.
34. S.R. Ovshinsky, in Proc. of the Intern. Ion Engineering Congress, ISIAT '83 & IPAT '83 (Kyoto, Japan, 1983).

THIN FILM SILICON POWER MODULES:
CHALLENGES AND OPPORTUNITIES FOR MATERIALS SCIENCE

DON L. MOREL
ARCO Solar, Inc., P.O. Box 2105, Chatsworth, CA 91313

ABSTRACT

A survey of the materials issues involved in the development of thin film Si:H power modules is presented. Though there are significant opportunities for advancement of Si:H itself, especially in terms of new alloys, the range of opportunities is extended dramatically when all aspects of the product -- from gaseous fuels through encapsulants -- are included. Examples from each of these areas are presented to indicate the current state of development as well as the needs for the future. Viable products exist today, but the extent of greater success depends critically on materials innovation.

INTRODUCTION

Consumer electronics products based upon thin film Si:H (TFS) have been in the marketplace for several years, and power modules employing TFS have been introduced within the past year. These products are the result of years of R&D effort on TFS, and reams of reports discussing the results of these efforts can be found throughout the scientific literature. While TFS, as the photoactive element, is the heart of these products, it is only one of several materials that collectively make up the product and determine its success. In photovoltaic power modules for example, the additional materials include glass, transparent conducting oxides, metal electrodes, conductive pastes, solders, encapsulants, frame materials, wire leads, etc. Each plays a vital role, and performance is measured as the product of the individual performances, not as the sum. Failure of one component usually means complete product failure, not partial failure.

The discussion which follows briefly surveys the types of materials issues being confronted in developing TFS power modules. The discussion addresses many of the component materials that make up the product, not just the TFS. In fact, our particular experience is that TFS is significantly ahead of many of the other component materials in development. However, many exciting developmental opportunities still lie ahead for TFS and even more so for the entire module package. These opportunities will be delineated in the discussion below as we begin with fuel and systematically work through a sequence of materials issues which parallels the buildup of modules.

FUELS

Though TFS can be deposited by sputtering from silicon targets, by electrodeposition from solutions, and by other techniques, the predominant deposition process has been glow discharge using silane (SiH_4) gas as the fuel. Gaseous phosphine (PH_3) and diborane (B_2H_6) are used to add dopants to form n and p layers respectively in building up the successful p i n junction structure. While much is known about the effect of impurities on crystalline Si, this is not yet the case for TFS. To resolve this issue, fuel suppliers need to provide increasingly purer gases so we may find where thresholds might lie. As seen in Table I, the gas industry has been working on the problem. These data cover only a two-year period, but it is apparent that progress is being made. Early entries have been lowering impurity

Table I. Impurities in silane. Levels indicated as "0" usually mean below 1 ppm; those with "-" were not measured.

Impurity	Mfr.	1983 4Q	1984 1Q	2Q	3Q	4Q	1985 1Q
N	A			1		1	1
	B	132	16				9
	C	100					9
	D						139
CO2	A			0		6	5
	B	40	0				6
	C	0					6
	D						7
Silox-	A			–		–	5
anes	B	–	–				25
	C	–					70
	D						400
Other	A			0		0	0
	B	158	3				12
	C	1					20
	D						100

levels, and new entries apply further downward pressure. Though other impurities may also be operative, those listed seem to be the most problematical. Oxygen is also suspect, but all manufacturers seem able to control it below the 1 ppm level.

The effect of these impurities on film and device properties is still the subject of much investigation. Recent results by the Xerox group indicate that impurities such as oxygen and nitrogen are harmful only at levels of about 10^{20} per cm^3 or greater [1]. Other researchers have seen conflicting results. For our particular reactors, as shown in Fig. 1, we have seen a dependence of device fill factor on oxygen levels down to about 10 ppm. We have achieved fill factors of up to 0.77, but have not been able to correlate these high values to low oxygen or other impurity levels.

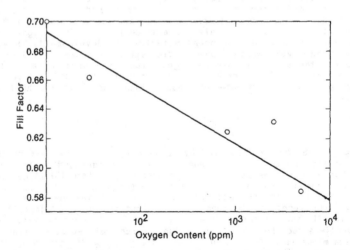

Fig. 1. Effect of oxygen on fill factor of PV device.

These types of conflicting results are not unusual in this field. TFS is not a specific material, but is rather more generic in nature with properties that are very much a function of local content. In several cases we have seen an applied perturbation appear ineffective only because its influence was secondary to another dominating factor. Only when this dominating factor was accounted for could the real performance of the perturbation be seen. Gas suppliers have responded to our request for pure fuels. We must request their forbearance as we struggle further with the difficult issue of determining what impurities are important and at what levels.

REACTORS

The main feature that distinguishes glow discharge reactors from standard vacuum processing units is the glow discharge itself. Depending on the particular experiment under way and the bent of the experimenter, all manner of radicals, charged and excited state species of C, O, N, Si, B, P, F, H, He, Ar, etc. can be present and in contact with the inner walls, electrodes, and other internal components of the reactor. Some of these species are sufficiently long-lived to reach pumping ports and the pumps themselves. Special care must be exercised in choosing reactor materials in order to minimize corrosive effects and to prevent effusion of agents from surfaces which become incorporated as harmful impurities in growing films. In reality, for the latter problem, this is an issue primarily only for runs done in a clean chamber. Shortly into the first such run, the inside components are already coated with TFS; this is the exposed surface during subsequent deposition. Indeed, in attempts to develop better internal component materials, e.g., using carbon instead of stainless steel, we have observed negligible differences in film quality once initial coating has occurred.

A more critical activity is cleaning, once depositions become adversely affected by film buildup. Physical abrasion and plasma etching are two commonly used techniques. In either case, the component materials need to survive the process without loss of structural or compositional integrity. Plasma etching is the more efficient of the two, but the most severe for non-metallic components such as wire insulation. Development of improved options here would facilitate the use of this technique.

SUBSTRATES

The two principal classes of substrates in use today are glass and metal. Some success is also being realized with plastic substrates such as Kapton which can withstand the typical TFS deposition temperatures of 200-300°C. Glass substrate devices typically employ a transparent conductor such as tin oxide as the light exposed electrode onto which a p i n structure is deposited. Since the p layer is only 100-200Å thick and is the principal junction former, the region including the transparent conductor and the p layer is critical. This is discussed further below.

The issue to focus on here is the significant potential for stress to occur among these various layers. The deposition process itself can give rise to stress, and it has recently been shown that defect levels can be caused by stress [2]. In addition to process-induced stress, the potential for thermal stress is present, depending upon the particular set of materials used to build up a given device structure. Table II lists commonly used components for these devices along with their coefficients of thermal expansion and typical thickness. A popular structure is soda lime glass/tin oxide/TFS/aluminum. The TFS/Al interface, with a factor of 6 difference in thermal expansion coefficient, is perhaps the most vulnerable and requires appropriate scrutiny. Of additional concern is the potential

Table II. Expansion coefficients in device components.

Material	Use	Coefficient of Thermal Expansion ($\times 10^{-6}$/°K)	Typical Thickness (Microns)
TFS	Photoactive Layer	4.1	0.5
Borosilicate Glass	Substrate	3.1	10^3
Soda Lime Glass	Substrate	8.0	10^3
Stainless Steel	Substrate	14.7	10^3
Aluminum	Electrode	23.0	2
Nickel	Electrode	12.8	2
Tin Oxide	Transparent Electrode	3.8	0.5

for Na diffusion from the soda lime glass into the tin oxide and perhaps TFS, and oxidation at the TFS/Al interface. All of these problems have been solved to a reasonable extent because, in fact, very good devices with this structure are in use today. Further resolution of these issues can, however, still result in significant improvements in this performance.

TIN OXIDE (TO)

As mentioned above, the tin oxide/p layer interface is critical. TO is usually evaluted in terms of transparency and sheet resistance. However, this is not enough. Shown in Fig. 2 are two samples with nominally identical transmission and sheet resistance. A slight variation in run conditions, however, results in the observed difference in surface morphology while not affecting the above two parameters. And, because of the sensitivity of the TO/p interface, this difference in surface morphology is also sufficient to result in significant differences in device performance.

Another factor that must be considered is crystallite size and orientation. As shown in Fig. 3, variation of the freon flow rate during TO deposition can cause a shift in crystalline orientation. Thus, while the dopant level is being changed, other subtle changes are occurring in the TO. These effects need to be separated and accounted for in order to optimize TO and device performance.

The best TO is made at substrate temperatures that exceed the softening point of glass. More exotic and expensive types of glass can be used to

Fig. 2. Difference in surface morphology of TO results in significant differences in performance due to sensitivity of the p/TO interface.

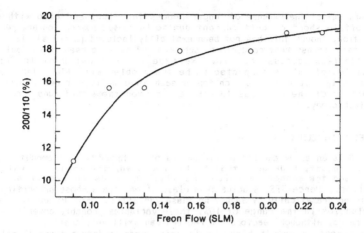

Fig. 3. Shift in crystalline orientation due to change in freon flow rate during TO deposition.

circumvent this problem in the laboratory, but not on the production line. TO is usually grown from a CVD process using either $SnCl_4$ or $Sn(CH_3)_4$ as a fuel. In addition to the high temperature problem, these fuels, especially $Sn(CH_3)_4$, are highly toxic and require significant care in handling. New and better procedures for making transparent conductors would be well received by the industry.

The importance of transparent conductors to device performance is demonstrated in Fig. 4, which shows a simulation of fill factor and efficiency vs. sheet resistance. The data points were measured on a G100 module. The curve simulates the effect of lower or higher sheet resistance on performance. Lower sheet resistance can be achieved by making the TO

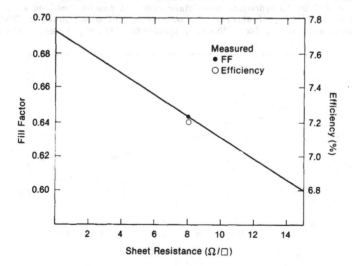

Fig. 4. Effect of sheet resistance on device performance.

thicker, but at the expense of longer processing time, and also with some drop-off in short circuit current due to increased absorption and perhaps reflection, though this has not been explicitly included in the simulation.

Other transparent conductors such as ITO have also been tried, but thus far with less success than TO. Release and diffusion of In during deposition of TFS is suspected to be the problem with ITO. The need to successfully coat thin layers on top of each other while not destroying the properties of the previous layers is a common theme that runs throughout this technology.

THIN FILM SILICON (TFS)

TFS is an alloy and not an element as often thought. The amount of the alloying agent, hydrogen, that is incorporated, the manner in which it bonds, and the manner in which the silicon bonds to itself are all variables. Hence TFS is actually a class of materials whose properties can be altered dramatically by adjusting these variables. For PV devices with efficiencies in the range of 10%, these variables probably cover a rather small range, although reactors and recipes can still vary greatly.

Progress on growth kinetics is being made, although there is still a great deal of controversy over precursors. This controversy extends to hydrogen content since films apparently contain unaccounted for excessive hydrogen. Part of the reason for the longevity of this issue is the difficulty in measuring hydrogen properly. In a paper by a colleague in this conference [3], a comparison of five techniques for profiling hydrogen is made, including four independent sets of results for one of the techniques (SIMS). There is a great deal of disagreement among the techniques, which renders model development difficult according to such data. The dependence of hydrogen content on film thickness by investigation of IR modes from this study is shown in Fig. 5. Because the general tendency of the other four techniques is to suggest no thickness dependence, this result has to be interpreted as a change in oscillator strength rather than in hydrogen content. This suggests a change in hydrogen environment with film thickness which might be expected to cause thickness-dependent electronic properties. This issue is being studied further.

In addition to hydrogen, many other elements can be combined with Si to form an entire family of alloys. The most prominent among these is C which, when doped with boron, forms the very successful Si/C:H p layer. Si/N:H has

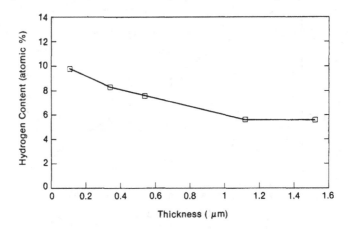

Fig. 5. Hydrogen content vs. thickness as measured by IR.

also been promoted for this role [4]. The results shown in Fig. 6 [5] suggest that while Si/N:H can be made p type, it prefers to be n type. In fact, we have successfully used it to make large gap n type contacts in 8% solar cells [5].

Si/Ge:H is another important class of alloys in that it represents the current best opportunity for making a successful smaller band gap thin film device. Progress to date in this area has been good although, in the author's opinion, somewhat slowed by the unavailability of pure GeH$_4$ or GeF$_4$. The best films seem to contain 40% or less Ge. As seen in Fig. 7 [6], the low energy PL tail increases strongly as Ge content beyond these levels is exceeded. An intensive study of the performance of these alloys and the opportunities they offer is under way throughout the world.

MODULE FABRICATION AND PERFORMANCE

A schematic of the ARCO Solar Genesis module, which is 30x30 cm and delivers 5 Wp of power, is shown in Fig. 8. All of the discussion to this point has been in reference to the cell circuit which consists of glass/TO/TFS/Al. Formidable materials issues also exist for the remainder of the package. Some of the options that were considered, the choices that were made, and the reasons for the choices are summarized in Table III [7]. The factors dictating these choices are manufacturability, low cost, and durability. These modules are just the beginning of the changeover to thin

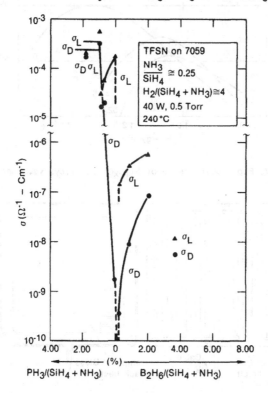

Fig. 6. Conductivity vs. dopant level in Si/N:H.

film technology. They are intended for limited consumer use and thus carry a standard one-year warranty. As sizes and efficiencies are increased to meet the large-scale power market requirements, life expectancies of 20-30 years are necessary to meet life cycle cost goals. The performance of conventional (i.e., crystalline Si) modules has been studied extensively, primarily through JPL as the lead center for such activity. The results of this study are summarized in Table IV' [8]. In essence, encapsulant performance and stress failures at joints significantly exceed problems with the cells themselves. This is expected to be the case for thin film technology as well. We do a good job on the photoconductor and take the rest of the package for granted. The opportunities for materials innovation here are extensive.

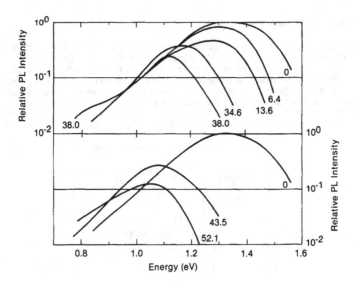

Fig. 7. Photoluminescence of $Si_{1-x}Ge_x$:H alloys vs. energy.

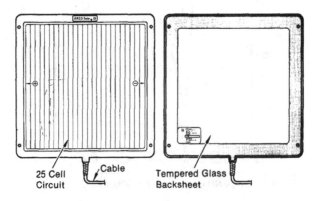

Fig. 8. Genesis TFS module.

As is commonly known, TFS has a performance loss-related problem referred to as the Staebler-Wronski effect [9]. Its manifestation depends upon the details of device fabrication and hence can exhibit a significant range. Our experience has been that there is an initial loss in device output of 10-15% followed by stabilization. We have in fact had devices (Fig. 9) maintain efficiencies of 7% for over two years of outdoor exposure

Table III. Options and choices for module materials.

	Pottants	Backsheets	Edge Seals	Frames	Terminations
Options	EVA*	Plastic	Tape	Injection molded plastic*	J-box
	PVB	Coated sheet metal	Gasket	Aluminum extrusions	Pigtail*
	Acrylic adhesives	Tempered glass*	RTV*	Roll formed sheet metal	
	Silicones			Gasket	
Choices & Reasons	*EVA	*Tempered glass	*RTV	*Injection molded plastic	*Pigtail
	• Weatherable	• Strength	• Low cost	• Low cost	• Low cost
	• Low cost	• Low cost	• Weatherable	• Aesthetics	• Weatherable insulation
	• Vacuum lamination processing	• Aesthetics	• Easy application		
	• Hi-pot/ voltage isolation	• Weatherable			
		• Hi-pot/voltage isolation			

Table IV. JPL summary of crystalline silicon module reliability problems.

Module Element	Function	Problem Areas	Comments
Encapsulant system	Structural support, electrical isolation	Soiling, yellowing, cracking, delamination, accelerated corrosion, voltage breakdown, processing stresses, differential expansion stresses	Greatest problem source
Circuit (inter-connection)	Energy delivery	Fatague failure due to differential expansion stress; poor solder joints	Next-greatest problem area
Cells	Energy conversion	Cell cracking; metallization adherence and series resistance; AR coating durability	Intrinsically reliable

274

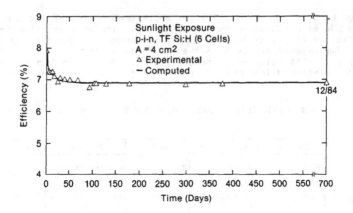

Fig. 9. Stability of device performance after two years.

under open circuit voltage conditions. Since the loss occurs almost immediately and then stops, we prefer to think of the problem as one of lost initial efficiency rather than instability. In fact, we often see improvement in output during summer months attributable to thermal reversibility of this loss [10]. We have learned how to somewhat circumvent this problem by, for example, fabricating tandem structures which are thinner and absorb less light in each layer. However, we have not yet agreed upon the responsible microscopic mechanisms or discovered a cure. This is one of the significant challenges of the field and a significant opportunity for materials innovation.

REFERENCES

1. C.C. Tsai, M. Stutzmann and W.B. Jackson, AIP Conference Proceedings No. 120, p. 242, American Institute of Physics.
2. M. Stutzmann, SERI Amorphous Silicon Subcontractors Review Meeting, Washington, D.C., March 1985.
3. G.A. Pollock, "Hydrogen Measurement of Thin Film Silicon:Hydrogen Alloy Films: Techniques Comparison," this conference.
4. (a) H. Kurata et al, Jpn. J. Appl. Phys., 21 (1982), Suppl. 21-2, 205-08. (b) M. Hirose, JARECT Vol. 6, Amorphous Semiconductor Technologies & Devices, (1983), Y. Hamakawa (ed.), 173-80. (c) I. Chambouleyron et al, Photovoltaic Solar Energy Conference, Athens, Greece (1983). (d) F. Alverez et al, Appl. Phys. Lett., 44 (1) (1984), 116-18.
5. B. Wong, D.L. Morel and V.G. Grosvenor, Proceedings of the First International Photovoltaic Science and Engineering Conference, Kobe, Japan, Nov. 1984, p. 433.
6. P.C. Taylor, SERI Amorphous Silicon Subcontractors Review Meeting, Washington, D.C., March 1985.
7. T. Jester, Proceedings of the Reliability and Engineering of Thin Film Photovoltaic Modules JPL Research Forum, Washington, D.C., March 1985.
8. R. Ross, Proceedings of the Reliability and Engineering of Thin Film Photovoltaic Modules JPL Research Forum, Washington, D.C., March 1985.
9. D. Staebler and C. Wronski, Appl. Phys. Lett., 31, (1977) 292.
10. H.S. Ullal, D.L. Morel, D.R. Willet, D. Kanani, P.C. Taylor and C. Lee, Proceedings of the Seventeenth IEEE Photovoltaic Specialists Conference, May 1984, Kissimmee, Fla.

HIGH PERFORMANCE a-Si SOLAR CELLS AND NARROW BANDGAP MATERIALS

SHOICHI NAKANO, YASUO KISHI*, MICHITOSHI OHNISHI, SHINYA TSUDA,
HISASHI SHIBUYA*, NOBORU NAKAMURA, YOSHIHIRO HISHIKAWA, HISAKI TARUI,
TSUYOSHI TAKAHAMA AND YUKINORI KUWANO

Research Center, SANYO Electric Co., Ltd.
Hashiridani, Hirakata City, Osaka, Japan
*Applied Research Center, SANYO Electric Co., Ltd.
Moriguchi City, Osaka, Japan

ABSTRACT

High performance a-Si solar cells were developed. A conversion efficiency of 11.5% was achieved for a textured TCO/p-SiC/in/Ag structure with a size of 1 cm^2 using the high quality i-layer fabricated by a new consecutive, separated reaction chamber apparatus. A conversion efficiency of 9.0% was obtained with a size of 10cm x 10cm. A high quality a-SiGe:H:F, which is a new narrow bandgap material for a-Si solar cells, was fabricated by a glow discharge decomposition of SiF_4 + GeF_4 + H_2.

A photo-CVD method was investigated in order to improve the interface properties of a-Si solar cells. A conversion efficiency of 11.0% was obtained with a solar cell in which the p-layer is fabricated by the photo-CVD method. a-SiGe:H films were fabricated by the photo-CVD method for the first time as a narrow bandgap material for multi-bandgap a-Si solar cells.

INTRODUCTION

Research and development of amorphous silicon (a-Si) materials are proceeding at a rapid rate. Their application to solar cells is also rapidly expanding. One of the main requirements for a-Si solar cells is to achieve higher conversion efficiency. Recently, their conversion efficiencies have been improved to a level of more than 11%[1] by improvements in a-Si materials and cell structures.

In this paper, first, we report an improvement of the conversion efficiency of a-Si solar cells by a new cell structure and high quality a-Si layers. Next, we discuss a-SiGe:H:F, which is a high quality new narrow bandgap material. Finally, we describe a-Si solar cells and a-SiGe:H films fabricated by a photo-CVD method, which is an attractive new method for fabricating a-Si materials.

HIGH PERFORMANCE a-Si SOLAR CELLS

Conversion efficiencies of a-Si solar cells have been increased owing to the development of new materials and new cell structures[2]. At present one of the key points for improving conversion efficiencies is the effective utilization of the wide spectrum of sunlight.

We achieved a conversion efficiency of 11.5% for a

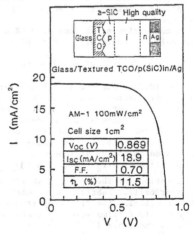

Fig. 1   Illuminated I-V characteristics of an a-Si solar cell

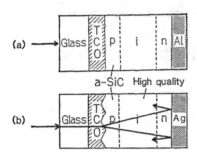

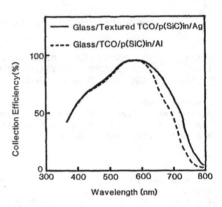

Fig. 2 Structures of a-Si solar cells
(a) Conventional structure
(b) New structure for high efficiency

Fig. 3 Collection efficiency spectra of a-Si solar cells

textured TCO/p-SiC/in/Ag structure with a size of 1 cm^2 by improving the film quality of the i-layer (Fig. 1). The new cell structure is shown in Fig. 2 together with a conventional structure. A textured transparent electrode is employed at the front side of the cell and silver is used for the back electrode (Fig. 2(b)).

The textured transparent electrode reduces the reflection of the incident light at the front side of the cell. It also makes the light path in the i-layer longer by a scattering effect. Furthermore, the silver back electrode effectively reflects the long wavelength light which comes through the cell, and sends it back to the active region. Thus this new structure has an 'optical confinement effect' in particular for the long wavelength light. The quality of the i-layer is improved by reducing impurity concentrations and by optimizing fabrication conditions, in order to collect photo-generated carriers most efficiently.

Collection efficiency spectra of the cells are shown in Fig. 3. The collection efficiency in the long wavelength region is greatly improved by the optical confinement effect of the textured TCO and the silver back electrode.

A conversion efficiency of 9.0% was obtained for an integrated type a-Si solar cell submodule with a size of 10cm x 10cm.

PREPARATION AND PROPERTIES OF a-SiGe:H:F

In order to achieve much higher efficiencies, it is essential to develop high quality narrow optical bandgap (Eopt) materials. The optimum values of the Eopt are about 1.4eV to 1.5eV. Conventional a-SiGe:H films have been fabricated by a glow discharge decomposition of SiH_4 + GeH_4. However, their electrical and optical properties were not satisfactory.

Recently, Shimizu et al. have suggested a fabrication of hydrogenated fluorinated amorphous silicon germanium (a-SiGe:H:F) by a glow discharge decomposition of SiF_4 + GeF_4 + H_2[3]. In this paper we report a detailed investigation of the optimum fabrication conditions to form a-SiGe:H:F films for solar cells. Fabrication conditions were investigated in wide ranges in order to optimize the film properties.

a-SiGe:H:F films were fabricated with the apparatus and conditions shown in Fig. 4 and Table I. The amounts of Si and Ge atoms were determined by ESCA. Fig. 5 shows variations of the properties of a-SiGe:H:F films as a function of the flow rate ratio of GeF_4 ($R_{Ge} = GeF_4/(SiF_4 + GeF_4)$). They show

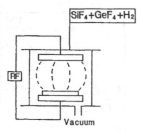

Fig. 4 Schematic diagram
of the fabrication
of a-SiGe:H:F

Table I Typical fabrication
conditions for a-SiGe:H:F

Ts		$150 \sim 250\,°C$
RF Power		$30 \sim 100\,W$
Pressure		$10 \sim 100\,Pa$
Flow rate	SiF₄	$50 \sim 250\,SCCM$
	GeF₄	$0.5 \sim 2.0\,SCCM$
	H₂	$10 \sim 50\,SCCM$

good properties when R_{Ge} is less than 0.7%. When R_{Ge} is more than 0.7%, the
dark conductivity (σd) becomes large and the value of $\sigma ph/\sigma d$ approaches unity.
Fig. 6 shows Raman spectra of (a) a film of $R_{Ge} > 0.7\%$, (b) a film of
$R_{Ge} < 0.7\%$ and (c) a conventional a-SiGe:H film. Spectrum (b) is
almost the same as (c). However, spectrum (a) has only one peak at $305\,cm^{-1}$,
which originates from microcrystalline germanium (μc-Ge). Thus it was found
that films of $R_{Ge} > 0.7\%$ are μc-Ge rather than a-SiGe:H:F. Therefore, it is
necessary to precisely control the flow rate of material gases.
    Other fabrication conditions were investigated in order
to attain higher film qualities. Fig. 7 shows properties of a-SiGe:H:F films
as a function of the substrate temperature (T_s). It was found that excellent
film properties were obtained when Ts is between 200°C and 250°C ($\sigma ph = 3 \times 10^{-4}\ \Omega^{-1}\ cm^{-1}$, $\sigma d = 1 \times 10^{-7}\ \Omega^{-1}\ cm^{-1}$ when Eopt = 1.5 eV).
    The Eopt and the Ge content ($C_{Ge} = Ge/(Si + Ge)$) are linearly related as
follows:

$$Eopt = -0.8 \times Cge + 1.75 \qquad \cdots\cdots\cdots\cdots\cdots\cdots (1)$$

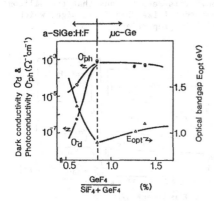

Fig. 5 Properties of a-SiGe:H:F films
as a function of the gas flow
rate ratio of GeF₄

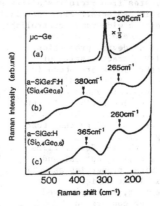

Fig. 6 Raman spectra of
Si-Ge alloys
(a) Film of $R_{Ge} > 0.7\%$
(b) Film of $R_{Ge} < 0.7\%$
(c) Conventional a-SiGe:H
($R_{Ge} = GeF_4 /(SiF_4 + GeF_4)$)

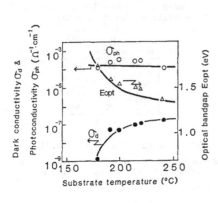

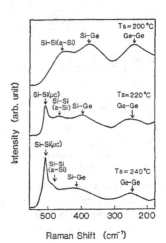

Fig. 7 Properties of a-SiGe:H:F films as a function of the substrate temperature

Fig. 8 Raman spectra of a-SiGe:H:F films as a function of the substrate temperature

This relation is qualitatively the same as that of conventional a-SiGe:H [4].

New structural information for a-SiGe:H:F films was obtained using Raman spectroscopy. Fig. 8 shows the Raman spectra of three a-SiGe:H:F films fabricated at different T_s. All the spectra show three broad peaks which originate from Si-Si bonds (480 cm^{-1}), Si-Ge bonds (400 cm^{-1}) and Ge-Ge bonds (250 cm^{-1}) in a-SiGe:H:F. In addition to these peaks, the spectra of the films of T_s = 220°C and T_s = 240°C show a narrow peak at 500 cm^{-1}, which originates from microcrystalline silicon. This result means that the silicon atoms in a-SiGe:H:F films partially microcrystallize when T_s is high, which suggests a wide-range controllability of film properties.

a-SiGe:H:F also shows excellent device properties. The I-V characteristics of Schottky diodes are shown in Fig. 9. A Schottky diode using a-SiGe:H:F has a rectification ratio of 10^7 at a bias of 1 volt, which is greater by 2 orders than that using a-SiGe:H.

As the a-SiGe:H:F films have excellent optical and electrical properties as well as excellent device properties, it is promised that high conversion efficiencies are obtained for solar cells using them.

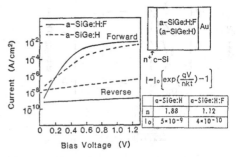

Fig. 9 I-V characteristics of Schottky diodes using a-SiGe:H:F and a-SiGe:H

Table II Typical reaction conditions
of the photo-CVD method

Gas	Si_2H_6 C_2H_2 B_2H_6	Si_2H_6 $SiH_2(CH_3)_2$ B_2H_6
Total Flow Rate	.10 - 100sccm	
Substrate Temp.	200 - 350°C	
Pressure	10 - 400Pa	

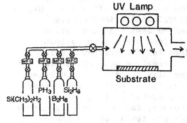

Fig. 10 Schematic diagram of the
photo-CVD method

PROPERTIES OF a-Si ALLOYS AND SOLAR CELLS FABRICATED BY A PHOTO-CVD METHOD

A photo-CVD method, in which material gases are decomposed by ultraviolet rays, has recently attracted much attention as a new fabrication method[1]. The photo-CVD method is expected to form high quality interfaces because there is no damage to a-Si films caused by ion bombardment.

The schematic reaction apparatus and typical fabrication conditions of the photo-CVD method are shown in Fig. 10 and Table II. Low-pressure mercury lamps were used as a light source.

Fig. 11 shows IMA depth profiles of In (indium) atoms near the TCO/p and p/i interfaces of a-Si solar cells. When the p-layer is fabricated by the plasma CVD method, In atoms in the ITO layer diffuse into the p and i layers by ion bombardment during the fabrication process (broken line). When the p-layer is fabricated by the photo-CVD method, however, In atoms don't diffuse at all (solid line). The interface properties of a-Si solar cells fabricated by the photo-CVD method are excellent, thereby improving the collection efficiency in the short wavelength region. At present we have obtained a conversion efficiency of 11.0 % for an a-Si solar cell in which

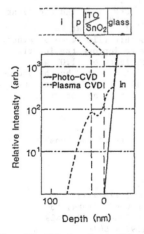

Fig. 11 IMA depth profiles
of In atoms in a-Si
solar cells

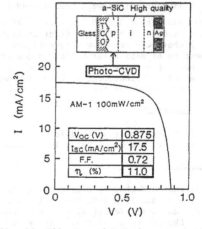

Fig. 12 Illuminated I-V characteristics
of an a-Si solar cell using the
photo-CVD method

the p-layer was fabricated by the photo-CVD method (Fig. 12). Higher conversion efficiencies are promised when the fabrication conditions are set to optimum values for the photo-CVD method.

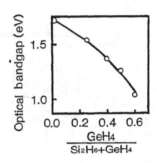

Interface properties are especially important for multi-bandgap solar cells because they have a number of interfaces. Therefore, fabrication of a narrow bandgap material by the photo-CVD method is desirable for multi-bandgap solar cells. We have fabricated a-SiGe:H films by the photo-CVD method for the first time and controlled their Eopt. Si_2H_6 + GeH_4 + H_2 gases were used as material gases. The Eopt could be controlled between 1.7 eV and 1.05

Fig. 13 Optical bandgaps of a-SiGe:H films fabricated by the photo-CVD method as a function of the flow rate ratio of material gases

eV by varying the flow rate ratio of material gases (Fig. 13). The Eopt and the Ge content are again linearly related by Eq.1.

Higher interface properties and higher conversion efficiencies are expected when this material is applied to solar cells.

CONCLUSION

Improvements in conversion efficiencies, new narrow bandgap materials, and improvements in interface properties by a photo-CVD method were discussed in view of their application to a-Si solar cells.

A conversion efficiency of 11.5% was achieved for a textured TCO/p-SiC/in/Ag structure with a size of 1 cm^2 by improving the film quality of the i-layer. A conversion efficiency of 9.0% was obtained for an integrated type a-Si solar cell submodule with a size of 10cm x 10cm.

a-SiGe:H:F films were fabricated from SiF_4 + GeF_4 + H_2 and excellent properties (σph = 3 x $10^{-4} \Omega^{-1} cm^{-1}$, σd = 1 x $10^{-7} \Omega^{-1} cm^{-1}$ at Eopt = 1.5 eV) were obtained. Partially microcrystallized a-SiGe:H:F films were obtained.

A photo-CVD method was investigated in order to improve the interface properties of a-Si solar cells. A conversion efficiency of 11.0% was obtained for a solar cell in which the p-layer was fabricated by the photo-CVD method. Furthermore, high quality a-SiGe:H films were fabricated by the photo-CVD method for the first time as a new narrow bandgap material.

ACKNOWLEDGEMENT

This work was supported in part by NEDO as a part of the Sunshine Project under the Ministry of International Trade and Industry.

REFERENCES

[1] Y. Kuwano:Technical Digest of the Int' 1 PVSEC-1 (1984)13
[2] Y. Hamakawa:17th IEEE Photovoltaic Specialists Conf. (1984) 63
[3] S. Oda et al.:Technical Digest of the Int' 1 PVSEC-1 (1984) 429
[4] J. Chevallier et al.:Solid State Commun. 24 (1977) 867

RESEARCH PROGRESS IN THE DOE/SERI AMORPHOUS SILICON RESEARCH PROJECT

E. Sabisky, W. Wallace, B. Stafford, K. Sadlon, and W. Luft

Solar Energy Research Institute, Golden, CO 80401

ABSTRACT

The Amorphous Silicon Research Project (ASRP), established at the Solar Energy Research Institute (SERI) in 1983, is responsible for all U.S. DOE government supported research activities in the field of amorphous silicon photovoltaics. The objectives and research directions of the project have been established by a Five Year Research Plan developed at SERI in cooperation with the Department of Energy in 1984. In order to accomplish project goals, research is performed by a combination of i) multi-year programs consisting of multi-disciplinary research teams based on strong government/industry partnerships and ii) basic research performed in university, government and industrial laboratories. A summary of recent research progress in the ASRP program is presented.

INTRODUCTION

The ASRP project is organized into two primary and five secondary research activities. The two primary activities are divided into research on high efficiency single-junction and high efficiency multi-junction stacked solar cells. A key feature of the primary activities is that single-junction solar cell research directly provides the technology base and expertise necessary for the development of multi-junction stacked cells, which are expected to achieve the longer term, high effiency goals of 20% and above called for in the DOE National Photovoltaics Program. The major goals for the applied research program are outlined in the Five Year Research Plan for the Amorphous Silicon Research Project [1] and are summarized in Table 1.

Table 1. Primary Long Term Goals of the Amorphous Silicon Research Project

	FY 86	FY 88
Single-Junction Cell R&D	12% (1 cm^2)	13% (1cm^2)
Single-Junction Submodule R&D	8% (1000 cm^2)	12% (100 cm^2)
Multi-Junction Cell R&D		18% (1cm^2)

In order to achieve these goals a strong reliance is placed on government/industry partnerships established in 1984 having the following characteristics: i) research is performed under multi-year subcontracts in which progress toward aggressive long-term objectives is measured on the achievement of significant milestones, ii) research is performed by highly interacting multidisciplinary teams capable of addressing all aspects of difficult projects and iii) significant cost sharing (a minimum of 30%) is required to leverage limited resources and allow the protection of proprietary industrial information.

Critical basic research is performed under the five secondary activities in order to address key fundamental issues, support the achievement of long range basic research objectives and provide a mechanism for investigating newer options. The five activities relate to: i) light induced effects, ii) material deposition rate, iii) alternate deposition methods, iv) device testing and reliability, and v) supporting research, including theory, plasma kinetics, and alloy materials.

MULTI-YEAR PROGRAMS

Four multi-year subcontracts, each three years in duration,were awarded during the first half of FY 1984 establishing research teams at Chronar Corporation, 3M Company, Solarex Thin Films Division, and Spire Corporation. Valued at a total of 18.6 million dollars, of which 30% is cost shared by industry, these programs are directed toward the objective of understanding and improving the performance of amorphous silicon single-junction cells and submodules based on material grown by the plasma assisted chemical vapor deposition (glow discharge) method, and the incorporation of amorphous silicon alloy materials into high efficiency multi-junction, stacked solar cells. Chronar, 3M, and Solarex are performing single-junction research and Spire is performing multi-junction research under the present program structure. To improve the probability of success in achieving overall objectives, each organization is pursuing a technical option that draws upon the internal strengths and technical expertise previously existing within the company. An overview of these technical options is given in Figure 1.

Single Junction Cell/Submodule Research

The primary goals of the single-junction three year programs are to demonstrate p-i-n solar cells of 12% in efficiency, area of 1 cm^2, and to demonstrate series interconnected submodules of 8% in efficiency, area of 1000 cm^2 by FY 1986. All efficiency goals have specific stability criteria associated with them. As shown in Figure 1, three technical approaches are being used in single-junction research [2,3]. Both Chronar and Solarex use tin oxide coated glass as the substrate and aluminum as a back metal contact. For deposition of amorphous silicon, Chronar uses two-electrode RF glow discharge and Solarex uses three-electrode DC glow discharge.3M uses two-electrode RF glow discharge deposition to deposit films on metal coated polyimide flexible polymer substrates and an ITO front contact material developed in-house. In order to minimize cross-contamination problems, multi-chamber deposition systems will be constructed in each program for cell and submodule research based on: i) the consecutive reactor design employing horizontal electrodes developed at RCA (Solarex), ii) a reactor design using vertical electrodes (Chronar), and iii) a continuous roll to roll system design employing a polyimide web substrate (3M). Laser scribing technology is used to interconnect cells in submodules. In all three organizations the processing capability exists for all steps involved in complete submodule fabrication.

During Phase I of the single-junction program, all subcontractors were on track in meeting individual milestones. SERI has measured efficiencies of 7-8% on Chronar p-i-n cells up to 1 cm^2 in area and Chronar has recently reported cell efficiencies of 9-10% on 0.1 cm^2 cells. SERI has measured efficiencies between 8 and 9% on Solarex cells up to 1 cm^2 and Solarex has recently reported efficiencies of 9-10.5% on cells up to 1 cm^2. SERI has measured efficiencies for Chronar submodules of 5 to 5.5% over active areas up to 720 cm^2. For Solarex submodules SERI has measured efficiencies of 5-6% over active areas of 80 cm^2 and Solarex has recently reported efficiencies of 6-7%. 3M has proceeded in optimizing p-i-n cells on polyimide substrates and has reported efficiencies over 7% on small area cells. In all organizations complete diagnostic and measurements capabilities have been established. Submodule research is being performed on one square foot glass substrates at Chronar and Solarex.

Multi-Junction Cell Research

The long-term goal of the multi-junction program is to demonstrate an efficiency of 18% by FY 1988 for cells based on amorphous silicon alloy materials. In order to achieve this objective, Spire is performing research based on a realistic design for a multi-junction, tandem cell consisting of an a-Si:H top cell and an a-(Si,Ge):H bottom cell. The two-cell structure will be optimized for bandgaps of 1.7 and 1.45 eV to achieve practical conversion efficiencies in the range of 17-20% [4]. It is perceived

Technical Approaches of New Multi-Year Initiatives in Amorphous Silicon

Three year program goals are to demonstrate stable p-i-n solar cells of at least 12% efficiency, area of 1 cm²; demonstrate stable submodule of at least 8% efficiency, area of 1000 cm²; and establish technology base in amorphous silicon alloys leading to 18% efficiency in 1988.

Company	Approach	A-Si Deposition Method

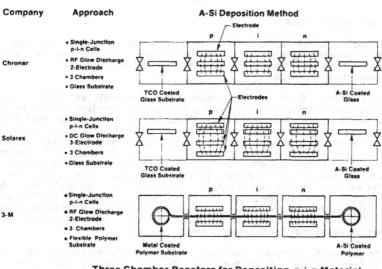

Chronar
- Single-Junction p-i-n Cells
- RF Glow Discharge 2-Electrode
- 3 Chambers
- Glass Substrate

Solarex
- Single-Junction p-i-n Cells
- DC Glow Discharge 3-Electrode
- 3 Chambers
- Glass Substrate

3-M
- Single-Junction p-i-n Cells
- RF Glow Discharge 2-Electrode
- 3 Chambers
- Flexible Polymer Substrate

Three Chamber Reactors for Depositing p-i-n Material

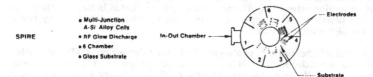

SPIRE
- Multi-Junction A-Si Alloy Cells
- RF Glow Discharge
- 6 Chamber
- Glass Substrate

Six Chamber Reactor for Depositing Material for Multi-Junction Cells

Figure 1. Schematic Overview of the Technical Approaches Being Persued in the ASRP Project for Single-Junction and Multi-Junction Research Programs. Chronar and SPIRE use Vertical Electrodes and Solarex and 3M use Horizontal Electrodes in Multi-Chamber Deposition Systems.

that this cell structure will have advantages in terms of stability as well.

During Phase I of the multi-junction program, Spire has concentrated on the construction of a six sector reactor consisting of six consecutive RF glow discharge reaction chambers for preventing cross-contamination in the deposition of intrinsic and doped alloy material layers. To date, Spire has reported small area single-junction a-Si:H cell efficiencies up to 7.8%, small area single-junction a-(Si,Ge):H cell efficiencies up to 5.8%, and small area a-Si:H/a-(Si, Ge):H tandem cell efficiencies up to 6.7%. The six sector reactor is presently operational in a manual mode.

BASIC RESEARCH PROGRAM

In addition to the cell/submodule activities covered in the previous section, a large basic research program is being conducted in support of the achievement of long-range goals. These basic research activities are outlined in Figure 2 and have recently been extensively reviewed [5].

Light-induced effects are being studied at Xerox, MIT, and Oregon State University. The effect of light soaking in a-Si:H samples deposited in an ultra-high vacuum (UHV) system or a conventional deposition system has been studied by measuring the change in dangling bond density using electron spin resonance (ESR). Although UHV-deposited samples have one to two orders of magnitude lower impurity levels than conventional glow discharge samples, the light soaking effect is nearly identical. This evidence indicates that impurities play a minimum role in generating light-induced effects. Two models for light induced degradation in amorphous silicon are being examined based on specific defect sites in amorphous materials leading to charge trapping. Both models are self-limiting in terms of long-term stability and predict primarily short term effects. Capacitance profiling studies indicate that light-induced effects may be the result of competing fundamental processes, complicating the interpretation of experimental observations.

The glow discharge deposition of silane and higher order silanes at high deposition rates is being investigated at Brookhaven National Laboratory and Vactronics Corporation. Brookhaven has reported a 6.2% cell efficiency for an intrinsic amorphous silicon layer deposited at 17 $\overset{\circ}{A}$/second and is investigating doped and intrinsic film properties at deposition rates of 20 $\overset{\circ}{A}$/second using disilane diluted in He. Vactronics is correlating film properties with deposition rate using monosilane diluted in hydrogen and deposited at high power densities.

The potential of thermal and photo-chemical vapor deposition (CVD and photo-CVD) as alternate deposition methods is being evaluated at Harvard University, Chronar Corporation and the Institute of Energy Conversion (IEC). Primary emphasis is being placed on photo-CVD due to its potential for producing high quality amorphous silicon films and minimizing the material damage encountered in the use of high energy plasmas. Using mercury sensitized photo-CVD of disilane, Chronar has reported the achievement of a 5.9% efficiency for an all photo-CVD cell. IEC has also recently initiated a study of mercury sensitized photo-CVD films from disilane. Using atmospheric pressure chemical vapor deposition, Harvard has investigated amorphous silicon films deposited at rates up to 90 $\overset{\circ}{A}$/second. The preparation of a-(Si,Ge):H films by photo-CVD will be studied at IEC under a recently awarded subcontract in the SERI New Ideas Program.

Device testing and reliability measurements activities are being pursued at SERI and JPL. A strong effort is now being pursued by SERI to standardize efficiency measurements in the amorphous silicon field in coordination with the photovoltaic community and various international standards committees. An Amorphous Silicon Measurements Task Force was recently established by SERI consisting of university/industry/government representatives, and a workshop was conducted in February, 1985. As a result of this workshop, recommendations were issued, which are

Outline of Fundamental Studies and
Supporting Research Activities

Light-Induced Effects
XEROX-PARC, R. Street   – Optical and electronic properties of
                                amorphous silicon alloy materials
MIT, D. Adler   – Theory of light induced effects
Oregon State U.,   – DLTS measurements
  J.D. Cohen
SERI, T. McMahon   – Controlled stability measurements of
                                a-Si:H materials and devices

Material Deposition Rate
Brookhaven, P. Vanier   – Efficient a-Si cells at high
                                deposition rates.
Vactronics, B. Wiesmann   – Investigation of properties of
                                materials deposited at high deposition rates.

Alternative Deposition Methods
Harvard, R. Gordon   – CVD of a-Si:H; SnO_x:F, and diffusion
                                barriers
Chronar, A. Delahoy   – Photo–CVD of higher order silanes
Institute of Energy   – Photo–CVD of disilane
  Conversion, B. Baron

Device Testing and Reliability
SERI, R. Hulstrom, L.   – Cell and submodule indoor and
  Kazmerski, and   outdoor standard measurements,
  D. Deblasio   resource assessment
JPL, R. Ross   – Submodule reliability measurements

Supporting Research
SERI, H. Mahan   – Amorphous silicon alloy studies
                                (C,Ge, and Sn)
Harvard, W. Paul   – Glow discharge of a-Si,Ge:H:F alloys
North Carolina State U.   – Dual magnetron sputtering of
  G. Lucovsky   a-Si:Ge:H alloys
Rockwell, R. Haak   – Electrochemical Photocapacitance of
                                a-Si:H materials and devices.
NBS, Univ. of Colorado,   – Glow discharge silane plasma
  A. Gallagher   characterization
NRL, W. Carlos   – ESR and NMR studies
JPL, D. Bickler   – Glow discharge reactor studies

Figure 2.   Breakdown of the Basic Research Activities of the Amorphous Silicon
    Research Project

currently under review, for indoor/outdoor cell and submodule measurement procedures. The reliability of amorphous silicon cells and submodules is under investigation at JPL in terms of encapsulation issues and long-term stability.

In support of multi-junction cell research, fundamental amorphous silicon alloy studies are being performed at North Carolina State University, Harvard University, Xerox, and SERI. Using dual magnetron sputter deposition, NCSU has been able to control preferential hydrogen attachment to Si and Ge in a-(Si,Ge):H films by employing independent Si and Ge hydrogen plasmas. Harvard has found that the electronic properties of a-(Si,Ge):H films are affected by structural modifications in the material as the Ge content increases, as well as by band structure changes. Harvard has also recently initiated a study of fluorinated a-(Si,Ge):H materials by glow discharge deposition. a-(Si,Sn):H has been investigated extensively at SERI with the conclusion that poor electronic properties are due intrinsically to Sn.

Other supporting research activities focus on the measurement of material optical and structural parameters. NMR, ESR, and Photoluminescence studies at the Naval Research Laboratory are directed toward probing the existence of molecular hydrogen in amorphous silicon and characterizing the spatial distribution of defect states in films. Xerox has also correlated defect state density with stress parameters in amorphous silicon. The technique of Electrochemical Photocapacitance Spectroscopy (EPS) is being used at Rockwell to correlate reversible changes in defect state densities with light induced effects. Detailed plasma diagnostic studies at the University of Colorado have established the important contribution of surface catalized reactions to the gas phase chemistry in glow discharge plasmas. JPL is investigating the design and scale-up of glow discharge deposition systems.

CONCLUSION

The Amorphous Silicon Research Project at SERI is pursuing an aggressive R&D program incorporating highly interacting and multidisciplinary teams in government/industry partnerships supported by a strong basic research program in university/industry/government laboratories. It is expected that the present ASRP program will achieve 12% efficiencies for small area cells by FY 1986. The next step in the evolutionary development of thin-film amorphous silicon technology will be the challenge associated with multi-junction devices that should achieve conversion efficiencies of 20% in the long term. Building on our present knowledge of single-junction cells and alloy materials, we march toward that goal.

ACKNOWLEDGEMENT

This work was supported by the U.S. Department of Energy under Contract No. DE-AC02-83CH10093.

REFERENCES

1. Amorphous Silicon Research Project - Five Year Research Plan 1984-1988, SERI/SP-211-2350 (May, 1984).
2. E. Sabisky, Proc. of the 1st International PV Science and Engineering Conference, (Kobe, Japan), p. 505, 1984.
3. E. Sabisky, W. Wallace, A. Mikhail, H. Mahan, and S. Tsuo, 17th IEEE PV Spec. Conf., p. 217, 1984.
4. V.L. Dalal, 17th IEEE PV Spec. Conf., p. 86, 1984.
5. Amorphous Silicon Subcontractors Annual Review Meeting, SERI/CP-211-2654, (March, 1985).

TUNNELING CONTRIBUTIONS TO NIP AND PIN
HYDROGENATED AMORPHOUS SILICON DEVICES

T.J. McMAHON* AND A. MADAN**
*   Solar Energy Research Institute, 1617 Cole Blvd., Golden,
CO 80401
**   Glasstech Solar, Inc., P. O. Box 52, Wheat Ridge, CO 80034

INTRODUCTION

It is well known that the performance of hydrogenated amorphous silicon
(a-Si) p-i-n type devices is determined by the sequence of deposition. For
instance, the stainless steel/p-i-n configuration generally shows a larger
value of the open circuit voltage (up to 200mV) compared to the n-i-p
sequence of deposition [1,2]. Explanations of this phenomena such as the
Dember potential [1], self field effect [2], residual doping [3], hydrogen
effusion effects associated with the p^+ layer deposition process [4,5] are
unable to satisfactorily explain the difference in performance.

We have performed a systematic study to examine the difference between
the   two   configurations   and   used   dark   and   light   current-voltage
characteristics, SIMS, junction recovery and device modelling.   After
pointing out flaws in previous studies, we present a plausible alternative
explanation which relies on tunneling at the p^+-i interface to explain some
of the observed differences in the performance of these devices[6].

SAMPLE PREPARATION AND SOME PROPERTIES

The p-i-n and the n-i-p type diodes were fabricated using the radio
frequency glow discharge in SiH_4 gas with gas mixtures of 1% B_2H_6 and 1% PH_3
for the deposition of the p^+ and the n^+ layers. The p^+ layer (200-400 Å)
was deposited onto a stainless steel substrate.   Next, an approximately
~5000 Å thick undoped intrinsic layer was deposited and followed by the
deposition of a thin n^+ layer (~100-300 Å).   The top contact was either a
thermally evaporated semitransparent Al-layer or a transparent conducting
oxide of 600 Å in thickness.   A space charge limited current (SCLC)
measurement was performed using a sample in an n^+-i-n^+ configuration
resulting in the density of states, $g(E) = 2 \times 10^{15}$ cm^{-3} eV^{-1} at an energy
of $E_c - E_f = 0.90$ eV and $g(E) \sim 10^{16}$ cm^{-3} eV^{-1} at $E_c - E_f = 0.60$ eV where E_c
and E_f correspond to the conduction band edge and Fermi level energies
respectively.

COMPOSITIONAL ANALYSIS

In this study we have used diodes in the p-i-n and n-i(B)-p configurations with the i(B)-layer doped with Boron in the range 0 to 15 vppm.   In
Fig 1, B and P concentrations are shown for the n-i(B)-p configuration;
these concentrations were determined by the use of SIMS at Charles Evans
Associates.   It should be noted that the Boron concentration increases from
4×10^{16} atm/cm^3 to 8×10^{18} atm/cm^3 when the B_2H_6 in the gas phase is increased
from 0 to 15 vppm.   An important feature in the P concentration is the
initial decrease with B_2H_6 gas flow (up to 2-3 vppm) followed by a rapid
increase to the mid 10^{18} atm/cm^3 range which is virtually equal to the Boron
concentration.   As yet we do not understand the reason for this high level
of phosphorous, but our studies are restricted to the lower doping ranges
and therefore, the higher levels are of no concern.   Similar observations

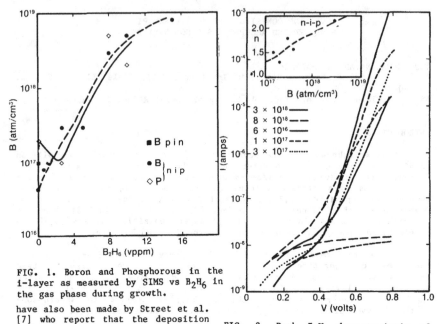

FIG. 1. Boron and Phosphorous in the i-layer as measured by SIMS vs B_2H_6 in the gas phase during growth.

have also been made by Street et al. [7] who report that the deposition process has a tendency to equalize the dopant concentration in the films and conclude that this could be due to B-P pair formation.

FIG. 2. Dark I-V characteristic of several n-i-p diodes with B concentration in the i-layer as noted. Inset shows diode quality factor n of each curve.

RESULTS

In Fig. 2, we show the dark log I-V (current-voltage) characteristics of n-i(B)-p devices with B concentrations up to 8×10^{18} atm/cm^3 in the i-layer. A well defined region in the forward bias can be observed from which the diode quality factor, n, can be extracted using

$$I = I_o \left[\exp \frac{qV}{nkT} - 1 \right] \qquad (1)$$

The inset of this Figure shows the diode quality factor n as a function of doping for the different diodes. We note that for a nominally undoped i-layer, n = 1.2 which is quite low when compared to a value of 1.6 usually observed for an undoped pin device [8]. Such low values of n (=1.2) are normally observed in Schottky Barrier and MIS type devices where the transport is dominated by majority carriers (electrons) [9]. We shall show in the junction recovery experiment that the transport in a n-i-p type device can become dominated by electrons. When the i-layer is doped with Boron, n increases until at a value of 8×10^{18} atm/cm^3 it reaches 2.5.

In Fig. 3, log I_{sc} versus V_{oc} results of n-i(B)-p diodes with B ranging from 4×10^{16} to 8×10^{18} are presented. It should be noted that over the illumination range investigated, n=1.21 for the lowest B concentration. With an increasing B concentration (4×10^{17} atm/cm^3) two regions become apparent: n=1.56 at low V_{oc} values and n=1.04 at higher V_{oc} values. With a further increase in B (8×10^{18} atm/cm^3), n exceeds 2 and at any current level, V_{oc} has decreased substantially from the maximum value that can be

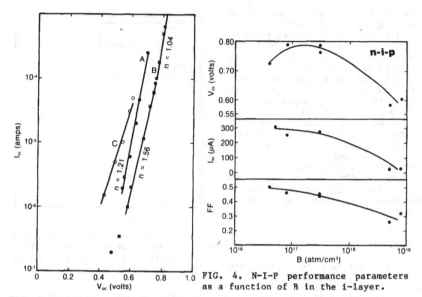

FIG. 3. Log I_{sc} vs V_{oc} for n-i(B)-P diodes: A-4x10^{16} atm/cm^3, B-4x10^{17} atm/cm^3 and C-8x10^{18} atm/cm^3.

FIG. 4. N-I-P performance parameters as a function of B in the i-layer.

obtained for a device with B of 4x10^{17} atm/cm^3.

In Fig. 4, we show the open circuit voltage, V_{oc}, the short circuit current, I_{sc}, and the fill factor, FF, of n-i(B)-p devices as a function of B concentration within the i-layer. V_{oc} increases by about 80mV with the addition of about 1-4x10^{17} B atm/cm^3 but beyond this level there is a rapid decrease in V_{oc}. Larger amounts of B degrade the performance of the device and is a direct consequence of a large defect state density, as well as a low $\mu\tau$ product for holes, where μ and τ are the mobility and the recombination lifetime of the minority carriers. Since this has been dealt with adequately in the literature [10] our discussion will be restricted to the basic differences between the two configurations and why the V_{oc} for the n-i(B)-p configuration can increase with B inclusion.

JUNCTION RECOVERY

Diodes of each configuration were studied with the junction recovery technique in which excess carriers are injected into the device using a far forward bias voltage pulse. After switching into reverse bias within 100 ns, the RC time of the circuitry, the recovery current transient was recorded and the recovered charge was determined by integration over the time scale t = 10 μs. In an earlier study it was concluded that the recovered charge for a-Si:H p-i-n junctions were holes due to the much broader valence bandtail [11].

In the inset of Fig. 5, a typical recovery transient is shown and in the main figure the recovered charge, Q_o, is shown for a p-i-n cell as well as for the n-i(B)-p cells. Almost no charge was collected with a B concentration of 4×10^{16} atm/cm^3 in the i-layer, just as in the Schottky barrier case. From this we conclude the n-i-p device is basically controlled by majority carriers (electrons) with little or no injection of holes into the i-layer from the p$^+$-layer. However, as the i-layer was doped with an increasing amount of Boron, the recovered charge increased as shown in Fig. 5 and exceeded that obtained for the p-i-n cell. The increase in the recovered charge can be attributed to the reduced potential barrier at the p$^+$/i interface —a direct result of the boron moving the Fermi level towards E_v in the i-layer.

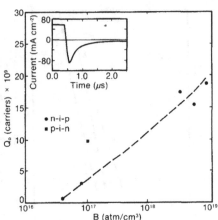

FIG. 5. Charge recovery for (B) n-i-p and p-i-n diodes as a function of B in the i-layer. Inset shows typical current transient.

DISCUSSION

Past attempts to explain the difference in V_{oc} between p-i-n and n-i-p diodes include Dember potential [1], self field effect [2], residual dopants [3] and hydrogen effusion effects associated with the p$^+$-layer growth [4]. We briefly discuss each of these in turn.

The Dember potential [1] and self field [2] explanation can be quickly tested by illuminating (a) glass/TCO/p-i-n/TCO and (b) glass/TCO/n-i-p/TCO cells from either direction. We found that similar voltages result if the illumination is adjusted so that identical current is generated by illuminating either through the p$^+$-layer or the n$^+$-layer. Since a directional dependence of V_{oc} for structures (a) and (b) is absent neither of these reasons can be valid. Nevertheless a difference of about 80 mV exists between the two with configuration (a) showing a higher V_{oc} than configuration (b). Results reported by Konagai et al. [3] have also shown that for each configuration there is virtually no difference in V_{oc} when the junctions are illuminated either through the n$^+$ or the p$^+$ layers; the small difference that did exist can be attributed to the slightly different amounts of generation rate. However a difference in V_{oc} of up to 180 mV did exist between the two configurations. Unlike Konagai et al [3] we do not find a large shift in the V_{oc}-I_{sc} plot from the forward dark I-V. They point out that the V_{oc} on the p-i(B)-n/substrate cell may be increased initially by the addition of B which compensates the residual P but do not explain why the V_{oc} is depressed without the addition of B.

Another reason cited for the V_{oc} difference is that the growth mechanism of the p$^+$-layer on a substrate could be somewhat different from when it is deposited onto an i-layer [4]. More specifically hydrogen evolution from the surface p$^+$-layer during the final stages of the deposition process is thought to result in a slight back barrier formation for the n-i-p type junction. Evidence for asymmetry in the junction formation was provided by the fabrication of a p$^+$-i-p$^+$ type junction and noting that the structure had led to V_{oc}=300 mV [4] upon illumination. However, careful examination of the hydrogen evolution results of heavily doped Boron a-Si:H layers reveals that at a deposition temperature of 250 C, the effusion of hydrogen is insignificant [12]. On several occasions we have compared the performance of our n-i-p type junctions when they were completed with metals of extremely different work functions (Al and Pd) and obtained identical performance in terms of V_{oc}. We therefore conclude that the p^+-layer provides a good ohmic contact and the cause for the different V_{oc}'s has to be sought elsewhere.

TUNNELLING

In Fig. 3 (curve B for B=4×10^{17} atm/cm^3) two regions in the log J_{sc}-V_{oc} characteristic exist from which the diode quality factor can be extracted; (a) a region n=1.6 which is representative of space charge recombination and (b) at higher V_{oc}, n=1 which is indicative of electron injection. By extrapolating the latter region to V_{oc}=0 it is clear that the built in potential thus derived would be somewhat higher in comparison with curve A, corresponding to B=4×10^{16} atm/cm^3 in the i-layer.

We explain this increase in V_{bi} in the compensated n-i-p cells by noting the work of Jackson and Amer [13] who report that the subband gap absorption of films with P increases (as would be expected from the residual P in the n-i-p cell). States below E_F must be the origin of this absorption and it is these states at a level of 2×10^{17} atm/cm^3 that we propose as the source of the reduced V_{bi} due to tunneling at the p/i interface. With compensation, there is a reduced sub band gap absorption [13] and an increase in photoluminescence [7]. The implied reduction in the density of states, especially below E_f, leads to a reduction in the field at the p^+/i interface. This in turn reduces the tunnelling component with the consequence of an increase in V_{bi}.

Changes in V_{oc} with the density of states can be determined from the potential profile at the p^+/i interface. For simplicity, we use T = 0K statistics and consider the density of localized states composed of donors and acceptors as shown in the inset of Fig. 6. The characteristic energies for the segments of the multi-exponential conduction and valence band tails are shown. Using this DOS spectra, it is straight forward to calculate the field and hence the potential profile by the use of Poisson's equation. In Fig. 6, we show the potential profile (curve A) calculated for the i-layer with residual phosphorous having a steep slope at the p^+/i interface with a correspondingly high field of F = 1.1×10^5 V cm^{-1}. The charge contained between E_f and the interface potential is 2×10^{17} cm^{-3} which we purposely fixed to correspond to the P contamination as shown in Fig. 1. In Fig. 6, curve B is the potential profile which couldoccur in the case of compensation. In this $E_c - E_f$ = 1.025 eV and the integrated charge is now reduced to 2×10^{16} cm^{-3}. The potential profile becomes softer in nature as is to be expected. Band bending at the p^+/i interface is now

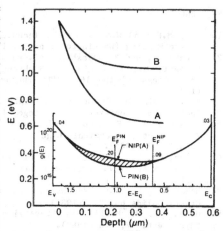

FIG. 6. Calculated potential as a function of distance from the p/i inter- face. Inset shows densities of states used for profile A (NIP) and profile B (PIN)

0.375 eV and the field was calculated to be only 3.6×10^4 Volt/cm.

Because of the high fields at the p^+/i interface (curve A) and the consequent thin barrier, tunnelling of electrons can occur lowering the built-in potential for electron transport and hence promotes the electron injection current. Since tunnelling distances are usually of the order of $\delta = 35$ Å, then for case A, Fig. 6, tunnelling would occur at $\Delta V = \delta x F = 38.5$ mV below the top of the barrier and for case B, Fig. 6, tunnelling occurs at 12.6 mV. The difference in the two translates to an increase of about 26 mV in barrier height for the electron injection current; this then translates into an enhancement of V_{oc} with B doping for the n-i(B)-p configuration.

CONCLUSION

We have cited some of the reasons previously given for differences in performance found between the pin and nip configuration and given arguments showing why they are not valid. The charge recovery data and J–V data suggest strongly that a reduction in V_{bi} has occurred and that tunnelling is part of the reason for different performance in the nip cell. Preliminary modelling on field profiles is presented explaining some of the difference suggesting qualitatively that tunnelling is a plausible explanation. If multiple donor like defects were produced per phosphorous atom, then such a model would quantatively explain such differences.

ACKNOWLEDGEMENTS

We would like to thank B. Nelson and C. Evans for sample preparation and K. Sadlon for performing the charge recovery experiments.

This work is supported under DOE Contract No. DE-AC02-83CH10093.

REFERENCES

1. Han, M.K. and W.A. Anderson, Technical Digest of International Electron Device Meeting (Washington), p. 34, 1981.
2. Sakata I., and Y. Hayashi, Appl. Phys. Lett. 42, 279 (1983).
3. Konogai, M., K.S. Lim, P. Sichanugrist, K. Komori and K. Takahashi, Proceedings of the 16th IEEE Photovoltaic Conference, San Diego, 1321 (1982).
4. Müller, G., M. Simon and G. Winterling, Proceedings of the 16th IEEE Photovoltaic Conference, San Diego, 1129 (1982).
5. Nakamura, G., K. Sato, T. Ishihara, M. Usui, H. Sasaki, K. Okaniwa and Y. Uykimoto, Technical Digest of the International PVSEC-1, (Kobe, Japan) 587 (1984).
6. Madan, A., T.J. McMahon and W. Pickin, submitted for publication.
7. Street R.A., D.K. Bregelsen and J.C. Knights Phys. Rev. B. 24, 969 (1981).
8. McMahon T.J., B.G. Yacobi, K. Sadlon, J. Dick and A. Madan, J. Non Cryst. Solids 66, 375 (1984).
9. Madan A., J. McGill, W. Czubatyj, J. Yang and S.R. Ovshinsky, Appl. Phys. Lett. 37, 826 (1980).
10. See for instance Moustakas T.D., R. Friedman and B.R. Weinberger, Appl. Phys. Lett. 40, 587 (1982).
11. Könenkamp R., A.M. Hermann and A. Madan, Appl. Phys. Lett., 46(4), 405 (1985).
12. Beyer W. and H. Wagner, J. de Physique 42, 783 (1981).
13. Jackson W.B. and N.M. Amer, Phys Rev B 25, 5559 (1982).

PART X

Stability

REVERSIBLE, LIGHT INDUCED CHANGES IN A-SI:H FILMS AND SOLAR CELLS

C. R. Wronski

Corporate Research Laboratories
Exxon Research and Engineering Co.
Annandale, New Jersey 08801

Continuous progress is being made in the conversion efficiencies of a-Si:H solar cells and efficiencies in excess of 11% have been achieved. Because of these advances and the development of a-Si:H cell technologies there is an increased interest in the long term performance of a-Si:H cells and the mechanisms responsible for their degradation. The reversible light-induced changes in a-Si:H solar cells are generally associated with the Staebler-Wronski effect (SWE) (1). This effect has been studied on a wide range of a-Si:H materials using a variety of different experimental techniques and this talk reviews the results that have been obtained on a-Si:H films and solar cells (2). It discusses in greater detail recent studies on a-Si:H solar cell structures in which simultanous measurements have been made on the changes in both the photovoltaic properties as well as their electronic properties and densities of gap states. In particular it focuses on several results obtained with semitransparent metal-undoped a-Si:H Schottky barrier solar cell structures (3).

In these studies the photovoltaic properties of the cell structures were evaluated and carrier collection measurements were carried out to determine the changes in deep gap states and their effect on the hole lifetimes (4). An example of the carrier collection efficiencies in State A and State B obtained after 145 hours of 100 mw/cm^2 of white light (connected for the transmission of the Pt films) are shown in Figure 1.

The spectral dependences of the carrier collection efficiencies in Figure 1 were obtained under short-circuit conditions and under an illumination of 100 mW/cm^2 of white light. The points are the experimental results and the solid lines were computed using experimentally determined values of the built-in potential, the model of Gutkowicz-Krusin (5) where N_0, the density of deep lying gap states, and L_p, the hole diffusion length, were the best fit parameters. Using N_0 as the density of recombination centers the hole capture cross-section, Sp, can be obtained from the ambipolar hole lifetimes deduced from L_p (4). There appears to be a small systematic shift in the hole capture cross-sections from $1-1.5 \times 10^{-15}$ cm^{-2} in State A to $2-2.5 \times 10^{-15}$ cm^{-2} in State B. Both these capture cross-sections correspond to atomic dimensions which suggests that in solar cells the hole recombination, in both State A and State B, is due to neutral centers, possibly dangling bonds (6).

The photoconductivities and the electron lifetimes of the bulk undoped a-Si:H in the cell structures were studies under far forward bias where the photocurrents are ohmic and are limited by series resistance of the bulk. By investigating these photocurrents in State A and State B over a wide range of volume generation rates it was possible to study the effect of the metastable defects on the photoconductive electron lifetimes and their recombination kinetics. In addition over the same range of generation efficiencies dual beam photoconductivity measurements were used to determine the optical absorbtion for photon energies, E, from 0.8 to 1.7

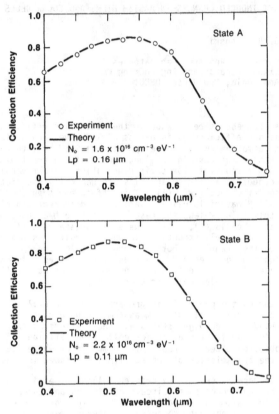

Fig. 1   Spectral dependence of carrier collection efficiency for a
Pt/a-Si:H Schottky structure before (State A) and after (State B)
light exposure.

eV. (7). The results for $\alpha(E)$ show an exponential region for E between 1.7
and 1.4 eV and a well defined shoulder for E below 1.4 eV similar to those
found with photothermal deflection spectroscopy (8). The results of α
versus photon energy, E, obtained for two films in State A and State B are
shown in Figure 2 for a bias red light illumination of 1mW/cm^2.

The distinct differences between the shoulders in State A and State B
in Figure 2 reflect the changes in the densities of state resulting from
the introduction of light-induced defects near midgap. The densities of
states for the region of energies correponding to the shoulders in $\alpha(E)$
were interpreted in terms of a band of gap states having a constant density
N_r cm^{-3} eV^{-1} which is centered at E_r from the conduction band and has a
half-width Δ (3). The values of E_r obtained on the films in State A are
from ~0.85 to 0.95 eV and decrease by ~0.2 eV for State B. In contrast the
corresponding shifts in the values of $E_r - \Delta$ were significantly smaller
implying that the edges of these bands of states do not move as much as
E_r. The values of N_r, just as the electron lifetimes, exhibit a dependence

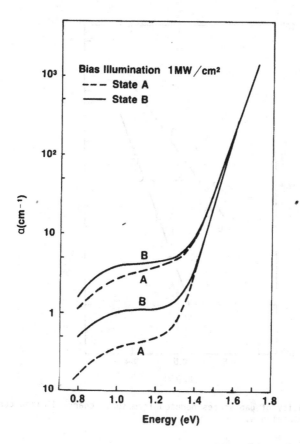

Fig. 2   Absorption coefficient α derived from bias photoconductivity
measurements.

on the volume generation rates where in State B they range from ~1 to 3 x
10^{16} cm^{-3} eV^{-1}. However it was not possible to correlate the electron
lifetimes with a constant electron capture cross-section as can be done
with films which have higher values of N_r (films primarily deposited on
quartz)(3). This large change in recombination kinetics found with lower
densities of N_r centers indicates that other states in the gap also have a
major role in determining the photoconductive electron lifetimes.

Evidence was found in these films that the transition from the State A
to State B is accompanied by the introduction of ~ 1 x 10^{16} cm^{-3} eV^{-1}
metastable defects well above midgap. This was obtained from results on
the dark currents under far forward bias where space charge limited
currents could be observed (9,10). These dark currents exhibit significant
differences between State A and State B and using the method described in
reference 9 the space charge limited currents were used to measure the
densities of states over the region of ~ 0.6 to 0.4 eV from the conduction

298

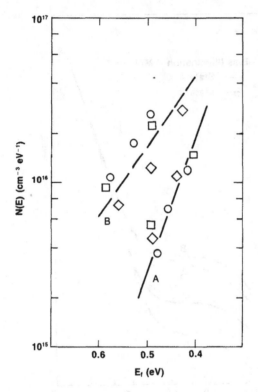

Fig. 3   Density of gap states deduced from space charge limited current measurements.

band. The results obtained for three ~ 1 μm thick films in State A and State B are shown in Figure 3. In Figure 3 the densities of states, $N(E)$ cm^{-3} eV^{-1}, are plotted as a function of E_f, the distance of the Fermi level from the conduction band. These results indicate that in these undoped a-Si:H films having low densities of gap states there is also a induced increase in the densities of gap states at ~ 0.5 eV from the conduction band.

The significance of the results on the performance of a-Si:H solar cells is reviewed in the talk.

REFERENCES

1. D. L. Staebler and C. R. Wronski, Appl. Phys. Lett., 31, 292, 1977; J. Appl. Phys. 51, 3262, 1980.
2. C. R. Wronski, Semiconductor and Semimetals, Vol. 21, Part C, J. Pankove Ed., Academic Press Inc., 1984, p. 347.
3. C. R. Wronski, J. Non-Cryst. Solids, 59 & 60; 401, 1983.
4. B. Abeles, C. R. Wronski, Y. Goldstein and G. D. Cody, Solid State Comm., 41, 251, 1982.
5. D. Gutkowicz-Krusin, J. Appl. Phys., 52, 5370, 1981.
6. R. A. Street and D. K. Biegelsen, Solid State Comm., 33, 1195, 1980).
7. C. R. Wronski, B. Abeles, T. Tiedje and G. D. Cody; Solid State Comm., 44, 1423, 1982.
8. W. B. Jackson and N. M. Amer, Phys. Rev., B 35, 5321, 1982. C. B. Roxlo, B. Abeles, C. R. Wronski, G. D. Cody and T. Tiedje, Solid State Comm., 47, 985, 1983.
9. A. Rose, Phys. Rev., 97, 1538 (1958).
10. R. Plattner, W. Kmuhler, H. Pjleider, H. Kausche and M. Muller, Tech. Digest, 1st Photovoltaic Science and Engineeing Conference, Kobe, Japan, Nov. 1984, p. 895.

LIGHT-INDUCED METASTABLE DEFECTS IN a-Si:H : TOWARDS AN UNDERSTANDING

MARTIN STUTZMANN, WARREN B. JACKSON, AND CHUANG CHUANG TSAI
Xerox Palo Alto Research Center, Palo Alto, CA 94304

ABSTRACT

The dependence of the creation and the annealing of metastable dangling bonds in hydrogenated amorphous silicon on various material parameters will be discussed in the context of a recently proposed model. After a brief review of the kinetic behaviour governing defect creation and annealing in undoped a-Si:H, a number of special cases will be analyzed: the influence of alloying with O, N, C, and Ge, changes introduced by doping and compensation, and the role of mechanical stress. Finally, possibilities to increase the stability of a-Si:H based devices will be examined.

INTRODUCTION

Light-induced reversible changes in the electronic and optical properties of amorphous semiconductors ("photo-darkening" of chalcolgenide glasses [1], "Staebler-Wronski effect" (SWE) in a-Si:H [2]) have attracted a considerable amount of attention in the solid state community. The reasons for this interest are widespread. From a basic physics point of view, the reversible changes are interesting because they appear to be connected with the overall metastable character of the amorphous phase. For technological applications, the negative influences of these phenomena on device stability have to be minimized without loosing the inherent advantages of amorphous solids.

In the case of hydrogenated amorphous silicon (a-Si:H), the metastable changes have been traced back to a reversible increase in the density of dangling bond defects [3,4]. The creation of these metastable dangling bonds during prolonged illumination and their annealing in the dark at elevated temperatures has been studied recently in detail, using electron spin resonance (ESR) and photoconductivity (PC) as a microscopic and macroscopic probe, respectively [5,6]. We will start by briefly reviewing the implications of these experiments for the possible defect creation and annealing processes.

The model used to explain the creation of metastable dangling bonds is shown schematically in Fig.1. Electrons and holes are created during the illumination with a generation rate G. The carriers thermalize into band tail states, where they contribute to the steady state concentrations, n and p, of localized excess electrons and holes, respectively. At room temperature, recombination of these carriers takes place predominantly through nonradiative processes involving the dangling bond states with density N_s near midgap as recombination centers. This gives rise to a monomolecular dependence of n and p on light intensity and

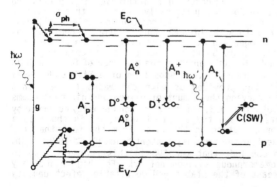

Fig. 1: Schematic model for the creation of metastable dangling bonds in a-Si:H (see text for details).

defect density:

$$n,p \propto G/N_s \tag{1}$$

New dangling bonds can be created by bond breaking following a direct nonradiative tail-to-tail recombination between an electron in a weak Si-Si antibonding state and a hole in a weak bonding state as discussed elsewhere in more detail [6]. The increase of the number of dangling bonds during illumination is then given by:

$$dN_s/dt = C\,n\,p = C'\,G^2/N_s^2 \quad . \tag{2}$$

This equation can be integrated to obtain the increase of N_s as a function of illumination time, t, and generation rate, G. The predicted behaviour is in excellent agreement with the experimental results obtained by ESR and PC [5,6].

A major conclusion from the study of the creation kinetics is that the dangling bonds already existing in a sample will quench the tail-to-tail recombination generating new defects. Thus, the SWE is self-limiting; once a critical dangling bond density has been reached, the electron and hole lifetimes have become so short that the defect inducing tail-to-tail transitions are highly improbable. Or, put differently, a significant light-induced degradation is the mark of high quality a-Si:H. If the initial photoresponse of a sample is small, the observed degradation will also be small.

Interesting information about the metastable dangling bond states created during illumination can be obtained by analyzing the thermal annealing process. We have shown recently [6] that the decay of the light-induced spin density, N_{ind}, at high temperatures follows monomolecular kinetics:

$$dN_{ind}/dt = -\nu N_{ind} \quad , \tag{3}$$

with a thermally activated decay constant

$$\nu = \nu_0 \exp(-E_A/kT) \quad . \tag{4}$$

The activation energy, E_A, reflects the energy barrier separating the metastable state from the stable ground state. In a given sample, E_A is not well defined; rather, a broad distribution of metastable states with different values for E_A exists. For a-Si:H deposited in an UHV system [7], this distribution is shown in Fig.2a. When such a sample is subjected to a large number of illumination-annealing cycles, the distribution shifts slightly to higher energies (Fig.2b).

THE INFLUENCE OF ALLOYING AND DOPING ON THE SWE

So far, we have discussed the SWE as observed in intrinsic, high purity and high quality a-Si:H. It has been noticed quite early, however, that magnitude and character of the metastable changes can be altered considerably by alloying (with N, O, C, Ge) or doping (with B, P) of amorphous silicon. Especially the question as to the role of nitrogen and oxygen in the degradation process has been the subject of some controversy. So it has been proposed repeatedly that the SWE is not intrinsic to a-Si:H, but linked to the presence of O and/or N in the material [8,9]. Recent progress in the deposition of a-Si:H under well controled UHV conditions [7] has provided us with the means to test this hypothesis by systematically varying the amount of oxygen and nitrogen atoms incorporated into a sample [10]. In Fig.3, the dangling bond density observed by ESR in the illuminated state, B, and the annealed state, A, is shown as a function of the impurity concentration in the film. From the fact that the degradation becomes independent of the impurity content below [O] $\approx 10^{20}$ cm^{-3}, [N] $\approx 10^{19}$ cm^{-3}, it is safe to conclude that the SWE in this regime is due to intrinsic properties of the amorphous silicon network. For higher impurity contents, however, the increase of the stable and metastable defect density

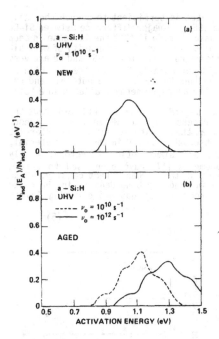

Fig.2:Normalized distribution of activation energies, E_A, for the annealing of metastable defects in a new UHV deposited sample (a), and in a sample subjected to a large number of annealing-illumination cycles (b). ν_0 is the prefactor in Eq. (4).

Fig.3:Dependence of the ESR spin density in the annealed state (A) and the light-soaked state (B) on the concentration of nitrogen (upper part) and oxygen (lower part) in a-Si:H.

points towards an enhancing effect of oxygen and nitrogen on the light-induced degradation. Concomitant with this increase we observe a change of the metastable state distribution in the annealing experiments. According to Fig.4, at high O or N concentrations the symmetric distribution obtained in UHV a-Si:H is replaced by a distribution rising continuosly towards higher activation energies. This is consistent with the widening of the optical band gap in these samples, allowing the creation of higher energy defects by tail-to-tail recombination. Moreover, the incorporation of group V and VI atoms into a tetrahedrally bonded network will increase the bonding disorder [11,12] and consequently, the density of tail states that can break during a nonradiative tail-to-tail recombination event.

Consistent with the above interpretation, alloying of a-Si:H with group IV elements, i.e. carbon and germanium, has quite different effects on the SWE. In the case of alloying with carbon a significant lowering of the energy barrier for thermal annealing of the metastable defects has been reported [13] , in contrast to the increase of E_A in O and N rich samples. In a-Si$_x$Ge$_{1-x}$:H alloys, a complete disappearance of the reversible changes for Si contents x < 0.8 is commonly observed in conductivity experiments [14]. Corresponding results obtained by monitoring the Si dangling bond density with ESR are shown in Fig.5. The main reason for the disappearance of the SWE with increasing Ge alloying is the strong increase of Ge dangling bond states. About 10^{17} cm^{-3} Ge dangling bonds are present in a Si-Ge alloy sample with a Ge content of 20% or more. As argued above, such a large density of midgap states is sufficient to quench the tail-to-tail transitions responsible for the SWE. The major difference between Si-Ge alloys and alloying with O and N as discussed before is that the increase in deep gap states is no longer balanced by a similar increase in the density of weak Si-Si bonds (tail states), since the bonding characteristics for Si and Ge are very similar [15]. A second factor that could potentially inhibit the creation of metastable defects in a-SiGe:H alloys is the decrease of the optical band gap with increasing Ge content. This lowers the average energy available in a tail-to-tail recombination event.

As in the case of the various alloys of a-Si:H, very little is known about the kinetic behaviour and the microscopic processes of the SWE in doped amorphous silicon. This is partly due to the fact that the metastable defects of interest, i.e. the dangling bonds, are no longer observable by ESR in the doped material, since the charged states, D^+ and D^-, are diamagnetic. Therefore, mostly transport data [16] or optical measurements [17] have been used to obtain information about the reversible changes of the midgap defect density. These experiments indicate that the increase in dangling bond density in the doped samples is proportional to the square root of the gas phase doping level:

$$N_{ind} \propto [PH_3]^{1/2} , [B_2H_6]^{1/2} \qquad (5)$$

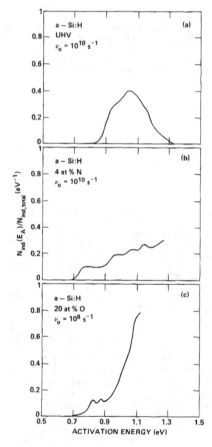

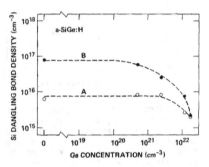

Fig.5: Density of silicon dangling bonds in a-SiGe:H-alloys as a function of the Ge concentration in the annealed (A) and illuminated state (B).

Fig.4: Distribution of activation energies for annealing of light-induced dangling bonds in high purity a-Si:H (a), a sample with 4 at% N(b), and a-Si:H with 20 at% O (c).

For high doping levels, N_{ind} reaches values around 10^{18} cm^{-3} which is about an order of magnitude higher than typical values measured in light-soaked undoped a-Si:H. Possible reasons for this increased absolute degradation are discussed in ref. 17, but no conclusive explanation exists so far. It is interesting to note, however, that the stable dangling bond density in doped a-Si:H obeys the same dependence on doping level as given by Eq.(5) [18], suggesting that the light-induced and the doping-induced increase in the midgap density of states in doped a-Si:H seem to be related in origin. A positive side effect due to the same dependence of the stable and metastable dangling bond density on the dopant concentration is that the degradation of the electronic properties of a-Si:H becomes less noticeable with increasing doping level. The relative increase of the dangling bond density following light-soaking is typically a factor of ten in undoped a-Si:H, but less than a factor of two in doped material.

A special case concerning the doping dependence of the SWE is the light-induced degradation in compensated a-Si:H, i.e. amorphous silicon doped with equal concentrations of phosphine and diborane in the discharge plasma. It has been reported first by the Berkeley group [17] that nominally compensated samples exhibit a much smaller light-induced degradation than undoped samples, although the position of the Fermi level in both cases is near midgap. Since compensated a-Si:H samples have roughly the same density of stable dangling bonds as undoped samples, this experimental result seems to be in variance with the creation kinetics as proposed in the first part of this paper. However, the inhibited degradation in compensated samples has been confirmed by the ESR measurements shown in Fig.6. This apparent inconsistency with the kinetic model leads to one of three possible schemes to reduce the SWE in a-Si:H which will be discussed in the next section.

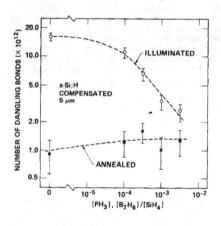

Fig.6:Influence of compensation level on the light-induced degradation of nominally compensated a-Si:H (sample size 1cm x 0.5cm x 5μm; 2 x 10^{13} spins correspond to a dangling bond density of app. 10^{17} cm^{-3}).

POSSIBLE WAYS TO MINIMIZE LIGHT-INDUCED DEGRADATION

According to the previous section, the key to the elimination of the negative consequences of light-soaking on the properties of amorphous silicon lies in the question how to minimize the degradation effects in high quality, undoped a-Si:H. Looking at Eq. (2), two possibilities are evident. The first way to minimize dN_s/dt is by trying to reduce the "creation efficiency constant", C. The second possibility is given by a reduction of the n p product, although this means simultaneously an often undesirable reduction of the photoresponse of the material. The constant C is determined by a number of terms, e.g.the densities of initial and final states for those nonradiative transitions that can result in defect creation, the recombination energy, and the electron-phonon coupling. For example, in undoped a-Si:H the initial and final states for a defect creating transition most likely are the antibonding and the bonding orbitals of a weak Si-Si bond [6]. An electronic transition between these states can be expected to have a strong electron-phonon coupling and, at the same time, minimizes the number of phonons that have to be emitted, since both states are deep tail states.

We can use this picture to explain the smaller SWE in compensated a-Si:H. To understand this, we show in Fig. 7 the occupancy of tail states under illumination as determined by light-induced spin resonance (LESR) for compensated samples with increasing doping level. In undoped a-Si:H the usual distribution is observed: holes occupy weak Si-Si bonding states in the valence band tail (vb, g=2.011), and the same density of electrons is localized in the weak antibonding states of the conduction band tail (cb g=2.0044). As the compensation level is increased, the distribution of holes remains unchanged. The density of electrons in cb-states, however, decreases continuously by more than two orders of magnitude. Instead, most of the electrons required by charge neutrality are now localized in different shallow states, which are denoted by "hf" in Fig. 7. These states can also be observed in phosphorus doped a-Si:H and are easily distinguishable from the usual cb-states because of a large ESR hyperfine splitting. Recently, we have shown that these hyperfine split states are due to electrons at fourfold coordinated phosphorus sites (neutral donors, P_4^0) [19]. As far as the magnitude of the SWE in compensated a-Si:H is concerned, the replacement of the usual Si-Si antibonding states by the P_4^0 states as the predominant shallow electron traps has two important consequences. First, the change in bonding character of the P_4^0 states most likely decreases the electron-phonon coupling, and secondly, P_4^0 sites will be spatially separated from the weak bonding states, which still are the major hole traps. Thus, the probability of nonradiative transitions between P_4^0- and vb-states will be much smaller than for transitions between cb- and vb-states. Indeed, a comparison between Figs. 6 and 7 shows quite clearly that the decrease of the light-induced degradation occurs parallel to the decrease in cb-state density seen in LESR, although the total n p product remains unchanged for all samples investigated.

The increased stability of compensated a-Si:H has interesting implications for the application of this material. Since according to Fig. 6 the degradation of a-Si:H compensated with $[PH_3]=[B_2H_6] \geq 10^{-3}$ is reduced by nearly a factor of ten without any increase in the stable defect density, devices using compensated

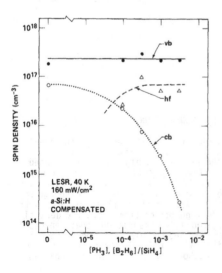

Fig.7: Spin density in compensated samples observed by low temperature light-induced electron spin resonance (LESR) as a function of compensation level. vb, cb, and hf refer to holes in the valence band tail (g=2.011), electrons in the conduction band tail (g=2.0044), and electrons in hyperfine split neutral donor states, respectively.

instead of intrinsic layers in some cases may turn out to have a superior performance in the long run. Note also that the use of compensated material will reduce the cross-contamination problem, e.g. in p-i-n diodes, and therefore, may allow simplified mass production techniques. As an additional point, it might be useful to look for impurities that form nondoping shallow states in a-Si:H, since such impurities could enhance the stability similar to compensation by quenching nonradiative tail-to-tail transitions without increasing the density of deep recombination centers.

A second possibility for improvement of a-Si:H device stability has recently been proposed by us in connection with mechanical stress measurements in undoped a-Si:H samples [20]. The existence of large compressive stresses in this material has been known for quite a while, and the magnitude of the stress can be measured quite easily using substrate curvature techniques [21]. Qualitatively, the curvature, 1/R, of a substrate is proportional to the average stress, σ, in a thin film of a-Si:H deposited on this substrate. By examining samples with different thicknesses on Corning 7059 glass substrates, it has been found that the stress is largest ($\approx$2kbar) near the substrate/sample-interface and vanishes at a distance of about 2-3 μm from this interface. The surprising result is that the number of metastable dangling bonds, N_{ind}, induced by light-soaking does not increase linearly with sample thickness, but instead with the average stress (i.e., the substrate curvature) [6]. This suggests that metastable defects are created predominantly in the stressed regions of an undoped sample, a concept easily understandable in the microscopic picture described above (i.e., recombination induced breaking of weak bonds). The correlation between substrate curvature and N_{ind} is shown in Fig.8 and is valid

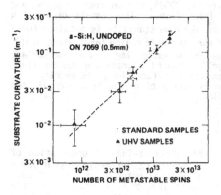

Fig.8:Correlation between the stress-induced substrate curvature and the number of light-induced metastable dangling bonds for undoped a-Si:H samples deposited on Corning 7059 glass substrates. Open symbols: samples deposited in a standard glow discharge system. Solid symbols: high purity a-Si:H obtained from an UHV system.

for all the samples examined. The important role of the mechanical stress for the SWE can be shown, also, by changing the stress for a fixed sample thickness and monitoring the maximum number of defects in the light-soaked state. This is demonstrated in Fig. 9, where the average stress in a 0.5 μm thick sample has been altered by external deformation or by deposition onto a thin Al-substrate, which will relax the stress through plastic flow already at small stress levels in the sample. Again a roughly linear dependence of N_{ind} on the average stress is observed. The implications of these results for the application of a-Si:H are quite obvious: care has to be taken not only in the optimization of the electronic properties of a-Si:H, but also of its mechanical properties. Preferentially, substrate materials and deposition conditions should be used which will reduce stress build-up at the substrate/sample interface.

The final example of how the stability of a-Si:H against light-soaking can be improved deals with the influence of band bending. It is known, e.g., from time-of-flight experiments [22], that high quality a-Si:H films show a considerable amount of band bending near the free surface and the substrate interface. It is conceivable that in the resulting space charge regions the generation rate of

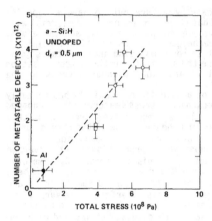

Fig.9:Dependence of the number of light-induced defects in a 0.5 μm thick sample on the estimated total stress (internal plus external). The data point denoted "Al" refers to a thin aluminum substrate, all other points to a thick 7059 glass substrate.

metastable defects can be affected by changes in the n p product and/or the Fermi level position (cf. the higher metastable defect density in doped a-Si:H). To test this idea, we have investigated the degradation of a series of thick (2 μm) intrinsic a-Si:H samples sandwiched between two n- or p-type layers. By changing the thickness of the doped layers (0-2000Å) one can control amount and sign of the band bending in the intrinsic layer. The results of the ESR measurements are summarized in Fig. 10. Optimum stability is obtained for a 750Å thick p-type layer on both sides of the intrinsic region (equivalent to 10^{15} positive surface charges per cm^2). Reducing or increasing this p-type surface layer leads quite rapidly to a largely enhanced degradation during light soaking. Moreover, the fact that a positive surface charge is necessary to reach flat band conditions implies the existence of an electron accumulation layer in the isolated intrinsic sample.

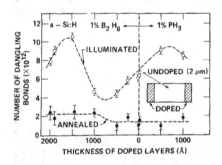

Fig.10:Light-induced degradation of a 2μm thick intrinsic a-Si:H layer sandwiched between two thin, highly doped layers as a function of the thickness of these layers. The doped layers are either both n-type (1% PH_3) or both p-type (1% B_2H_6).

In summary, we hope to have shown that the model developed to describe the metastable defect creation and annealing in high quality undoped a-Si:H can be applied at least qualitatively to a large variety of cases that are of potential interest for the commercial use of amorphous silicon. These ideas could provide a guideline for future efforts to enhance and model device stability. As examples, we have discussed the role of shallow trap states, mechanical stress, and band bending in the Staebler-Wronski effect.

ACKNOWLEDGEMENTS

This work was supported by the Solar Energy Research Institute, Golden, Colorado.

REFERENCES

[1]   J.P. de Neufville, in Optical Properties of Solids - New Developments, edited by B.O. Seraphin (North-Holland, Amsterdam, 1976)

[2]   D.L. Staebler, C.R. Wronski, Appl. Phys. Lett. 31, 292 (1977)

[3]   K. Morigaki, I. Hirabayashi, M. Nakayama, S. Nitta, K. Shimakawa, Sol. State Commun. 33, 851 (1980)

[4]   H. Dersch, J. Stuke, J. Beichler, Appl. Phys. Lett. 38, 456 (1980)

[5]   M. Stutzmann, W.B. Jackson, C.C. Tsai, Appl. Phys. Lett. 45, 1075 (1984)

[6]   M. Stutzmann, W.B. Jackson, C.C. Tsai, Phys. Rev. B, to be published

[7]   C.C. Tsai, J.C. Knights, R.A. Lujan, B. Wacker, B.L. Stafford, M.J. Thompson, J. Non-Cryst. Solids 59,60, 73 (1983)

[8]   R.S. Crandall, Phys. Rev. B24, 7457 (1981)

[9]   D.E. Carlson, J. Vac. Sci. Technol. 20, 290 (1982)

[10]   C.C. Tsai, M. Stutzmann, W.B. Jackson, AIP Conference Proceedings No. 120, p. 242 (American Institute of Physics, New York, 1984)

[11]   E. Holzenkämpfer, F.-W. Richter, J. Stuke, U. Voget-Grote, J. Non-Cryst. Solids 32, 327 (1979)

[12]   J.C. Knights, R.A. Street, G. Lucovsky, J. Non-Cryst. Solids 35,36, 279 (1980)

[13]   R.S. Crandall, D.E. Carlson, A. Catalano, H.A. Weakliem, Appl. Phys. Lett. 44, 200 (1984)

[14]   G. Nakamura, K. Sato, Y. Yukimoto, Solar Cells 9, 75 (1983)

[15]   M. Stutzmann, J. Stuke, Sol. State Commun. 47, 635 (1983)

[16]   M.H. Tanielian, N.B. Goodman, H. Fritzsche, J. Physique 42, C4-375 (1981)

[17]   A. Skumanich, N.M. Amer, W.B. Jackson, Phys. Rev. B31, 2263 (1985)

[18]   R.A. Street, Phys. Rev. Lett. 49, 1487 (1982)

[19]   M. Stutzmann, R.A. Street, to be published

[20]   M. Stutzmann, Appl. Phys. Lett., in print

[21]   D.S. Campbell, in Handbook of Thin Film Technology, edited by L.I. Maissel and R. Glang (McGraw-Hill, New York, 1970)

[22]   R.A. Street, M.J. Thompson, N.M. Johnson, Phil. Mag. B51, 1 (1985)

CAPACITANCE STUDIES OF LIGHT-INDUCED EFFECTS IN
UNDOPED HYDROGENATED AMORPHOUS SILICON

K. Zellama[+], J.D. Cohen[*], and J.P. Harbison[**]
+Present address; Universite Paris VII, 75521 Paris Cedex 05, France
*University of Oregon, Department of Physics, Eugene, Oregon 97403
**Bell Communications Research Inc., Murray Hill, New Jersey 07974

ABSTRACT

The effects of light saturation on the properties of undoped a-Si:H films were studied by a new capacitance profiling technique which can be used to directly determine changes in the dangling bond density of states near mid-gap. Coplanar conductivity and capacitance vs. temperature measurements save the changes in activation energies for electrical conductivity. These studies indicate that, while substantial increases in the dangling bond densities are observed for most samples, the detailed behavior of the light induced changes in these films are inconsistent with the creation of such defects by breaking weak valence band tail states.

INTRODUCTION

It is widely acknowledged that prolonged light exposure alters the properties of amorphous silicon (a-Si:H) films through the creation of metastable defects in the mobility gap. Numerous electron spin resonance (ESR) studies indicate that the dominant such defect is the silicon dangling bond (DB) [1,2]. However, capacitance transient measurements on n-type doped films seemed to be inconsistent with this conclusion [3]. In the present study, capacitance methods were employed to examine changes in the DB defect densities induced by light exposure in undoped films. Auxillary conductivity and ESR measurements were also carried out. While the capacitance measurements verify a marked increase in the DB density for these samples, the data do not appear consistent with a model [1,2] suggesting that these DB defects arise from the breaking of weak valence-band tail states. Furthermore, the behavior in one sample appears more consistent with the earlier capacitance transient results [3] and indicates a dominant role for a second type of metastable defect.

CHARACTERIZATION MEASUREMENTS

A series of undoped glow discharge samples were grown on p+ crystalline Si substrates with growth conditions as listed in Table 1. Samples 1-3 were co-deposited on SiO_2 substrates for the coplanar conductivity and ESR measurements. Samples were prepared in State B by exposure to Tungsten-Halogen unfiltered white light (at 1.1 W/cm^2) for 4-15 hours at room temperature or, alternatively, by exposure to 6100 Å light from a tunable flashlamp pumped dye laser (at an average power of 1.5 W/cm^2) for 2 1/2 hours. For the samples studied by capacitance the actual light intensity into the film was one tenth as large due to the semitransparent top metal barrier.

Measurements of the coplanar conductivity versus temperature for both state A and state B gave the activation energies summarized in Table 2 for samples 1-3. We also employed the capacitance-temperature slope method [4] for which the conductivity is determined in a sandwich configuration and which, because it is ac measurement, is greatly insensitive to surface barrier effects. Figure 1 shows the results of such measurements for one sample. As discussed previously [3,4], an Arrhenius plot of log ω vs. the inverse temperature of the intercepts of Fig. 1(b) gives the activation energy of the bulk conductivity. These results are also tabulated in Table 2. Agreement between the two methods is seen to be fairly good. Some of the

TABLE 1. Deposition Growth Parameters of Undoped Glow Discharge Amorphous Silicon Films. Sample 1 was grown at Tektronix and samples 2 – 5 were grown at AT&T Bell Laboratories.

Sample	SiH_4/Ar Ratio (%)	Substrate Temperature ($^{\circ}C$)	Power (mW/cm^2)	Film Thickness (microns)
1	50	255	85	1.5
2	30	260	240	1.3
3	100	260	230	1.55
4	30	250	250	1.5
5	10	250	250	2.0

differences can be attributed to the fact that if g(E) increases below E_F, this C-T analysis will yield slightly too large an activation energy.

A second important piece of information obtained from the slope of the fitted lines in Fig. 1(b) is the density of states at E_F. These values are listed in Table 3 for each film in both states A and B. Because the D_3^- defect band is very nearly a Gaussian distribution with an energy width parameter of 100-150 meV [4], the density of D_3^0 is given approximately by the product of $g(E_F)$ and this energy width. In all cases this density of D_3^0 states is estimated to be less than $3 \times 10^{15} cm^{-3}$ which is below the detectable limit for ESR spin densities for 1-2 micron thick films. Indeed, we failed to observe any clear DB spin density in our samples. This can now be attributed to the very low defect densities in these films.

TABLE 2. Activation Energy for Conductivities in State A and State B Determined by Coplanar Conductivity Measurements and the Capacitance-Temperature Slope Method.

Sample	$E_C - E_F$ (State A)		$E_C - E_F$ (State B)		
	Coplanar conductivity	C - T Slope Method	Coplanar conductivity	C - T Slope Method	
1	0.506	0.545	0.733	0.782	(Laser)
2	0.716	(?)	-----	0.782	(Laser)
3	0.801	0.82	0.60	(?)	(Laser)
			0.65	0.65	(W-Hal.)
4	-----	0.875	-----	0.871	(Laser)
5	-----	0.80	-----	0.83	(Laser)
			-----	0.748	(W-Hal.)

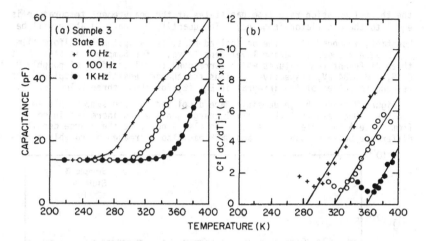

FIG. 1. (a) Variation of junction capacitance of Sample 3 with temperature at three ac frequencies as indicated.

(b) Plot of $C^2/(dC/dT)$ vs. temperature to determine the activation energy of the conductivity and the density of states at E_F. The slope of the lines fitted to these data gives $g(E_F) = 1.8 \times 10^{16}$ cm^{-3} eV^{-1}.

DRIVE LEVEL PROFILING RESULTS

We employ the "drive-level capacitance profiling" method [5] to determine the charge density in the depletion region of our samples. This method is much more easily interpretated than the standard C-V profiling method [6] for a-Si:H and is also essentially free from any influence of interface effects. In this method the capacitance at each value of applied dc bias is determined as a function of the peak-to-peak alternating voltage amplitude or drive level, δV.

In the drive level profiling method one plots the value of the quantity

$$N_{DL} = -C_o^3/(2q\varepsilon A^2 C_1) = \rho_e/q \quad (1)$$

vs. the spatial value $W = \varepsilon A/C_o$. Here C_o denotes the capacitance for infinitesimal drive amplitude and C_1 is the slope of the capacitance vs. δV dependence. The value of ρ_e is given simply by the integral over the density of states

$$\rho_e = \int_{E_c - E_e}^{E_F^o} g(E)dE \quad (2)$$

where E_F^o is the position of the neutral bulk Fermi level, E_c is the energy of conduction band mobility edge, and where E_e is the energy depth for which

TABLE 3. Densities of States at E_F Determined by the Capacitance-Temperature Slope Method.

Sample (State)	Density of States at E_F (cm^{-3} eV^{-1})
Sample 1 (A)	7.4×10^{15}
(B: Laser)	1.7×10^{16}
(B: W-Hal.)	1.05×10^{16}
Sample 2 (A)	2.6×10^{15}
(B: Laser)	1.0×10^{16}
(B: W-Hal.)	1.5×10^{16}
Sample 3 (A)	3.5×10^{15}
(B: Laser)	1.8×10^{16}
(B: W-Hal.)	7.15×10^{15}
Sample 4 (A)	4.2×10^{15}
(B: Laser)	4.6×10^{15}
(B: W-Hal.)	4.55×10^{15}
Sample 5 (A)	1.15×10^{15}
(B: Laser)	1.1×10^{15}
(B: W-Hal.)	1.2×10^{15}

the thermal emission rate from gap states at the measurement temperature is equal to the oscillator frequency. The quantity N_{DL} defined above will be constant, independent of the dc applied bias, for a spatially uniform film.

Figure 2 shows the drive level density vs. W for Sample 3 in state A at three different temperatures which correspond to values of E_e of roughly 0.77, 0.79, and 0.84 eV, respectively. Thus the charge densities displayed correspond to values of the integral in Eq. (2) for these three values of E_e.

In Figure 3 these charge densities are displayed for the same sample in state B. One sees that the overall charge density has increased in going from state A to state B; moreover, one also finds that the charge density difference between, say, 350K and 380K has also increased by roughly a fac-

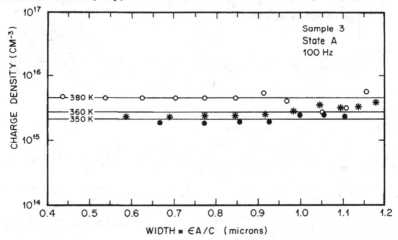

FIG. 2. Drive level profile for Sample 3 in State A at three different temperatures. The increase in the charge density value at higher temperatures is due to the contribution of gap state charge from deeper in the mobility gap.

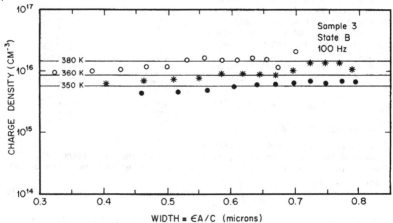

FIG. 3. Drive level profile for Sample 3 in State B at the same 3 temperatures

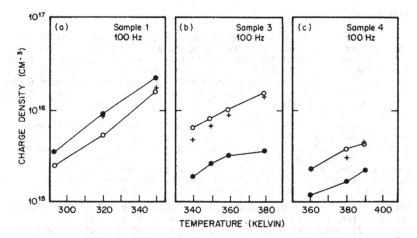

FIG. 4. Summary of drive level densities of states for three samples in both states A and B as a function of the measurement temperatures. The date symbols denote: (●) State A, (O) State B produced by exposure to 2 eV laser light, and (+) State B produced by exposure to unfiltered Tungsten-Halogen white light. Note the increase in the gap state charge density for state B for samples 3 and 4 and the decrease for sample 1.

tor of 5. This corresponds to an increase in the DB density of states by just that factor.

Figure 4 summarizes the results of such drive-level densities vs. temperature for states A and B for three of our samples. We see that Sample 1 shows a decrease in its absolute charge density and a modest decrease in the DB density of states. (Within experimental error one would probably deduce that this density of states was unchanged.) Sample 3 shows a marked increase in the DB density. Sample 4 shows a marked increase in overall charge density but a much more modest increase in the DB density (by perhaps a factor of 2). A summary of the changes in the density of dangling bond states due to light saturation for all five samples are listed in Table 4 along with the changes in the position of the dark Fermi level.

TABLE 4. Summary of Changes in Fermi Level Position and Dangling Bond Density as a Result of Long Term Light Exposure.

Sample	Dangling Bond Density		Fermi Level Position below E_c (eV)	
	State A (cm^{-3})	Factor of Change for State B	State A	Increase in Depth for State B
1	1.8×10^{16}	0.8	0.51	+ 0.2
2	2.5×10^{15}	7	0.72	+ 0.06
3	1.5×10^{15}	5	0.80	− 0.2
4	1.1×10^{15}	2	0.88	~ 0
5	1.0×10^{15}	1.5	0.80	~ 0

DISCUSSION

From the results of these measurements we can conclude, first of all, that none of the samples which show an increase in the DB density (Samples 2-5) are consistent with the model in which light exposure is thought to increase the DB density by breaking weak valence band tail states [5,6]. If this were the dominant mechanism, the Fermi level would move downward (creating more unpaired spins as observed) but the total number of doubly occupied (D_3^-) dangling bond states in the bulk of the film would not change. This follows directly from conservation of charge.

The value we obtain for N_{DL} at a particular temperature and frequency tells us the total number of occupied gap state electrons (in the neutral bulk material) down to the energy depth E_e below E_c. For the undoped films in our study, these electrons lie almost entirely in D_3^- centers. Since the total number of D_3^- centers for states A and B must be the same in the model in question, the state A and state B drive level values, N_{DL}^A and N_{DL}^B, should become equal for a value E_e near midgap. At smaller values of E_e (measurements made at lower temperatures) the value N_{DL}^A should be larger than N_{DL}^B due to the shallower Fermi level in state A (see Eq. 2), while the slope of N_{DL}^B vs. T should be greater (due to the larger DB density of states in state B). We see no indication of this kind of behavior in Fig. 4. Although the slope of N_{DL} vs. T is indeed greater for Samples 2-5, so is the absolute value of the charge density at each temperature.

Sample 1 was the only sample to exhibit essentially no change in the DB density within experimental error. Thus the motion of the Fermi level and decrease in conductivity of this sample cannot be attributed to the dangling bond defect but should instead be attributed to a light induced change in the density of some unknown defect. This behavior and conclusion are, in fact, quite consistent with earlier DLTS results on n-type doped samples [3]. Sample 3 presents a different situation where the DB density increases by a factor of 4-5 but where E_F exhibits an unusual upward motion of 0.2 eV. While this upward shift of E_F might be questioned on the basis of uncertainties in determining the Fermi level position accurately, we note that such Fermi level motion precisely preserves the total density of D_3^0 centers in this sample for the observed 5-fold increase of dangling bonds.

The remaining samples fall into neither of the two categories described above. Sample 2 shows a large increase in the DB density and only a slight motion of E_F downward. Samples 4 and 5 were significantly more stable against light induced effects since they exhibited only modest increases in the DB density and almost no motion of E_F. It is difficult at the present time to correlate these behaviors in detail with sample preparation conditions. Further studies now underway will hopefully help us determine which factors are responsible for these differences.

Work at the University of Oregon was supported in part by Solar Energy Research Institute subcontract XB-4-03101-1.

REFERENCES

1. H. Dersh, J. Stuke, and J. Beichler, Appl. Phys. Lett. 38, 456 (1981).
2. M. Stutzmann, W.B. Jackson, and C.C. Tsai, Appl. Phys. Lett. in press.
3. D.V. Lang, J.D. Cohen, J.P. Harbison, and A.M. Sergent, Appl. Phys. Lett. 40. 474 (1982).
4. D.V. Lang, J.D. Cohen, and J.P. Harbison, Phys. Rev. B25, 5285 (1982).
5. C.E. Michelson, A.V. Gelatos, and J.D. Cohen, submitted to Appl. Phys. Lett.
6. G.L. Miller, IEEE Trans. Electron. Devices ED. 19, 1103 (1972).

STABILITY OF P-I-N AMORPHOUS SILICON SOLAR CELLS
WITH BORON-DOPED AND UNDOPED I-LAYERS

ANTHONY CATALANO, RAJEEWA R. ARYA, AND RALPH C. KERNS
Solarex Corporation, Thin Film Division, Newtown, PA 18940

ABSTRACT

Boron-doping the i-layer in p-i-n amorphous silicon solar cells improves
the device performance when the density of impurities in the undoped i-layer
material is high (> $10^{20} cm^{-3}$). While this technique can boost the initial
device efficiencies for poor quality i-layer material, our devices degrade
faster than devices made with undoped, low impurity i-layer material. We
have measured the degradation of photovoltaic parameters as a function of
continuous AM1 exposure time for devices with and without B-doped i-layers.
For single junction p-i-n solar cells with comparable initial conversion
efficiencies (> 7%, area > $1 cm^2$) we find that our devices containing i-layers
deposited from gas mixtures containing 2-3 ppm diborane degrade faster than
devices containing undoped i-layers. Similar effects are observed when two-
junction stacked cells with B-doped i-layers are compared to two-junction
stacked cells with undoped i-layers.

INTRODUCTION

The purity of the i-layer material in amorphous silicon p-i-n solar cell
structures is of crucial importance for achieving high device performance.
The primary determinants of purity are pure starting materials and a leak-
tight, well out-gassed vacuum system. Achieving the necessary clean vacuum
in bell jar systems requires long pumpdown times at elevated temperatures.
This is a considerable handicap for both manufacturing and research environments.

Among the solutions to this problem, the chemical approach of passivating
the impurities in the solid material by doping the deposition gas mixture
has the major advantage of direct control over doping level. Favorable
results have been reported when the i-layer deposition mixture is lightly
doped with a few ppm of diborane in the i-layer gas mixture [1-4].

When the vacuum system is outgassed for a few hours before depositing
the a-Si:H layers, a pronounced optimum diborane concentration is observed
in the range 2-3 ppm [1]. Typical data are shown in Figure 1. This result
is encouraging since it suggests that relatively impure i-layer material,
normally capable of being fabricated into devices with efficiencies limited
to 5 to 6%, can be easily modified to achieve device efficiencies in excess
of 8%. Since the mechanism of this effect is not well understood, a
fundamental issue is raised as to whether the improvements resulting from
the low-level B-doping are stable. To address this issue, we compared the
light-induced degradation of devices with lightly B-doped relatively impure
i-layers to that of devices with undoped very pure i-layers.

Other authors have reported improved stability as well as improved
efficiency in cells with a B-doping profile in the i-layer [2-4]. They
attributed their results to boron counterdoping of phosphorus impurities [2]
and the effects of the dopant profile on the internal electric field [3,4].
Our study is qualitatively different in two respects. First, their device
structures differ from ours [2,3]. Second, they varied the boron concentration
through the i-layer [3,4]. We have used a uniform doping in the i-layer of
p-i-n solar cells in an attempt to compensate impurity states that arise
from residual gases in the vacuum system, not from unintentional phosphorus
contamination from a previously deposited n-layer. We thus hoped to produce

318

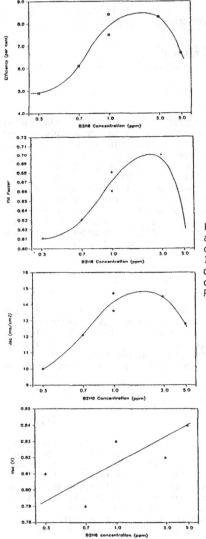

Figure 1. Solar cell device parameters as a function of the concentration of diborane in the deposition gas mixture. In these cases, the vacuum system was outgassed for 2.5 hours before depositing the a-Si:H layers. (From Reference 1.)

an improved intrinsic material with a lower density of states in the gap. In this study we focus on the stability of this lightly doped intrinsic material in solar cells.

EXPERIMENTAL

The amorphous silicon layers were deposited in a stainless steel vacuum system with a glass bell jar and Viton seals. The device structure was glass/CTO/p-i-n/metal. The substrates were soda lime glass coated with 3000 to 4000Å of low resistivity (< 20Ω/) tin oxide. The substrates were baked in situ under high vacuum at the deposition temperature (230-250°C) for either a few hours (high impurity B-doped i-layers) or overnight for 16 - 18 hours (low impurity undoped i-layers) prior to the deposition of the amorphous silicon layers. Deposition was from a DC glow discharge in undiluted silane flowing at a rate of 100sccm at a pressure of 0.5 torr. The deposition rate was 4 - 5Å/sec.

The two-junction stacked cells in this study had a glass/CTO/p-i-n-p-i-n/metal structure using the same materials as used for the single junction cells. In particular, the bandgap of the i-layer material was the same in the top and bottom cells, but the i-layer thicknesses were designed to produce identical currents in the top and bottom cells. The conversion efficiencies of these devices can be improved by low-level B-doping in the i-layers in the same way as for the single junction cells.

The solar cells that were selected for this study had fill factors of about 0.6 and initial conversion efficiencies of 5 to 6.5% for two-junction stacked cells and over 7% for single junction cells. The cells were exposed under open circuit conditions to continuous unfiltered AM1 illumination from sodium vapor lamps.

RESULTS

The normalized photovoltaic parameters for single junction solar cells are shown in Figures 2-4. From Figure 2 one can see that B-doping our i-layer is detrimental to device stability. The majority of the light-induced degradation takes place in the fill factor, as shown in Figures 3 and 4. The change in open circuit voltage is negligible.

All our two-junction stacked cells are markedly more stable than any of our single junction cells, as shown in Figures 5-7. Again, however, cells with B-doped i-layers suffer more rapid degradation than cells with purer undoped i-layers. Most of the degradation is again due to the degradation in the fill factor, while the open circuit voltage does not change appreciably.

DISCUSSION

Low level B-doping of the i-layers in p-i-n solar cells improves the initial conversion efficiencies of single junction and two-junction stacked solar cells if the i-layer material contains a relatively high concentration of impurities. Unfortunately, in our case we found that this improvement is not stable. As the cells are light-soaked, those with impure B-doped i-layers degrade much faster than those with purer undoped i-layers.

We believe our results can be understood in terms of recent work relating impurities in amorphous silicon films to solar cell degradation [5] and a new model of light-induced effects in amorphous silicon films [6]. Our working hypothesis is that, at low doping levels, boron forms impurity complexes, either with other impurities associated with silicon dangling bonds or with the dangling bonds themselves. Formation of these complexes reduces the concentration of recombination centers. Hence, the cells show an initial improved photovoltaic performance. However, the complexes must be weakly bonded so that the weak bonds are disrupted when an optically excited electron-hole pair recombines at a complex, thus creating more recombination centers. The cell performance degrades rapidly because we have increased the total concentration of impurities in the film. This last aspect of our hypothesis is supported by recent work showing that, for high purity i-layers, photovoltaic performance decreases monotonically as a function of low level B-doping [1].

All of our two-junction cells are much more stable than any of our

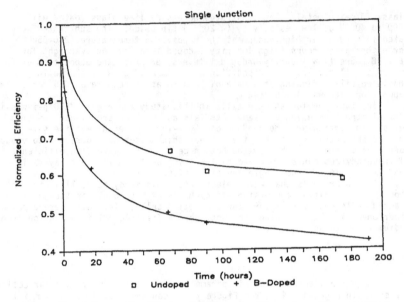

Figure 2. Solar cell efficiencies as a function of continuous AM1
exposure time for single junction cells with undoped and B-doped i-layers.

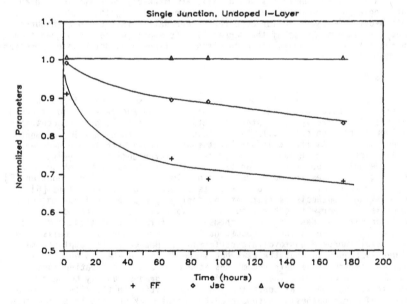

Figure 3. Solar cell parameters as a function of continuous AM1
exposure time for single junction solar cells with pure undoped i-layers.

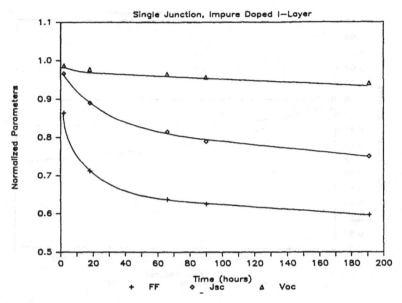

Figure 4. Solar cell parameters as a function of continuous AM1 exposure time for single junction solar cells with impure B-doped i-layers.

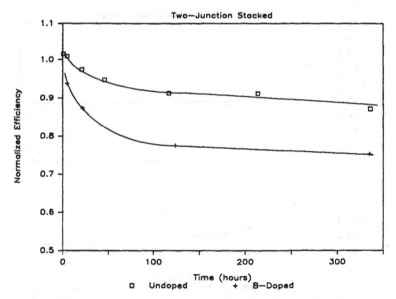

Figure 5. Solar cell efficiencies as a function of continuous AM1 exposure time for stacked two-junction solar cells with pure undoped and impure B-doped i-layers.

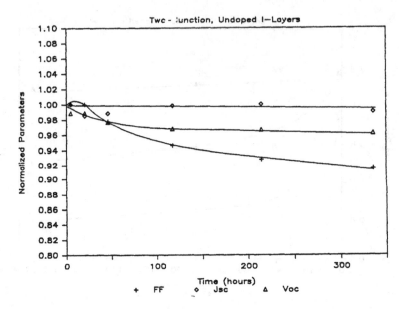

Figure 6. Solar cell parameters as a function of continuous AM1 exposure time for stacked two-junction solar cells with pure undoped i-layers.

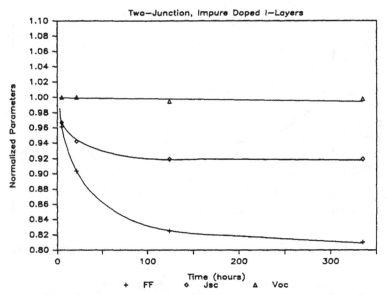

Figure 7. Solar cell parameters as a function of continuous AM1 exposure time for stacked two-junction solar cells with impure B-doped i-layers.

single junction cells in this study. This is easy to understand since each i-layer in the stacked cell is thinner than an i-layer in a single junction cell of comparable conversion efficiency but the open circuit voltage is very nearly twice that of a single junction cell. The internal fields are therefore higher in each i-layer of the stacked junction cell than they are in a normal single junction cell, so carrier collection is more efficient in stacked junction cells. It appears that stacked junction cells offer the most favorable route presently available for the simultaneous achievement of long-term device stability and high conversion efficiency in amorphous silicon solar cells.

CONCLUSION

While the conversion efficiency of p-i-n solar cells can be improved by uniform light boron doping in the i-layer introduced by adding 2-3 ppm of diborane to the i-layer deposition mixture, in the present case, such cells degrade at an accelerated rate compard to cells containing low-impurity i-layers. This is true for single junction cells and for two-junction stacked cells. Stacked cells degrade much more slowly than single junction cells.

REFERENCES

1. A. Catalano, B.W. Faughnan, and A.R. Moore, to be published.
2. H. Sakai, A. Asano, M. Nishiura, M. Kamiyama, Y. Uchida, and H. Haruki, Optical Effects in Amorphous Semiconductors, AIP Conf. Proc. 120 (AIP, New York, 1984), p. 318.
3. M. Moller, B. Rauscher, W. Kruhler, R. Plattner, and H. Pfleiderer, Proc. 16th IEEE Photovoltaic Specialists Conf. (IEEE, New York, 1982), p. 1376.
4. P. Sichanugrist, M. Konagai, and K. Takahashi, Solar Energy Mater., 11, 35 (1984).
5. M. Stutzmann, W.B. Jackson, and C.C. Tsai, in press.
6. N. Nakamura, S. Tsuda, T. Takahama, M. Nishikuni, K. Watanabe, M. Ohnishi, and Y. Kuwano, Optical Effects in Amorphous Semiconductors, AIP Conf. Proc. 120 (AIP, New York, 1984), p. 303.

ENHANCED STABILITY OF AMORPHOUS SILICON pin SOLAR CELLS BY DOPING PROFILES

M.MOELLER, H.KAUSCHE, E.GUENZEL, W.JUERGENS AND W.STETTER
Siemens Research Laboratories,
Otto-Hahn-Ring 6, D-8000 München 83, F.R.G.

ABSTRACT

pin cells with the light entering the n-layer (pinITO) or the p-layer (SnO_2p(C)in) were prepared taking into account the "basic" boron profile in the i-layer. Their efficiencies (up to 7.4% and 10.1%) show a light-induced degradation. This can be removed by doping the i-layer of pinITO cells with a decreasing boron profile. The initial efficiency is only slightly affected. Cells containing fluorine due to prior plasma-etching require higher amounts of boron for stability. For the cell type SnO_2p(C)in, phosphorus profiles, however increasing towards the n-layer, lead to reduced cell degradation without affecting the efficiency. The results of profiling were interpreted in terms of a recombination model.

INTRODUCTION

In the development of amorphous silicon pin solar cells structures were investigated with the light entering the n-layer or the p-layer. With the first type, glass/Ag/pinITO, conversion efficiencies of up to 7.4% were obtained (on $6mm^2$) /1/ and large area cells ($100 \ cm^2$) without an Ag back reflector showed an efficiency of about 5% /2/. With the second cell type, glass/SnO_2/p(C)in/Ag, a top efficiency of 10.1% was recently obtained on small specimens in our lab.
The Staebler-Wronski (SW) degradation of cells due to recombination or hole trapping /3/ in the i-layer which is influenced by dopants /1/ or impurities like oxygen or fluorine /4/ has been a subject of intense research. The objective of this study is to ascertain the effect of doping profiles on cell degradation, making use respectively of boron and phosphorus for the two types of cell and taking into account fluorine residues left from plasma-etching in the reactor.

PREPARATION OF pinITO CELLS WITH BORON PROFILES

The preparation of pinITO cells on stainless steel substrates was described elsewhere /1/. The boron concentration in the 5 nm p-layer amounts to 1%. The deposition sequence p-i-n leads to a boron profile in the i-layer which finally determines cell efficiency. The profile results from intermixing effects caused by the glow discharge plasma on the underlying p-layer as well as from boron outgassing from the internal chamber walls or the gas feeders. The shape of the profile and, consequently, the cell efficiency therefore depend on process parameters such as the rf power, intermediate flushing or the geometry of the reactor.

326

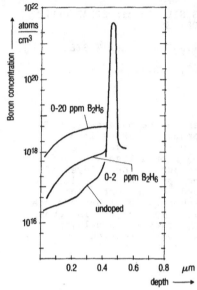

fig.1 SIMS boron profile
in pinITO cells

The pin cells were deposited in a small reactor and in a large two-chamber system with an electrode area of 50 x 50 cm², where the doped and undoped layers can be deposited in different chambers. Cleaning the large system after cell deposition is performed by plasma-etching with CF_4/O_2 or NF_3 /2/ while wet chemical etching with KOH is normally used to clean the small reactor.

In addition to preparing nominally undoped cells with the "basic" boron profile, cells were doped with a profile using a B_2H_6-concentration decreasing linearly with deposition time from the initial value at the p-i interface to zero at the i-n interface. The initial values were varied between 0 and 4 ppm in the small reactor and between 0 and 20 ppm in the large reactor.

The resulting boron profiles were measured by SIMS. For a nominally undoped cell prepared in the small reactor, the boron concentration decreases more or less exponentially with the distance from the p-layer, from about 2 x 10^{17} cm⁻³ at the p-i interface to 2 x 10^{16} cm⁻³ at the i-n interface (fig.1). This profile leads to an optimum electric field resulting in maximum cell efficiency.

The B_2H_6-concentration (of 2 ppm in the small and of 20 ppm in the large reactor) decreasing linearly with time during deposition of profiled i-layers corresponded to a boron content decreasing almost linearly with the distance from the p-layer in the SIMS profile (fig. 1). The concentration in the layer is enhanced by a factor of about three as compared with the concentration in the gas. Finally we observed that the boron content near the i-n interface increases with profiling, namely to about twice the value of a nominally undoped cell for profiling with 2 ppm. Controlling the B_2H_6 to very low concentrations is therefore inhibited by B_2H_6 remaining in the reactor.

EFFICIENCY AND STABILITY OF pinITO CELLS

The performance of pinITO cells of 6 mm² without an Ag back reflector and with various boron profiles was measured under AM1 illumination and the open-circuit voltage (V_{oc}), short-circuit current (J_{sc}), fill-factor (FF) and efficiency (η) were evaluated.

With the nominally undoped cell prepared in the small reactor the highest efficiency of 6.4% was obtained. An initial boron profile concentration of 4 ppm leads to a cell efficiency which is lower by only about 15% (fig. 2).

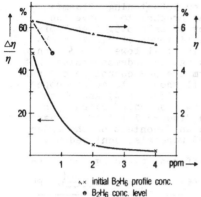

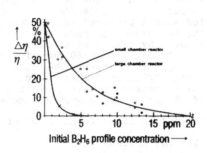

fig.2 η and $\Delta\eta/\eta$ after 16h
       AM1 for pinITO cells
       with a boron profile
       prepared in the small
       reactor

fig.3 $\Delta\eta/\eta$ after 16h AM1 for
       pinITO cells with a
       boron profile in
       different reactors

A constant level of boron throughout the i-layer of only
0.7 ppm, however, reduces cell efficiency by as much as
20%. For high efficiency a low boron content near the i-n
interface is therefore required. In practice, however, this
content increases with profiling due to diffusion effects,
as explained before. This may be the main reason for the
efficiency decrease with profiling which will be further
discussed in the last paragraph.

    After measuring the initial efficiency η , the cells
were exposed in open-circuit to AM1 illumination.The relative
degradation $\Delta\eta/\eta$, which occurs mainly within the first
16 hours, was determined after this time. As can be seen
from figs. 2 and 3 it decreases clearly with the increase
of the initial B_2H_6 profile concentration of cells prepared
in the small reactor, dropping to zero at a concentration
of 4 ppm. For cells deposited in the large system, however,
an initial profile concentration of 20 ppm is necessary
to reduce the relative degradation to zero (fig.3), which
will be discussed in the next paragraph.
    Another prerequisite for obtaining stable cells is
that the profile has to extend up to the n-layer. Profiling
only the first two thirds of the i-layer results in almost
the same relative degradation as in an unprofiled cell,
which demonstrates that the effects in the illuminated side
are mostly responsible for stability.

PLASMA-ETCHING AND CELL STABILITY

    The need to provide higher profile concentrations to
obtain stable cells in the large system probably arises
from the residual fluorine (F-)concentration due to plasma-
etching with CF_4/O_2. Therefore, cells were prepared in a

small system after either plasma- or KOH-etching. Plasma-
etching followed by a short glowing process to remove the
F-residues yields a cell efficiency of only 5.5% (see Table I).
After extended glowing, a higher efficiency of 6.28% is obtained. The fill-factor, however, is still lower by 4% than
for a cell prepared after KOH-etching. This demonstrates that
the F-residues degrade cell performance and cannot be entirely
removed by a glowing process. A cell whose i-layer was doped
intentionally with 60 ppm of fluorine (from NF_3) and which
was deposited after KOH-etching showed an only[3] slightly reduced efficiency. This indicates a "lower" limit of the harmful F-content. However, cells with an F-content on the order
of 1% and an efficiency of above 6% were also achieved/5/.

Table I. Cell performance after different cleaning processes

cleaning process	η (%)	V_{oc} (mV)	$J_{sc} (\frac{mA}{cm^2})$	FF
CF_4/O_2-etching,short glow	5.53	764	13.3	0.54
CF_4/O_2-etching,extended glow	6.28	804	13.8	0.57
KOH-etching	6.33	803	13.4	0.59
KOH-etching,60ppm F in i-layer	6.12	864	12.1	0.58

The effect of fluorine on cell stability can be characterized as follows:
Similarly as in /4/, an influence of the F-content on
the relative degradation of nominally undoped cells could
not be observed here, when comparing for example, the results
in the large and small reactors (fig. 3).
Fluorine, however, reduces the stabilizing effect of
boron profiling, since larger amounts of boron are necessary
to prepare stable cells in the large system where fluorine
is present. This may be due to the bonding of fluorine to
the incorporated boron, thereby inhibiting its interstitial
doping effect. The B-F bond is very strong, as previously
discussed in /6/.

$SnO_2 p(C)$in CELLS WITH DOPING PROFILES

Cells of the type $SnO_2 p(C)$in are deposited also in
the sequence p-i-n on glass substrates coated with tin oxide.
The silicon-carbon p(C) layer is doped with only about 0.1%
of boron, giving rise to a boron content in the i-layer
of typically $3 \times 10^{16} cm^{-3}$ which is almost constant over
the i-layer and is lower than that in pinITO cells. The
best performance recently obtained with such $SnO_2 p(C)$in
cells of 6 mm^2 was an efficiency of 10.1%, with V_{oc} = 905 mV,
J_{sc} = 17.05mA/cm and FF = 0.66.
Linear boron profiles decreasing from between 0.1 and
0.6 ppm to zero were tried, resulting in a decrease of the
cell efficiency by as much as 50%, with only 0.2 ppm as the
initial value (fig. 4). The relative degradation after 16 hours
AM1 illumination was determined to be 7% which is clearly
lower than the 18% obtained for an "unprofiled" cell.
Other cells were doped with phosphorus profiles, the
PH_3 concentration during deposition rising from zero to the
final value of 3 ppm and 6 ppm. This reduces cell efficiency
by only 5% and 10% (fig. 5). A PH_3 concentration level of

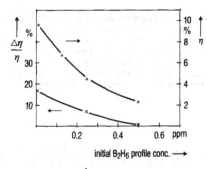

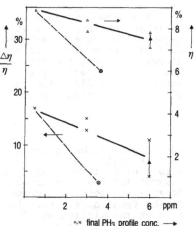

fig.4 η and $\frac{\Delta\eta}{\eta}$ after 16h AM1 for SnO$_2$p(C)in cells with a boron profile

only 3 ppm, however, reduces cell efficiency by 30%. The relative degradation for a cell with a 6 ppm final PH$_3$ -profile concentration proved to be less than 10% (fig.5) as against 18% for an unprofiled cell. This shows that increasing phosphorus profiles yield similar results with respect to efficiency and stability for this type of cell as do decreasing boron profiles for pinITO cells.

fig.5 η and $\frac{\Delta\eta}{\eta}$ after 16h AM1 for SnO$_2$p(C) in cells with a phosphorous profile

DISCUSSION

The study of boron and phosphorus profiles in pinITO and SnO$_2$p(C)in cells demonstrates that only in the case where the dopant concentration increases from the illuminated (front) side to the rear side of the cell, profiling will have no degrading effect on efficiency. The front side of the i-layer contains the highest concentration of both photo-generated electrons and holes. Doping in that part creates recombination centers causing enhanced recombination and reducing cell efficiency. In the rear part of the i-layer, either electrons or holes prevail so that recombination is hardly increased there by the creation of additional centers through doping.

Profiles, especially those increasing to the rear side, reduce cell degradation. This effect can be explained in two ways:
One is the increased density of defect centers by doping which leads to recombination via these centers. This process is alternative to band-band recombination creating the SW defects. Consequently the creation rate of light-induced defects should be limited by the pre-existing defects /7/, also if the defects are created by dopants. Additionally, the recombination occurs predominantly in the region of high center or dopant concentration, thus creating fewer SW defects in the region of low doping, i.e. the front side.

The other explanation is by an increase of the electric field in the front part of the i-layer which prevents the defect creation caused by recombination therein.Defects in that part would be especially deleterious to cell efficiency. A strong field extends from the front side into the i-layer especially in the case of a sharp rise in dopant concentration /1/ which therefore requires profiling very close to the front side.

CONCLUSION

Having used either a boron or a phosphorus profile in different pin cells, we conclude that both improve cell stability almost without reducing cell efficiency when they increase towards the rear (or dark) side. Recombination centers by dopants are more deleterious to cell efficiency if created in the front side. Profiling reduces cell degradation either through the limitation of defect creation by dopants or through the effect of a strong field in the front side preventing defect creation therein.

Fluorine incorporated in the cells as a result of plasma-etching can be tolerated in small amounts as far as cell efficiency is concerned. It requires, however, higher boron concentrations to ensure stability.

Acknowledgment
We are grateful to Mr.I. Weitzel for the SIMS measurements and to Dr. H. Pfleiderer for valuable discussions.

This work has been supported under the Technological Program of the Federal Department of Research and Technology of the FRG (sponsorship no. 03E-8430-A). The authors alone are responsible for the content.

REFERENCES

/1/ M.Möller, B.Rauscher, W.Krühler, R.Plättner, H.Pfleiderer:
Proc. of the 16th IEEE Photovoltaic Spec. Conf. in
San Diego, USA, 1982, 1376-1380
/2/ H.Kausche, M.Möller, R.Plättner:
Proc. of the 5th E.C. Photovoltaic Solar Energy Conf.
in Athens, Greece, 1983, 707-711
/3/ R.Plättner, W.Krühler, H.Pfleiderer, H.Kausche, M.Möller:
Proc. of the 1st Int. Photovoltaic Science and Engineering Conf. in Kobe, Japan, 1984, 695-698
/4/ D.E.Carlson, R.V.d'Aiello, C.R.Dickson, R.S.Oswald:
AIP Proc. of the Conf. on Optical Effects in Amorphous
Semiconductors in Snowbird, USA, 1984 No.120,234-241
/5/ A.Madan, J.McGill, W.Czubatyi, J.Yang, S.R.Ovshinsky:
Appl.Phys. Lett. 37, 826 (1980)
/6/ D.E.Carlson, R.W.Smith:
Proc. of the 15th IEEE Photovoltaic Spec. Conf. in
Orlando, USA, 1981, 694-697
/7/ M.Stutzmann, W.B.Jackson, C.C.Tsai:
Appl. Phys. Lett. 45 (10) (1984)

A CARRIER LIFETIME MODEL FOR THE OPTICAL DEGRADATION
OF AMORPHOUS SILICON SOLAR CELLS

Z E. SMITH[a] and S. WAGNER
Department of Electrical Engineering, Princeton University, Princeton, New Jersey 08544

ABSTRACT

The light-induced performance degradation of amorphous silicon solar cells is described well by a model in which the carrier lifetimes are determined by the dangling bond density. The kinetics of the defect generation follow the model in which band-to-band recombination provides the energy for the creation of dangling bonds, which in turn introduce gap states that reduce carrier lifetime. Degradation will be slower in solar cells operating at lower excess carrier concentrations. This is documented with a comparison of degradation data for cells of different i-layer thickness, cells operating at open circuit vs. load, and for single vs. cascade cells. The model also correctly predicts the relation between short circuit current and fill factor degradation. At sufficiently long times, the efficiency will decrease at approximately the same rate for all cell structures and dimensions, with an offset in time between different device types which can be calculated.

Introduction

When solar cells made of hydrogenated amorphous silicon (a-Si:H) are exposed to light, their performance characteristics degrade with time. The mechanism of this degradation is usually assumed to be associated with the thermally-reversible light-induced changes in photoconductivity of undoped a-Si:H films known as the Staebler-Wronski effect [1]. Many other materials properties change upon light exposure as well: the density of neutral dangling silicon bonds, as determined by electron spin resonance (ESR) measurements, rises [2]; the density of levels in the gap region, as determined by deep level transient spectroscopy, increases [3]; and since some of these levels have net charge in some of their occupation states, the density of fixed space charge rises as well [4]. These observations lead to the following possible mechanisms for the light-induced degradation of p-i-n a-Si:H solar cells:

(1) The increase in the density of dangling bonds results in reduced carrier lifetime, because the recombinations centers associated with the neutral dangling bond are efficient at reducing both electron and hole lifetime [5,6].

(2) The increase in other levels not associated with the dangling bond reduces carrier lifetime in a similar fashion.

(3) The increase in the density of fixed space charge distorts the electric field in the i-layer and results in a "dead" (low field) region in the center of the cell. In this region, flow is diffusion-limited instead of drift-limited as in the ideal, uniform field case [7].

(4) The increase in the density of deep levels promotes recombination at the p-i or i-n interface [8].

Postulate (4), interface recombination, seems at best a minor effect, since interface recombination would be expected to affect the reverse saturation current and hence reduce open circuit voltage (V_{oc}), an effect rarely observed. Recent numerical simulation of p-i-n devices by Sakata and Hayashi [9] also casts doubt on the importance of this degradation mechanism. They showed·that increasing the interface recombination rates reduces short circuit current density (J_{sc}) significantly but leaves fill factor (FF) unchanged; most experimental data show FF dropping significantly and J_{sc} affected less severely. Furthermore, the experimental studies of Guanghua et al. on the p-i interface [10] also minimize the importance of this effect. Postulate (2), deep levels not associated with dangling bonds, cannot be categorically ruled out; but the work of Street identifying the dominant recombination

a) AT&T Bell Laboratories Scholar

center in the light-soaked state with the dangling bond [6] would appear to minimize the importance of other levels. Postulate (3), field deformation due to increased space charge, has probably been the most widely accepted mechanism; but as the electric field profile in some high efficiency ($\eta > 10\%$) solar cells has been measured to be highly non-uniform while that in some lower efficiency cells is comparatively uniform [11], the waters here are muddied.

Instead, we will consider a model in which the dominant effect of light-soaking is an increase in the density of dangling silicon bonds; the recombination center associated with the neutral dangling bond reduces both electron and hole lifetime. Using simple, analytical formulae, we relate the increase in dangling bond density N_s to the commonly measured cell performance parameters FF and J_{sc}. By assuming that the generation of dangling bonds follows the kinetic model recently proposed by Stutzmann et al. [12,13], we derive expressions for FF and J_{sc} as a function of exposure time. Some approximations must be made along the way to avoid recourse to computer simulation, but we feel that this approach best illuminates the physics at work.

Degradation kinetics

Although over two hundred articles dealing with the Staebler-Wronski effect have been published since it was first described in 1977, a clear picture of the mechanism behind it has yet to emerge. Many theories have been proposed to explain the effect; most fall into one of two categories [14]:

(1) A fixed number of metastable defect sites are present at all times; light-soaking changes the occupation of these levels [15].

(2) Some fraction of all Si-Si bonds are sufficiently strained that a recombination event in their neighborhood can break them. If hydrogen is nearby, some reconfiguration of bonds is possible which prevents the Si-Si bond from reforming. Thus the number of defects changes with time [16].

Stutzmann et al. [12,13], proceeding on the second premise, proposed a model in which band-to-band (or bandtail-to-bandtail) recombination events provide the energy for bond breaking. If n and p are the free electron and hole concentrations, r_t the rate constant for (nonradiative) band-to-band transitions, and c_{sw} the fraction of these transitions which result in the creation of a Staebler-Wronski defect, then the time rate of change of dangling bond density is given by

$$\frac{dN_s}{dt} = c_{sw} r_t np \qquad (1)$$

where it is assumed that light soaking is proceeding at a sufficiently low temperature that annealing effects can be ignored. In an illuminated photoconductor in open circuit condition, $n = G\tau_n$ and $p = G\tau_p$, where G is the volume carrier generation rate, equal to the photon absorption rate. If the dominant carrier recombination center is associated with the neutral dangling bond [5,6] then carrier lifetimes are given by $\tau_n = 1/(r_n N_s)$ and $\tau_p = 1/(r_p N_s)$. Whether it is the dangling bond itself or a three-center bond required to stabilize the metastable defect [17] which actually acts as the recombination center does not affect this argument. Then Eq. (1) becomes

$$\frac{dN_s}{dt} = \frac{c_{sw} r_t}{r_n r_p} \frac{G^2}{N_s^2} \qquad (2)$$

which integrates to

$$N_s^3(t) - N_s^3(0) = \alpha G^2 t \qquad (3)$$

where $\alpha = (3c_{sw} r_t)/(r_n r_p)$. From the data presented in Fig. 1 of reference 12, we calculate $\alpha = 2100 \pm 200$ cm^{6}s^1. At sufficiently long times,

$$N_s(t) \cong \alpha^{1/3} G^{2/3} t^{1/3} \tag{4}$$

We have fit the photoconductivity data of Guha [18] and some of the ESR data of Lee et al. [19] to Eqs (4) and (5) as well. The quality of the fit gives us confidence for applying this kinetic model to a-Si:H films in general.

Collection length model

Having established a form for $N_s(t)$ which is derived using $\tau_n = 1/(r_n N_s(t))$ and $\tau_p = 1/(r_p N_s(t))$, we now seek to relate cell performance parameters to carrier lifetimes. We do so using the collection length model of Faughnan and Crandall [20].

In most solar cell structures, it is minority carrier transport which determines device behavior. Since undoped a-Si:H is usually weakly n-type, this has led some to assume that hole transport limits performance. In a p-i-n device structure, however, both electron and hole transport are important, because the assignation of majority and minority carrier changes over the i-layer [21]. This observation is used in the regional approximation [22] to simplify the relevant transport equations which are then joined at the crossover point (where the assignation of majority/minority carrier reverses). Crandall [23] finds that for a p-i-n solar cell with uniform electric field and uniformly absorbed light, the collected primary current is

$$J = e G l_c [1 - \exp(-d/l_c)] \tag{5}$$

which can also be expressed as the fraction f of the absorbed photocurrent $J_p = eGd$,

$$f = \frac{J}{J_p} = \frac{l_c}{d}[1 - \exp(-d/l_c)] \tag{6}$$

where d is the thickness of the i-layer and l_c is the carrier collection length; Crandall finds $l_c = l_n + l_p$, with the carrier drift lengths given by $l_n = \mu_n \tau_n E$, $l_p = \mu_p \tau_p E$. Under the uniform field assumption $E = (V_{bi} + V_a)/d$, where V_{bi} and V_a are the built-in and applied voltages. Thus scanning a range of applied voltages produces a normalized J-V characteristic, with the shape of the curve characterized by the parameter $l_{c0} = (\mu_n \tau_n + \mu_p \tau_p) V_{bi}/d$. Although the regional approximation has been extended to include non-uniform light absorption [23] and electric field [24], the fit of Eq. (6) to the measured J-V curve of most solar cells is so good that we retain this simple form. Eq. (6) is identical in form to the expression derived by Hecht in 1932 for collection of a single carrier type in alkali halides [25], with $l_c = \mu \tau V_a/d$, and so it has become common to refer to Eq. (6) as the Hecht expression, even though it was derived under different assumptions.

It should be noted that Eq.(6) is not, strictly speaking, a J-V relation but rather a small-signal response function; $f = 0$ when $V_a = -V_{bi}$, while $J = 0$ when $V_a = -V_{oc}$. By making the assumption that V_{bi} and V_{oc} are linearly related over some range [26], Faughnan and Crandall take the "fill factor" of Eq.(6) as a function of l_{c0}/d and compare the result with experiment. Their model and data are shown in Fig. 1. While the fit to the model is quite good, data are available only over the mid-region of the modelled curve. For our purposes, we simply fit the data to

$$FF\left[\frac{l_{c0}}{d}\right] = FF_1 + k_1 \log_{10}\left[\frac{l_{c0}}{d}\right] \tag{7}$$

with $FF_1 = 0.39$ and $k_1 = 0.30$.

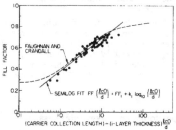

Figure 1: Fill Factor vs. Carrier Collection Length for p-i-n a-Si:H solar cells, after ref. 20, with semilog fit to data superimposed.

We note in passing that the identification $l_c = l_n + l_p$ is not universally accepted. The argument given is usually along the lines that in the limiting case in which one strongly dopes the i-layer, l_c should stay long because majority carrier lifetime would not be reduced by the doping; yet experimentally such cells do not work. The error in this reasoning is that while majority carrier lifetime might not be reduced, the resulting field distortion would greatly reduce the electric field over most of the i-layer, and hence, the majority carrier drift length. For those unconvinced, we stress that our use of the collection length concept only requires that l_c be some function of carrier lifetimes inversely proportional to dangling bond density. Thus our hypothesis for the form for (l_{c0}/d) to be used in Eq. (7) is

$$\frac{l_{c0}}{d} = \frac{\overline{\mu}}{\overline{\tau}}\frac{V_{bi}}{d^2}\frac{1}{N_s}$$
(8)

where $\overline{\mu}/\overline{\tau}$ is to be determined. The Crandall model predicts

$$\frac{\overline{\mu}}{\overline{\tau}} = \frac{\mu_n}{\tau_n} + \frac{\mu_p}{\tau_p}$$
(9)

Substituting Eqs. (8) and (4) into Eq. (7) yields an expression for the expected behavior of fill factor with exposure time and carrier generation rate for a p-i-n a-Si:H solar cell light-soaked in open circuit condition. This expression is valid once the light-generated spin density is large compared to the initial value:

$$FF(t) = FF_1 + k_1 \log_{10}\left[\frac{\overline{\mu} V_{bi}}{\overline{\tau}d^2\alpha^{1/3}G^{2/3}t_0^{1/3}}\right] + k_1\log_{10}\left[\frac{t^{-1/3}}{t_0^{-1/3}}\right]$$

$$= FF_2 - \frac{k_1}{3}\log_{10}\left[\frac{t}{t_0}\right]$$
(10)

where t_0 is any time sufficiently large that our requirement that $N_s(t_0) \gg N_s(0)$ is met. Figure 2 shows a plot of FF vs. t for a typical solar cell [27]. The fit of the data to Eq. (10) gives $FF_2 = 0.68$ and $k_1 = 0.25$; the agreement in the value of slope is within our uncertainty in choosing k_1 from Fig. 1. Plots of $FF(t)$ data recently published by Tawada [8] and Uchida [28] imply similar values for k_1.

The value of FF_2 implies $\overline{\mu}/\overline{\tau}$ $= 6\times10^{-8}\mathrm{V}^{-1}\mathrm{cm}^{-1}$. The measurements of Street [5,6] on undoped a-Si:H films yield μ_n/τ_n $= 4\times10^{-8}\mathrm{V}^{-1}\mathrm{cm}^{-1}$ and μ_p/τ_p $= 1\times10^{-8}\mathrm{V}^{-1}\mathrm{cm}^{-1}$; these values are consistent with the value for $\overline{\mu}/\overline{\tau}$ predicted by Eq. (9).

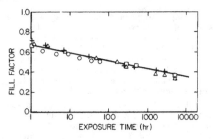

Figure 2: Degradation of Fill Factor with light exposure.

Dependence on i-layer thickness

Motivated by the space-charge model for cell degradation, some researchers have explored using thinner i-layers in solar cells to reduce the width of the low-field "dead" region. Results on cells of various i-layer thickness [27] are shown in Fig. 3. Stability is improved for thinner cells, but only in the sense of forestalling the time at which cell performance begins to plummet. This effect is predicted by the carrier lifetime model. Retaining the initial spin density term in Eq. 3,

$$\frac{l_{c0}}{d} = \left[\frac{\overline{\mu}}{\overline{\tau}}\frac{V_{bi}}{d^2}\right]\left[\alpha G^2 t + N_s^3(0)\right]^{-1/3} \quad (11)$$

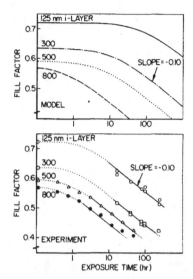

We substitute this form into Eq. (7). The "MODEL" curves in Figure 3 result. The value of G is adjusted according to the J_{sc} and thickness of each cell. The value of $N_s(0)$ was estimated from the initial FF using Eqs. (7) and (8); for example, $N_s(0) \cong 2\times10^{16} \mathrm{cm}^{-3}$ for the 800nm cell. The agreement between modelled and experimental slopes of $FF(t)$ is excellent; both find $k_1/3=0.10$, or $k_1=0.30$. These cells were grown on rough substrates to maximize optical absorption [29], and so the FF for the thinnest cell is reduced by shunting effects; since we have chosen for simplicity not to include these effects, we calculate an initial spin density value $N_s(0)$ for this cell which is artificially high (above $2\times10^{17}\mathrm{cm}^{-3}$); hence the turning point in $FF(t)$, the time at which light-generated defects become significant compared with $N_s(0)$, is somewhat overestimated by the model for very thin cells.

Figure 3: Effect of i-layer thickness on FF degradation.

Open circuit vs. load degradation

When solar cells are light soaked in conditions other than open circuit, degradation has been observed by many groups to proceed more slowly [8,30,31]. The effect of cell bias during light soaking enters the carrier lifetime model in the following way. Since the product of n and p appearing in Eq. (1) is lower when current is allowed to flow than in open circuit, the rate at which dangling bonds are generated will be reduced. Numerical modeling of p-i-n solar cells by Swartz [32] indicate that the np product is reduced by a factor of 10 to 100 at the maximum power load point compared with open circuit. The ratio $np(\mathrm{oc})/np(\mathrm{load})$ is spatially dependent, but for illustrative purposes let us define an effective value for this ratio to be $R_{oc/l}$. Then Eq. (11) becomes

$$\frac{l_{c0}}{d} = \left[\frac{\overline{\mu}}{\overline{\tau}}\frac{V_{bi}}{d^2}\right]\left[\frac{\alpha G^2 t}{R_{oc/l}} + N_s^3(0)\right]^{-1/3} \quad (12)$$

The effect of moving the operating point from open circuit towards short circuit is shown in Fig. 4.

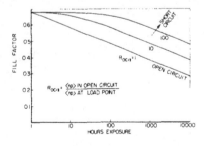

Figure 4: Effect of open circuit vs. load point operation on FF degradation.

Short circuit current degradation

By definition, l_{c0} is the collection length at zero applied bias, $i.e.$, short circuit. Thus Eqs. (5) and (6) imply

$$\frac{J_{sc}}{J_p} = \frac{l_{c0}}{d}[1-\exp(-d/l_{c0})] \quad (13)$$

The behavior of Eq. (13) is illustrated in Fig. 5(a), along with that of the FF predicted by Eq. (6) for comparison. The computer simulations of Sakata and Hayashi [9] of the effect of

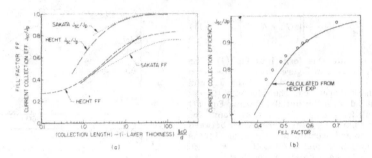

Figure 5: (a) Dependence of current collection efficiency (J_{sc}/J_p) and fill factor (FF) on l_{c0}/d, for various device models; (b) J_{sc}/J_p and FF plotted parametrically, Hecht model and experimental data.

varying lifetime (they assume $\tau = \tau_n = \tau_p$) are plotted as well, where we have used $l_{c0} = (\mu_n \tau_n + \mu_p \tau_p) V_{bi}/d$. The trends are the same. The predicted correlation of J_{sc}/J_p with FF is plotted in Fig. 5 (b), along with experimental data for a typical cell.

Once the initial value of FF is used to find l_{c0}/d, Eq. (13) can then be used with the initial J_{sc} to find a value for J_p, and then a more accurate estimate of the carrier generation rate, $G = J_p/(ed)$, can be made.

Single vs. cascade structure cells

Solar cells of the cascade or tandem structure are attractive because they allow better matching to the solar spectrum, and hence at least the possibility of higher efficiencies. Recently, a three-layer cascade cell was reported by Nakamura *et al.* [33] to have the added advantage of improved stability. The structure is shown in Figure 6(a). The initial perform-

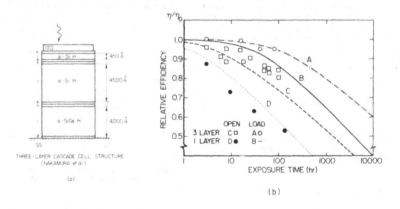

Figure 6: (a) Structure of the cascade cell structure of ref. 33; (b) relative efficiency degradation for the cascade cell and a single i-layer control cell. $R_{oc/i} = 20$.

ance characteristics of the cell were $J_{sc} = 6.4$ mA/cm^2, $V_{oc} = 2.2$V, and $FF = 0.62$. From the carrier lifetime model, the reasons for the improved stability are

(1) The top cell is so thin that l_{c0}/d stays high even for very short carrier lifetimes, and

(2) The middle cell, while approximately as thick as a normal 1-layer cell, has less than half the average carrier generation rate G, as.reflected in the lower J_{sc}, because of the spectrum-splitting of the cascade structure.

The bottom cell, with an a-SiGe:H i-layer, is assumed to be stable because the defect density is already so high in a-SiGe:H as presently produced [34]. In Figure 6(b) we compare the degradation of efficiency for single and three-layer cells predicted by our model and as measured by Nakamura *et al.*. After calculating the degradation of the top and middle cells independently, the top cell degradation is observed to be negligible, and so the "model" curves in Fig. 6(b) for the 3-layer cell are actually just the characteristic of the middle cell. Agreement between model and experiment is good so far as experimental data are available; unfortunately, no published data are available past 100 hours of light-soaking for cascade structure cells.

We emphasize that given the dimensions of the cell and and the initial FF (to find $N_s(0)$) and J_{sc} (to estimate G), there are no adjustable parameters in our model when light soaking is carried out under open circuit condition. The model introduces one adjustable parameter, $R_{sc/i}$, when operation at the maximum power point or short circuit is used.

Field deformation and boron profiling

Field deformation will become important, experimentally speaking, when the observed bias-dependent photoresponse curve no longer fits Eq. 6. For cells with good initial lifetimes, this appears to occur only when the cell performance is already so degraded [27] that it is of little economic interest.

On the other hand, if the undoped i-layer quality is such that initial lifetimes are low, then tailoring the electric field in the i-layer through boron profiling [28,35] may improve stability. Unfortunately the stability data in refs. 28 and 35 are reported only out to 4 hours exposure time, and as Figs. 3 and 6 suggest, one cannot extrapolate from short times with confidence. Moreover, the initial efficiencies are only 7-8%; recent work [36] indicates that introduction of boron into high-lifetime i-layer material involves a trade-off between loss in collection length due to boron reducing the electron lifetime [37] and gain in collection length due to improved field distribution.

Conclusions

We have shown that the degradation characteristics of p-i-n a-Si:H solar cells are well-characterized by a model in which carrier lifetime, not field deformation, is the dominant decay mechanism. Modifying cell structure can forestall performance degradation, but not eliminate it, given present materials. The model as presented here does not include annealing effects. In real-world environments, where cells may reach operating temperatures above 50 °C, the competition between generation and annealing. of defects will result in a saturation of defect density; the concomitant saturation of degradation in cell performance has already been reported [38]. In a future publication we will report the incorporation of annealing effects into the carrier lifetime model.

The amorphous silicon program at Princeton University is supported by the Electric Power Research Institute under contract RP 1193-6.

[1] D.L. Staebler and C.R. Wronski, Appl. Phys. Lett. **31**, 292 (1977).

[2] H. Dersch, J. Stuke and J. Beichler, Appl. Phys. Lett. **38**, 456 (1980).

[3] J.D. Cohen, D.V. Lang, J.P. Harbison and A.M. Sergent, Solar Cells **9**, 119 (1983).

[4] R.S. Crandall and D.L. Staebler, Solar Cells **9**, 63 (1983).

[5] R.A. Street, Appl. Phys. Lett. **41**, 1060 (1982).

[6] R.A. Street, Appl. Phys. Lett. **42**, 507 (1983).

[7] D.E. Carlson, A.R. Moore, D.J. Szostak, B Goldstein, R.W. Smith, P.J. Zanzucchi and W.R. Frenchu, Solar Cells **9**, 19 (1983).

[8] Y. Tawada, K. Nishimura, S. Nonomura, H. Okamato and Y. Hamakawa, Solar Cells **9**, 53 (1983).

[9] I. Sakata and Y. Hayashi, IEEE Trans. Electron Dev. **ED-32**, 551 (1985).

338

[10] C. Guanghua, Z. Fangqing and X. Xixiang, *Technical Digest of the International PVSEC-1* (Kobe, Japan, 1984) p. 421.

[11] D.J. Szostak and B. Goldstein, J. Appl. Phys. **56**, 522 (1984).

[12] M. Stutzmann, W.B. Jackson and C.C. Tsai, Appl. Phys. Lett. **45**, 1075 (1984).

[13] M. Stutzmann, W.B. Jackson and C.C. Tsai, in *Opt. Eff. in Amorph. Semicond. (AIP Conf. Proc. **120**, Snowbird, Utah, 1984)* p. 213.

[14] H. Fritzsche, in *Opt. Eff. in Amorph. Semicond. (AIP Conf. Proc. **120**, Snowbird, Utah, 1984)* p. 478.

[15] D. Adler, Solar Cells **9**, 133 (1983); D. Adler, J. Phys. (Paris), Colloq. C4 **42**, 3 (1981).

[16] J.I. Pankove and J.E. Berkeyheiser, Appl. Phys. Lett. **37**, 705 (1980).

[17] M.E. Eberhart, K.H. Johnson and D. Adler, Phys. Rev. B **26**, 3138 (1982).

[18] S. Guha, Appl. Phys. Lett. **45**, 569 (1984).

[19] Fig. 4 of C. Lee, W.D. Olsen, P.C. Taylor, H.S. Ullal and G.P. Ceasar, in *Opt. Eff. in Amorph. Semicond. (AIP Conf. Proc. **120**, Snowbird, Utah, 1984)* p. 205.

[20] B. Faughnan and R. Crandall, Appl. Phys. Lett. **44**, 537 (1984).

[21] S. Nonomura, H. Okamoto, H. Kida and Y. Hamakawa, Jap. J. Appl. Phys. **21**, Suppl. 21-2, 279 (1982).

[22] M.A. Lampert and R.B. Schilling, in *Semiconductors and Semimetals* **6**, edited by A. Beer (Academic Press, New York, 1970) p.1.

[23] R.S. Crandall, J. Appl. Phys. **54**, 7176 (1983).

[24] R.S. Crandall, J. Appl. Phys. **55**, 4418 (1984).

[25] K. Hecht, Z. Phys. **77**, 235 (1932).

[26] *Theory:* M. Hack, M. Shur, W. Czubatyj, J. Yang and J. McGill, J. Non-Cryst. Solids **59&60**, 1115 (1983); *Experiment:* H. Schade, Z E. Smith and A. Catalano, Solar Energy Mat. **10**, 317 (1984).

[27] B.W. Faughnan, RCA Laboratories, private communication.

[28] Y. Uchida, M. Nishiura, H. Sakai and H. Haruki, Solar Cells **9**, 3 (1983).

[29] H. Schade and Z E. Smith, J. Appl. Phys. **57**, 568 (1985).

[30] R. Williams, Inner Mongolia Electronics **2**, 1 (1984) [in Chinese].

[31] S. Tsuda, N. Nakamura, K. Watanabe, T. Takahama, H. Nishiwaki, M. Ohnishi and Y. Kuwano, Solar Cells **9**, 25 (1983).

[32] G.A. Swartz, J. Appl. Phys. **53**, 712 (1982).

[33] G. Nakamura, K. Sato, T. Ishihara, M. Usui, K. Okaniwa and Y. Yukimoto, J. Non-Cryst. Solids **59&60**, 1111 (1983).

[34] G. Nakamura, K. Sato and Y. Yukimoto, Solar Cells **9**, 75 (1983).

[35] M. Konagai, T. Matsushita, P. Sichanugrist, K. Takahashi and K. Komori, *Proc. of the 17th IEEE Photovoltaic Spec. Conf., FL* (IEEE Press, New York, 1984) p. 347.

[36] A. Catalano, B.W. Faughnan and A.R. Moore, *Technical Digest of the International PVSEC-1, Late News Section* (Kobe, Japan, 1984) p. 3.

[37] R.A. Street, J. Zesch and M.J. Thompson, Appl. Phys. Lett. **43**, 672 (1983).

[38] H.S. Ullal, D.L. Morel, D.R. Willet, D. Kanani, P.C. Taylor and C. Lee, *Proc. of the 17th IEEE Photovoltaic Spec. Conf., Kissimmee, FL* (IEEE Press, New York, 1984) p. 359.

PART XI

Thin-Film Transistors
and Switching Devices

AMORPHOUS SILICON THIN FILM TRANSISTORS AND MEMORY DEVICES

P.G. LECOMBER
Carnegie Laboratory of Physics, University of Dundee,
Dundee DD1 4HN, Scotland, U.K.

ABSTRACT

The preparation of amorphous silicon with a low density of defect states by the glow discharge decomposition of silane and the ability to control its electrical conductivity over many orders of magnitude by the addition of phosphine or diborane to the silane, stimulated a worldwide interest in this material and in its possible applications. This paper begins with a description of the preparation technique and a brief review of some of the important properties of the material. The fabrication and characteristics of a-Si thin-film field effect transistors will be described and followed by a discussion of the applications of these devices in large area liquid crystal displays, in simple logic circuits and in addressable image sensors. Finally, the use of a-Si in memory devices will be briefly described.

INTRODUCTION

During the last decade the use of amorphous silicon (a-Si) films in over twenty different applications has been proposed by groups working throughout the world [1,2]. At the present time five of these proposals have led to products that form part of a consumer item that can be purchased in at least one country. The origins of this rapidly growing industry can be traced back to two important fundamental developments during the 1970s. The first of these was the discovery [3-5] that a-Si films prepared by plasma-enhanced chemical vapour deposition (PECVD), also called the glow discharge process, possess a very low density of localised states in the mobility gap. This property is probably the single most important factor in the subsequent development of the material since it led directly to the development of the a-Si thin-film field effect transistor and also is an essential material requirement for the second development, namely that of controlling the conductivity of the a-Si over a very large range by substitutional doping from the gas phase [6,7]. This important breakthrough occurred in 1975 and was soon followed by the report that a-Si p-n junctions could be prepared by this process [8]. Preliminary measurements of the photoresponse of these devices showed the typical characteristics of photovoltaic cells. Shortly afterwards, Carlson and Wronski [9] reported more detailed photovoltaic characteristics of p-i-n structures. These discoveries were followed by a worldwide interest in the material and to its rapid application in photovoltaic cells for consumer products, in electrophotography, etc. [1,2].

This paper is concerned more specifically with the use of a-Si in two applications, namely in thin-film field effect transistors (FETs) and in memory devices. Although neither of these developments have presently resulted in a commercial product, both have considerable potential. In fact, a number of companies have produced demonstration prototypes of liquid crystal displays employing arrays of a-Si FETs and it is possible that these will reach the market place in the near future.

In the following the preparation of a-Si and a-Si FETs will first be described. The material requirements of FETs, the performance of these devices and their applications will then be discussed. Finally, the recent published work on a-Si memories will be reviewed.

Mat. Res. Soc. Symp. Proc. Vol. 49. ᶜ 1985 Materials Research Society

FILM PREPARATION

a-Si

Plasma techniques are not new: as long ago as 1857 Siemens [10] described a plasma method for the preparation of ozone. However, it was not until the 1960s that Sterling and his collaborators [11,12] applied the technique to the production of thin films of a-Si and a-Ge from silane and germane gases respectively. A schematic diagram of the preparation unit developed in our laboratories is shown in Fig. 1. Doping of the material can be achieved by mixing small but accurately determined amounts of phosphine (PH_3) or diborane (B_2H_6) to the silane (SiH_4) to produce n- and p-type layers respectively [7]. The gas flow rate is measured by the mass flow controllers F and decomposition takes place in the rf glow discharge between plates A and B of the chamber. The substrates S are heated during deposition to about 300°C. The pressure in the chamber is a few tenths of a torr and the rf power usually very small, typically 5 to 10 Watts for a 30 cm diameter system. Films prepared in this way are extremely hard and possess a remarkably low density of localised gap states [3-5, 13-15]. The films generally contain a few atomic percent of hydrogen, mainly in the form of Si-H bonds in samples deposited at 300°C. Their room temperature conductivity can be controlled over 10 orders of magnitude, from about $10^{-12}(\Omega cm)^{-1}$ to over $10^{-2}(\Omega cm)^{-1}$ with either n- or p-type doping [6,7,13].

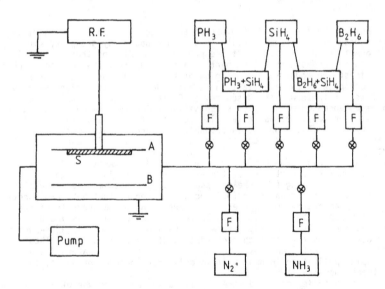

Fig. 1. Schematic diagram of the preparation unit
developed by the Dundee group (after ref.7)

a-Si FETs

The field effect experiments carried out in our laboratory since 1972 [3-5] demonstrated the remarkably low overall density of states of glow discharge amorphous silicon (a-Si) and also formed the basis for the subsequent development of the a-Si field effect transistor. The original aim of this work was to explore the possibility of using an array of thin film FETs in addressable liquid crystal (LC) displays [16-21] although attempts have also been made to explore their use in other applications [18,22-25].

a-Si FETs have been prepared in a number of configurations [26]. The structure most commonly used is shown in the insert to Fig. 2 and this is generally referred to as the "inverted-staggered" configuration. The inverted form of this, the normal staggered, is also used but less frequently [26] and recently a novel vertical structure has been proposed [27]. Double-gate structures have also been produced [28].

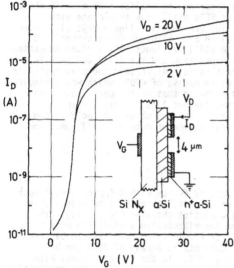

Fig. 2. Transfer characteristics of a-Si FET element [19].

The FET shown in Fig. 2 is prepared as follows. A metal electrode, typically 30μm wide, is evaporated onto a glass substrate to form the gate electrode. A thin insulating layer, typically about 0.3μm of amorphous silicon nitride (SiN$_x$), is then deposited by the PECVD technique from a mixture of SiH$_4$ and NH$_3$ to form the gate dielectric. An a-Si layer, also about 0.3μm thick, is next deposited onto the SiN$_x$ again by the PECVD technique. Both these depositions take place at about 300°C so that the process is compatible with low cost glass substrates. The final step consists of depositing the required pattern of source and drain contacts on to the a-Si surface. In devices made in other laboratories, alternative materials such as PECVD SiO$_x$ or CVD SiO$_x$ sometimes form the gate insulator. These and other factors associated with the a-Si FET fabrication are discussed in more detail in reference [26].

PERFORMANCE OF a-Si FETs

Material Requirements

In order to be of practical use the a-Si FETs have to satisfy a number of criteria:- (a) low OFF current; (b) high ON/OFF current ratio; (c) transition from OFF to ON at gate voltages V$_G$ compatible with modern integrated circuits, i.e. at V$_G$ ≤ 15V; (d) long term stability under operating conditions; (e) reproducible properties both within a given deposition and from a series of depositions; and (f) ability to produce arrays of devices, preferably over large areas. Of these, (a) - (c) are determined by the density of localised states in the a-Si and also by the number of interface states at the a-Si/gate dielectric insulator. The reproducibility of the FETs properties are governed by the control that can be achieved of the PECVD process. The long term stability of the devices is determined by the a-Si material itself, by the interface and by the gate dielectric. Finally, the need to produce (large area) arrays means that photolithographic techniques must be applicable and this is fortunately the case for the a-Si FETs [17].

We shall discuss the extent to which a-Si FETs meet the other requirements in the following but it is important to realise that even if these are all in the affirmative the a-Si FETs still have to compete with a number of other materials for use in, for example, liquid crystal displays. These are CdSe, Te and polycrystalline Si. Although the argument as to which of these is the most suitable still continues, it is often suggested that CdSe has problems related to deposition and stability and that Te does not have a sufficient ON/OFF current ratio. At present the polycrystalline Si technology involves temperatures in excess of 600°C which would require a costly quartz substrate. It is believed that of the many industrial companies throughout the world investigating liquid crystal display technology, the greater majority have converged on a-Si or polysilicon and it is therefore likely that the final choice of material will come from these.

dc Characteristics

The dc performance of an elementary device with a 4μm channel length and a 500μm channel width, produced by photolithographic techniques, is illustrated by the transfer characteristics shown in Fig. 2 [19]. The source-drain current I_D is plotted logarithmically against the gate voltage V_G for drain potentials V_D of 2V, 10V, and 20V. It can be seen that with +15V on the gate, drain currents in excess of 10μA can be achieved for drain voltages as low as 10V. In the OFF-condition, with $V_G = 0$, the current through the device drops below 10^{-11}A. The remarkable rise in I_D is caused by an electron accumulation layer formed at the Si/SiN_x interface, which produces an efficient current path between the source and drain electrodes.

As a measure of the performance of these a-Si FETs it is useful to use the field effect mobility μ_{fe} (see, for example, references [19,20 and 26]). This can be calculated from the gradient of the curve obtained by plotting $\sqrt{I_D}$ vs V_D as shown in Fig. 3 for one device [19]. At room temperature the data lead to $\mu_{fe} = 0.3$cm^{2}V^{-1}s^{-1} and the intercept gives a value for the threshold voltage of $V_T = 4$V. Powell [26] has collected together the data in the literature and concludes that although there are often significant differences in a-Si FET characteristics from different laboratories, most show the same general features and similar values of μ_{fe}. The best FETs are obtained in the inverted staggered configuration and with SiN_x as the gate dielectric [18,26]. Devices made with SiO_x dielectrics generally have higher threshold voltages than those utilising SiN_x [26].

Stability and Reproducibility

An important factor in any development of a-Si FETs will be the reproducibility and uniformity of the characteristics of arrays of elements.

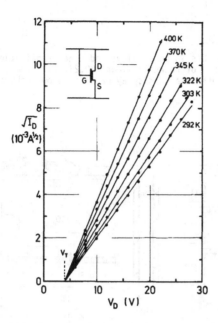

Fig. 3. Square root of source-drain current I_D plotted against source-drain voltage V_D, with gate and drain connected together, for various temperatures [19].

The evidence available at present looks encouraging and would not indicate that this should be a limiting factor [17-19]. As a check on the stability of the devices, the output from an a-Si FET was monitored in the period from September 1980 to September 1983, corresponding to about 8×10^9 switching operations [19]. During three years of continuous operation the peak voltage level changed by only 10 - 15%. This is encouraging, especially as this device was totally unpassivated and unencapsulated.

A more detailed discussion of the factors limiting the stability of the a-Si FETs has recently been given by Powell [26] who points out that the top passivation layer can also play a role.

The most extensive information on the uniformity of a-Si FETs is provided by the work of Powell, Stroomer and Chapman [29]. They investigated several thousand FETs and concluded that the small differences that were observed in the characteristics were mainly a result of variations in the threshold voltage of the order of 1V. These are unlikely to be a limiting factor in the applications described below.

Irradiation experiments on a-Si FETs with γ-ray doses of up to 5 Mrad (Si) have recently been reported [30]. Even at these levels, changes of less than 3V in threshold voltage and less than 10% in transconductance were observed. It was therefore concluded that the a-Si FETs are remarkably resistant to the effect of radiation, despite the relatively thick dielectric layer used at present in these devices.

APPLICATIONS OF a-Si FETs

Liquid Crystal Displays

The use of a-Si FETs in LC displays and other applications has been the subject of a number of recent reviews [20,21,26]. The general layout of a number of elements in a LC display panel is shown schematically in Fig. 4.

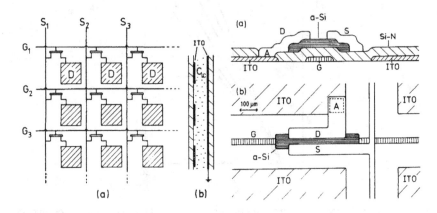

Fig. 4. Schematic layout of an addressable liquid crystal panel: (a) plan view, (b) side view. [17].

Fig. 5. Design of a-Si FET: (a) section through device, (b) FET in part of the matrix array [17].

A transistor is incorporated in a corner of each element of the array. The FETs are interconnected by means of X and Y buses, G_1, G_2 ... and S_1, S_2 ... linking gate and source contacts, respectively. The drain contact of each transistor is connected to the ITO squares, D. From the section through the panel in Fig. 4b it can be seen that the liquid crystal material is sandwiched between the substrate carrying the FETs and an ITO coated glass top plate which is normally returned to ground. The liquid crystal element is therefore in series with the drain circuit and behaves electrically as a capacitor C_{LC} with some leakage resistance. Fig. 5a shows a section through an individual a-Si device and Fig. 5b illustrates the design of the FET in part of the matrix array.

To simulate the operation of the FETs in an addressable array, the response of single elements to typical pulse voltages on the gate and source buses was investigated [17-19]. In these experiments the drain was connected to ground through a 10 pF capacitor to simulate the liquid crystal capacity C_{LC} of the 1 mm ITO elements (see Fig. 4). An important parameter, as far as multiplexing is concerned, is the rise time of the potential across the liquid crystal element. Even with a 10μm channel length a LC drive potential is reached in about 10μs. With a frame time of 25ms it is therefore possible with the present devices to scan more than 1000 lines of display in this time.

It has been estimated [31] that more than twenty independent electronics companies are actively developing a-Si FETs for LC displays and many of these have produced working prototypes. Of these the most impressive are the 5 inch colour TV display with 240 x 222 picture elements (240 x 666 FETs)

demonstrated by Sanyo [32], and a 7 inch colour graphics display with
325 x 325 picture elements produced by Hosiden [33]. The potential market
for large displays is enormous and the final choice of technology depends
upon many factors. However, in view of these impressive developments,
it would appear at the present stage that commercial production of LC
displays incorporating a-Si FETs is a real possibility in the not too
distant future.

Logic Circuits and Image Sensors

Matsumura, Hayama and co-workers at the Tokyo Institute of Technology
first showed that an integrated inverter circuit [22] and an image sensor
[23] could be made from a-Si FETs. Their work demonstrated the feasibility
of the application, but the comparatively poor characteristics of their
early FETs reflected adversely on the device performance. We have
published [18,24] the design of an integrated a-Si inverter consisting
of an a-Si FET in series with an a-Si load resistor. The output logic
of this circuit was clearly demonstrated: the output potential changing
from 14.5V to 2V as the input increased from about 5V to 15V. By
extending the above circuit, we have produced logic circuits such as NAND
and NOR gates, bistable multivibrators and also a shift register [18,24].
The latter is of particular interest as an integrated drive circuit for the
liquid crystal panels.

The factor that appears most likely to limit the use of a-Si FETs in
logic circuits is their frequency response. This problem will generally
be determined by the time required for the ON-current to charge circuit
capacitances. The FETs described above can charge a capacitance of 1 pF
to a few volts in under 1μs so that operation at frequencies of the order
of MHz would be feasible in certain applications. The maximum operating
frequency of an a-Si FET can be estimated from the "figure of merit" f_m,
given by the gain-bandwidth product [34]

$$f_m \approx \frac{\mu_{fe} V_G}{2L^2}$$

where L denotes the length between source and drain contacts. In large
area arrays it is unlikely that L will be less than a few μm so that with
V_G = 15V and $\mu_{fe} \sim 1cm^2 V^{-1} s^{-1}$ we find $f_m \sim 100MHz$. If this figure could
be approached in practical a-Si FETs then it is likely that these devices
could be used in commercial logic circuits.

The high photoconductivity of a-Si also makes it possible to integrate
light-sensitive elements into the circuitry. As an example, the performance
of addressable image sensing elements has been investigated [18,24,25]
and is shown in Fig. 6. These experiments demonstrated that the output
current of an integrated elementary device varied from about 10μA at a
white light intensity of 10^{17} photons $s^{-1} cm^{-2}$ to about 10nA at 10^{13}
photons $s^{-1} cm^{-2}$ [25]. This large dynamic range, providing an excellent
grey scale, could be achieved with read-times of about 20μs, suggesting
that in principle less than 100msec would be required to read the information from a 1000 line array. Further development work on this application
is being carried out at a number of laboratories throughout the world.

348

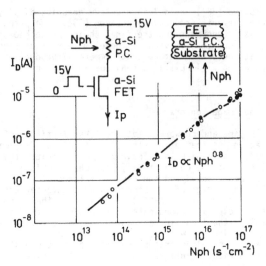

Fig. 6. Output current I_D from an elemental integrated image sensor
as a function of the incident photon flux N_{ph}. [25].

a-Si MEMORY JUNCTIONS

In this section we should like to discuss briefly a new a-Si memory
device [35-37] developed jointly by the Edinburgh and Dundee groups. It is
an electrically programmable, non-volatile memory element which in terms of
speed, retention time, operating voltages and stability compares very
favourably with crystalline MNOS (metal-nitride-oxide-semiconductor) or
FAMOS (floating-gate-avalanche-metal-oxide-semiconductor) devices currently
used for non-volatile programmable storage.

The most promising configuration investigated so far is of the p-n-i
type. After an initial forming step the structure operates as a non-volatile
memory. Immediately after forming the device is in its ON-state with a
resistance of a few hundred ohms. On increasing the reverse potential
(i.e. a negative voltage applied to the p-doped region) a reverse threshold
voltage V_{ThR} is reached beyond which the device switches to an OFF-state
with a resistance of the order of $1M\Omega$. The reverse threshold $V_{ThR} \approx 1V$
and the OFF-state is stable for voltage swings of $\pm 4V$. If the forward
potential is now increased beyond a value of V_{ThF}, the forward threshold
voltage, the device switches back into its high conductivity state, in some
cases through an intermediate state.

The above cycle has been repeated up to 10^5 times without observable
changes in characteristics or threshold voltages. Devices set in ON- and
OFF-states have been monitored for several months without detectable change
in either characteristic. Operation at temperatures of up to 180°C shows
little change in the threshold voltages.

The dynamic characteristics of the p-n-i devices has been investigated
with 10V forward and reverse pulses of 100ns duration. The experiments
demonstrate the important result that both the ON-OFF and OFF-ON transitions
are completed within 100ns and that there appears to be no observable time
delay in the switching response. It is estimated that the energy absorbed
during either transition is extremely low, typically in the range 10^{-6} to
10^{-8}J.

Although the mechanisms underlying the switching phenomena described above are unknown at present, we believe on the basis of the present experimental evidence it is likely that the ON-state of the a-Si memory devices involves the formation of a current filament [35].

CONCLUDING REMARKS

The last decade has been an extremely exciting and rewarding experience for those of us involved in the fundamental and applied work on a-Si. In that time the field has developed from a purely academic one to large scale commercial production of devices for consumer products. These applications have been made possible by the remarkable electronic properties of glow discharge Si, which had been clearly recognised in our fundamental work during the early 1970s.
Although a-Si FETs are only one solution to matrix addressing of LC displays, the impressive developments in recent years have made it possible that commercial displays based on this technology will be available in the future. The use of a-Si FETs in simple logic circuits may be limited by their speed of operation but this is unlikely to be a restriction in image sensing arrays. With the increasing number of laboratories working on a-Si and a-Si FETs it is likely that even further applications will arise in the future.

ACKNOWLEDGEMENTS

Our interest in a-Si FETs initially arose out of the work of W.E. Spear and myself on the density of states in a-Si, and from discussions with C. Hilsum and A.J. Hughes at RSRE. The subsequent development of these FETs with Walter Spear was carried out with a number of colleagues, particularly A.J. Snell and S. Kinmond, whose collaboration is gratefully acknowledged.

REFERENCES

[1] See, for example, Semiconductors and Semimetals, ed. by Willardson and Beer, Vol. 21, Part D, Academic Press (1984) and references cited therein.

[2] J. Non-Crystal Solids, 59 and 60 (1983) and references cited therein.

[3] W.E. Spear and P.G. LeComber, J. Non-Crystal Solids, 8-10, 727 (1972).

[4] A. Madan, P.G. LeComber and W.E. Spear, J. Non-Crystal Solids, 20, 239 (1976).

[5] A. Madan and P.G. LeComber, Proc. 7th Intern. Conf. on Amorphous and Liquid Semiconductors, Edinburgh, ed. by W.E. Spear (CICL, Univ. of Edinburgh, 1977), p. 377

[6] W.E. Spear and P.G. LeComber, Solid State Commun. 17, 1193 (1975).

[7] W.E. Spear and P.G. LeComber, Phil. Mag. 33, 935 (1976).

[8] W.E. Spear, P.G. LeComber, S. Kinmond and M.H. Brodsky, Appl. Phys. Lett. 28, 105 (1976).

[9] D.E. Carlson and C.R. Wronski, Appl. Phys. Lett. 28, 671 (1976).

[10] W. von Siemens, Pogg. Ann. 102, 120 (1857).

[11] H.F. Sterling, R.C.G. Swann, Solid State Electron. 8, 653, (1965).

[12] R.C. Chittick, J.H. Alexander and H.F. Sterling, J. Electronchem. Soc. 116, 77 (1969).

[13] P.G. LeComber and W.E. Spear, Topics in Appl. Physics, 36, Chpt 9, 251 (1979).

[14] W.E. Spear and P.G. LeComber, Topics in Appl. Physics, 55, 63 (1984).

[15] P.G. LeComber, Fundamental Physics of Amorphous Semiconductors, ed. by F. Yonezawa (Springer, 1981) p. 46.

[16] P.G. LeComber, W.E. Spear and A. Ghaith, Electron. Lett. 15, 179 (1979).

[17] A.J. Snell, K.D. Mackenzie, W.E. Spear, P.G. LeComber and A.J. Hughes, Appl. Phys. 24, 357 (1981).

[18] P.G. LeComber, A.J. Snell, K.D. Mackenzie and W.E. Spear, J. de Physique, 42, Supp. C4, 423 (1981).

[19] K.D. Mackenzie, A.J. Snell, I. French, P.G. LeComber and W.E. Spear, Appl. Phys. A31, 87 (1983).

[20] P.G. LeComber and W.E. Spear, Chp 6 of Semiconductors and Semimetals, 21D, ed. by Willardson and Beer, p 89 (Academic Press 1984).

[21] D.G. Ast, Chp. 7 of Semiconductors and Semimetals, ed. by Willardson and Beer, 21D, 115-138. (Academic Press 1984).

[22] M. Matsumura and H. Hayama, Proc. IEEE, 68, 1349 (1980).

[23] M. Matsumura, H. Hayama, Y. Nara and K. Ishibashi, IEEE, EDL-1, 182, (1980).

[24] A.J. Snell, W.E. Spear, P.G. LeComber and K.D. Mackenzie, Appl. Phys. A26, 83 (1981).

[25] A.J. Snell, A. Doghmane, P.G. LeComber and W.E. Spear, Appl. Phys. A34, 175 (1984).

[26] See, for example, M.J. Powell, MRS Symposium Proc. Vol 33 "Comparison of Thin Film Transistor and SOI Technologies", ed. by H.Y. Lam and M.J. Thompson, p. 258 (1984) and references cited therein.

[27] Y. Uchida, Y. Nara and M. Matsumura, IEEE Electron Device Letters, EDL-5, 105 (1984).

[28] H.C. Tuan, M.J. Thompson, N.M. Johnson and R.A. Lujan, IEEE Trans. Electron Device Letters, EDL-3, 357 (1982).

[29] M.J. Powell, M.V.C. Stroomer and J.A. Chapman, Proceedings of the SID Japan Display Conference, 1983 (at press).

[30] I.D. French, A.J. Snell, P.G. LeComber and J.H. Stephen, Appl. Phys. A31, 19 (1983).

[31] M.J. Powell, Proceedings of Fourth International Display Research Conference (EURODISPLAY), Paris (1984), p 131-136.

[32] M. Yamano, H. Takesada, M. Yamazaki, Y. Okita and H. Hado, Consumer Electronics Conference, Chicago (1984).

[33] Y. Ugai, Y. Murakami, J. Tamamura and S. Aoki, SID Symposium Digest, 308 (1984).

[34] S.M. Sze, Physics of Semiconductor Devices, p. 573, Wiley (Interscience), New York (1969).

[35] A.E. Owen, P.G. LeComber, G. Sarrabayrouse and W.E. Spear, IEE Proc., 129, 51 (1982)

[36] A.E. Owen, P.G. LeComber, W.E. Spear and J. Hajto, J. Non-Crystal Solids, 59 and 60, 1273 (1983).

[37] P.G. LeComber, A.E. Owen, W.E. Spear and J. Hajto, Chp. 15 of Semiconductors and Semimetals, 21D, ed. by Willardson and Beer, p. 275 (Academic Press 1984).

SHORT CHANNEL AMORPHOUS SILICON MOS STRUCTURES
WITH REDUCED CAPACITANCE

Z. YANIV, V. CANNELLA, G. HANSELL, and M. VIJAN
Ovonic Display Systems, Inc., 1896 Barrett Street, Troy, Michigan 48084

ABSTRACT

We report improvements in device structures by the reduction of
capacitance in short channel length thin film transistors of amorphous
silicon alloy materials. Employing techniques similar to those previously
reported [1,2], these MOS structures are fabricated with channel lengths
of 1 to 2 micrometers using standard photolithography with 10 micrometer
minimum feature size. Significant reductions in capacitance over earlier
reported device designs were achieved by improvements in device geometry
and innovative use of shadowing techniques utilizing oblique angle
deposition to minimize overlap between electrodes. Theses reduced
capacitance short channel length TFTs enhance the possibility of
fabricating on-board drivers for active matrix liquid crystal displays
using amorphous silicon alloy devices. Despite the relatively low
mobility of amorphous silicon (~ 1 cm^2/V-sec) these short channel length
TFTs can provide currents large enough for operation in the megahertz
regime when these reductions in capacitance are incorporated. The
noncritical photolithography assures that devices may be fabricated over
large area substrates (8" x 8") with acceptable yields. Computer
simulations predict that these TFTs will be able to provide the necessary
speed for on-substrate drivers. We will present experimental results from
the new TFT structures and describe modeling methods and results for
amorphous silicon TFT ring oscillators. We will discuss the significance
of these results as they pertain to drive circuitry for large area liquid
crystal displays.

INTRODUCTION

The use of a-Si alloys thin film transistors to serve as pixel
switching elements in active matrix addressed liquid crystal displays is
very attractive. The technology exists to produce uniform high quality
a-Si alloy materials over large areas very efficiently and economically
[3]. Even though the highest mobility achieved [2] in a-Si TFTs is ~ 1
cm^2/V-sec, the currents and speed available with 10-20μm channel lengths
is more than adequate for switching LCD pixels, and the low off-current
levels are ideal for holding pixel charge [4].
Nonetheless, if one is interested in large area, high density active
matrix LCDs, it is hoped that the TFTs will be used not only as the
switches at each pixel element, but also as the on-substrate drivers. The
incorporation of drivers directly onto the substrate with the display
elements will greatly reduce the external circuitry required to drive the
display in the TV mode and therefore will reduce the cost dramatically.
An example of the type of on-substrate circuitry required for a LCD is
shown in Figure 1. If one is driving a 640-column by 400-row display in
the fast update TV mode, then 400 rows must be addressed every 33.3 msec.
This implies that one has approximately 83 μsec to address an individual
row in the display. If the scheme in Figure 1 is utilized, then one must
be able to shift 640 bits of information into the shift register in the
allotted 83 μsec/row. This implies in a propagation delay of 0.13
μsec/bit. If one assumes that there are three transistor propagation
delays per film flop in the shift register and latch, then the transistor

must have approximately 23 MHz cut-off frequency.

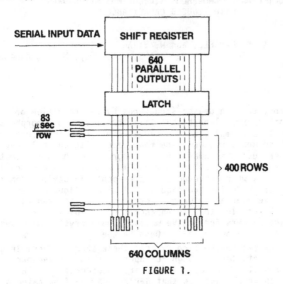

FIGURE 1.

An example of the type of on-substrate circuitry required to drive
a LCD in the TV mode. The shift register and latch can be built
and operated at sufficient speeds if one uses the a-Si TFTs
discussed in this paper.

This frequency level could be achieved with 10µm source-to-drain
spacing TFTs only if the mobility was in the range of 50-100 cm^2/V-sec and
if the interelectrode capacitance of the device was minimized.
Unfortunately, this high mobility is not achievable in a-Si materials at
present. There has been, however, progress in a-Si growth processes and
gate dielectric deposition techniques which have resulted in improved
mobility values. Field effect mobility values ranging from 0.1 to 0.3
cm^2/V-sec are typically found [5,6,7] and a value as high as 1.9 cm^2/V-sec
has been reported [8]. It is believed that the field effect mobilities
being achieved today are close to the best that are possible in a-Si
materials.

Consequently, research to improve field effect mobilities is unlikely
to provide major gains in TFT frequency response. Improvements in TFT
frequency response can, however, be quite dramatic if one reduces the gate
length in a device because increases in frequency proportional to the
inverse square of the gate length are possible. Thus, a reduction of gate
length from 10µm to 1µm can result in a factor of 100 increase in TFT
frequency of operation provided the transconductance increases and the
gate capacitance decreases as the gate length is reduced. For field
effect mobilities of approximately 1 cm^2/V-sec one must reduce the gate
length to ~ 1µm in order to obtain an a-Si TFT which can be operated at a
23 MHz frequency. In this paper, we will discuss a TFT configuration with
~ 1µm channel length which shows promise of approaching the cut-off
frequency required to do on-display drivers.

We have previously reported a successful 1µm channel length TFT
utilizing a vertical device structure to define the channel length [1].
This device had a threshold voltage V_{th} = 1.9V, a field effect mobility

of 1 cm^2/V-sec, and showed a dramatic increase in current over a 10 μm channel device. Moreover, this device was fabricated using 10μm feature size photolithography to preserve compatibility with large area processing. However, this device configuration had electrode overlap capacitances which were not optimized and which required reduction to achieve the necessary switching speeds.

MODEL

To determine the device parameters and capacitances which are compatible with driving circuits and ≳ 2 MHz eleven stage ring oscillators, an analysis of TFT-based ring oscillator has been conducted. The approach taken was to analyze a single-stage inverter with a capacitive load. Then a computer simulation was performed where many inverter stages were cascaded together to arrive at an estimate of the delay between successive stages in a ring oscillator [9]. The delay between alternate stages, 2Δ, represents one rise time and one fall time in a ring oscillator. Therefore, the frequency of operation may be calculated from

$$f = 1/n2\Delta$$

where n is the number of stages in the ring oscillator.

The analysis was based upon an inverter circuit in which the gate of the load transistor is directly coupled to its drain electrode. The analysis of a ring-oscillator in a small signal model predicts the node capacitance between two inverter stages is [10]

$$C = C_{gs}^L + C_{ds}^L + C_{ds}^D + C_{gs}^D + 2(1+A)C_{gd}^D$$

Where C_{gs}^L is the gate-source capacitance of the load, C_{ds}^L and C_{ds}^D are the drain-source capacitances of the load and driver respectively, C_{gs}^D is the gate-source capacitance of the driver, and $2C(1+A)C_{gd}^D$ is twice the Miller capacitance, $C_M = (1+A)C_{gd}^D$, between the gate and drain of the driver. A is the magnitude of the gain in the inverter stage. The Miller capacitance is counted twice since the gate-drain capacitance transitions are from a state where "0" is on the gate and a "1" is on the drain. Thus, the charge changes on the gate-drain capacitor by $Q \approx 2C_M V_{dd}$, where V_{dd} is the supply voltage.

A MOSFET model was used for the a-Si TFT and a computer simulation of numerous cascaded inverters was carried out where the input to the first inverter was a step function voltage, and the output of inverter i became the input to inverter i+1, and so on for a structure of m stages. After approximately ten inversions, every other waveform produced by inversion looked essentially, the same, but delayed in time. If n is the number of stages, then the output voltage $V_n(t) = V_{n+2}(t-2\Delta)$ where 2Δ is a constant time delay, $2\Delta = t_{rise} + t_{fall}$. Complete details of this analysis may be found in Reference 2. We have applied this analysis to determine the device parameters which are compatible with display driving circuits. A typical output from this simulation is presented in Figure 2. The inverter for the simulation in Figure 2 had a driver channel width of 250μm and ratio of driver to load widths of 3. The TFTs had channel lengths of 1μm, with mobility $\mu = 1 cm^2$/V-sec and $V_{th} = 2v$. The supply voltage used was 20 V. The node capacitance calculated was 0.2 pf. From this simulation, we conclude that a 2.5 MHz theoretical frequency in an eleven stage ring oscillator is possible for TFTs with $\mu=1$ cm^2/V-sec and

V_{th} = 2V, provided the total node capacitance is reduced to less than .2pf.

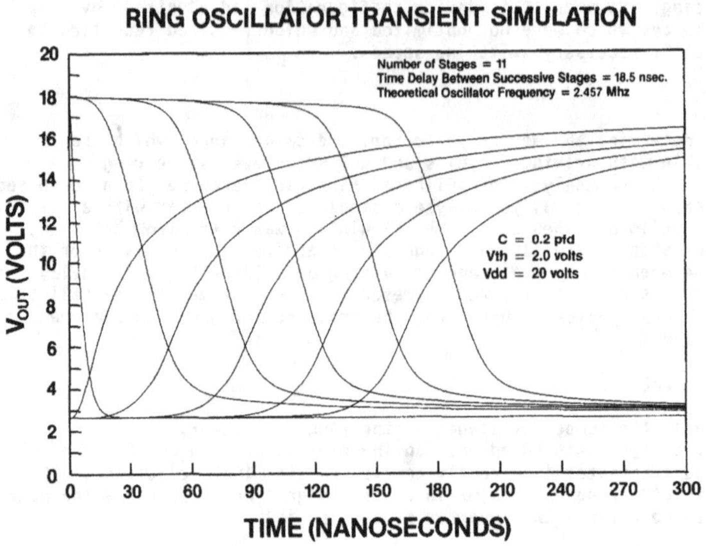

RING OSCILLATOR TRANSIENT SIMULATION

FIGURE 2.

Transient simulation of a ring oscillator composed of a-Si
transistors as described in the text.
The time response curve for each successive stage is shown.

EXPERIMENTAL DEVICE STRUCTURES

Figure 3a shows our original vertical channel TFT structure reported
in Reference 1. For this structure, the node capacitance from Eq.2 was
.45 pf. Examination of this structure suggested the need to reduce the
capacitance coupling between the gate and the upper source/drain
electrode. Our first attempt to reduce the interelectrode capacitance of
our vertical structure TFTs utilized an added dielectric layer between the
amorphous silicon layers and the upper of the source/drain electrodes as
shown in Fig. 3b. While this provided a 35% reduction in the node
capacitance, the new value of .33 pf was still too high to provide the
necessary frequency response. To reduce the capacitance further, we found
it necessary to go to a completely different arrangement of the device
electrodes as is shown in Figure 3c. In this device, it is the gate metal
which provides the vertical structure and the source and drain electrodes
are deposited in a single step by means of an angled or shadowed
evaporation. For this structure C_{sd}^D and C_{sd}^L became negligible. To
achieve this effect we used an electronic-beam evaporation which was
incident up on the structure at an angle indicated by the arrow in Figure
3c. The slight curvature of the gate metal provided by an isotropic etch
was helpful in achieving the correct structure.
The amorphous-silicon alloys in these devices were 0.3μm thick and
were deposited by a glow discharge decomposition of SiH_4 and SiF_4.

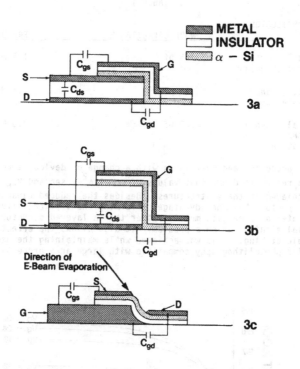

FIGURE 3. Various vertical channel TFT structures where S,
D, G label the source, drain, and gate respectively.
(a) Initial vertical TFT structure reported in Reference 1.
(b) Lower capacitance vertical channel TFT structure with additional
insulator layer. (c) Optimized low capacitance vertical channel TFT
structure using shadow evaporation of source and drain contacts.

The various gate insulators were also .3μm thick and were deposited by
glow discharge decomposition of SiH_4, N_2O, and/or NH_3 at substrate
temperatures below 400°C. Glass substrates were used, and the metals used
for the gate, source, and drain were Cr and MoTa. The vertical structure
allowed short channel lengths to be formed without using critical 1-μm
photolithography. Table I provides the total capacitances calculated for
each of the three structures according to our model, and the frequency of
an eleven stage ring oscillator utilizing each device as obtained from our
computer analysis. The use of this final structure with the shadowed
evaporation technique allows us to reduce the total device capacitance to
about .20 pf. Such a device can achieve a delay time per stage Δ = 18
nsec and a ring oscillator frequency of ≥ 2MHz.
    We have succeeded using this shadow evaporation technique to fabricate
a 2μm channel length device with this reduced capacitance structure. The
I vs V curves for this device is shown in Figure 4. The field effect
mobility achieved for this process was lower (.01 cm^2/V-sec) than that in
Reference 1 and V_{th} was higher (10V). This reflects the need to

TABLE 1

11 STAGE OSCILLATOR

DEVICE	NODE CAPACITANCE	FREQUENCY
Original Vertical Channel	.45 pf	1.1 MHz
Vertical Channel with Added Oxide	.33 pf	1.5 MHz
Vertical Channel with Shadow Evaporation	.20 pf	2.5 MHz

reoptimize processes and interfaces with a change of device structure, but we find no reason to doubt that values of $\mu = 1cm^2/V\text{-sec}$ and $V_{th}=2V$ can also be achieved in these structures. Besides the reduced capacitance, this new structure has the advantage of forming the vertical structure in a single material (the gate metal) rather than a layered structure where differential etching can be a problem. It also allows the effective use of isotropic etching. This was achieved while maintaining the use of 10μm feature size photolithography compatible with large area processing.

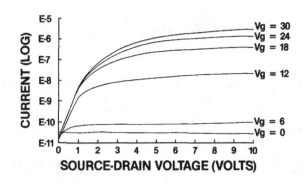

FIGURE 4. Current-Voltage characteristics for a 2 μm vertical channel length TFT using shadow evaporation of source and drain contacts.

CONCLUSIONS

We have designed and fabricated a 1 μm channel length TFT structure with low enough capacitance to allow clocking at low megahertz rates. The fabrication processes utilize 10 μm photolithography and are suitable for large area processing. This is an important step towards the incorporation of amorphous silicon TFT drivers on the substrate of an active matrix liquid crystal display. With this technique, we expect in the near future to have displays with both switches and drivers operating on the same substrate using amorphous silicon alloy technology.

ACKNOWLEDGMENTS

We would like to acknowledge S. R. Ovshinsky for his continued support and encouragement of this work. We are also grateful to Chris Willner for his computer calculations and to Walter Chapelle for helpful discussions.

REFERENCES

1. Z. Yaniv, G. Hansell, M. Vijan, V. Cannella, Mat. Res. Soc. Sym. Proc., Vol. 33, North Holland, 1984, pp. 293-296.
2. Z. Yaniv, V. Cannella, G. Hansell, C. Willner, M. Vijan, Mol. Cryst. Liq. Cryst., Vol. 124, 1985.
3. S.R. Ovshinsky, Proc. Symposium on Glass Science and Technology, Vienna, Austria, July, 1984.
4. D.G. Ast, IEEE Trans. on Elec. Dev., ED-30, 5d32 (1983).
5. M.J. Thompson, J. Vac. Sci. Technol. B2(4) , 827-834, 1984.
6. K.D. MacKenzie, A.J. Snell, I. French, P.G. LeComber, and W.E. Spear, Appl. Phys. A, 31, 87, (1983).
7. A.I. Lakatos, Conference Records of 1982 Int. Display Conf., SID, 146 (1982).
8. M. Matsumura, US-Japan Joint Seminar on Technological Applications of Tetrahedral Amorphous Solids, Palo Alto, CA, (1982).
9. R.S.C. Cobbold, Theory and Applications of Field-Effect Transistors, N.Y., Wiley-Interscience (1970).
10. C. Mead and L. Conway, Introduction to VLSI Systems, Reading, Massachusetts, Addison-Wesley (1980).

AMBIPOLAR FIELD EFFECT TRANSISTOR

H. PFLEIDERER AND W. KUSIAN
Siemens Research Laboratories, Otto-Hahn-Ring 6, 8000 München 83
Federal Republic of Germany

ABSTRACT

The characteristics of a thin-film transistor using an amorphous-silicon film are presented. The appearance of electron and hole channels is made possible by ohmic source and drain contacts. A theoretical model explains the phenomena.

INTRODUCTION

Thin-film transistors (TFTs) on the basis of amorphous silicon (a-Si) can be operated in n-channel and p-channel modes [1,2]. The n-channel mode is preferable for applications, but it would appear meaningful to consider the anomalous features of an "ambipolar" field-effect transistor (FET). We first present an ideal concept and then the characteristics of an a-Si TFT structure. The interplay between theory and experiment will characterize the ambipolar FET.

AMBIPOLAR FET MODEL

Fig.1 shows the considered TFT structure. An xy coordinate system describes the semiconducting thin film. The source (x = 0) is grounded. The drain voltage U_D is applied to the drain (x = a). The potential at the front surface (y = b) is determined by the gate and drain voltages U_G and U_D. At the rear surface (y = 0) we simply assume flat bands. Holes and electrons will be treated on an equal footing because both are current-carrying. Thus thermal equilibrium with respect to the y coordinate may be assumed. We accordingly introduce a Fermi potential $\phi(x)$ common for electrons and holes [1]; its boundary values are $\phi(0) = 0$ and $\phi(a) = U_D$. The electric potential is $\psi(x,y)$. The local field effect induces the transverse potential

$$\chi(x,y) = \psi(x,y) - \phi(x).$$

The gradual channel approximation yields $\phi(x)$ and the drain current I_D.

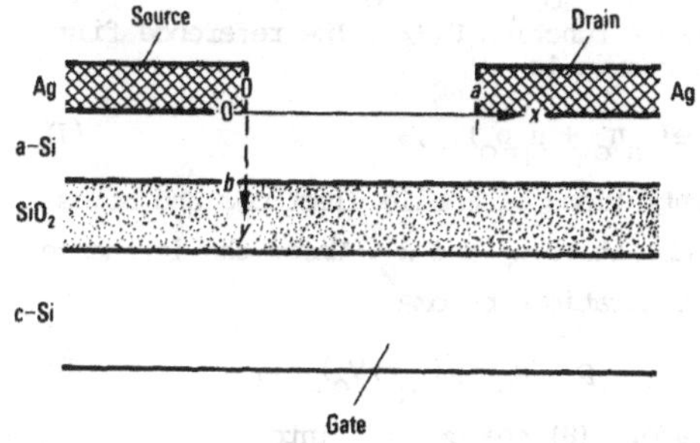

Fig.1.
Experimental TFT structure: crystalline-silicon (c-Si) substrate, thermally grown oxide layer (SiO$_2$), glow-discharge deposited amorphous-silicon film (a-Si), sputtered silver contacts (Ag). Channel length $a = 1.1 \cdot 10^{-2}$ cm, depth $b = 5.5 \cdot 10^{-5}$ cm, width $c = 1.0$ cm. Thickness of SiO$_2$ layer $3.2 \cdot 10^{-5}$ cm

We start by assuming an arbitrary bulk trap density $N_t(E)$. The gap energy E is limited by the valence- and conduction-band edges E_V and E_C. Fermi level E_F and elementary charge e define the Fermi function $f(E, E_F + e\chi)$. The concentration of trapped electrons,

$$n_t(\chi) = \int_{E_V}^{E_C} N_t(E) f(\chi) dE \quad , \tag{1}$$

determines the space charge, and the one-dimensional Poisson equation becomes

$$\chi'' = \frac{e}{\varepsilon}\left[n_t(\chi) - n_t(0)\right] = \int_0^{\chi} k^2 d\chi \quad , \tag{2}$$

with $\chi' = d\chi/dy$, semiconductor permittivity ε, and Debye length $1/k$. In (2) we assume

$$k^2 \equiv d\chi''/d\chi = e^2 N/\varepsilon \tag{3}$$

and define the effective density of gap states

$$N(\chi) = \frac{1}{e}\int_{E_V}^{E_C} N_t(df/d\chi) dE \quad . \tag{4}$$

The assumption of an homogeneous film allows a first integration of (2) as

$$\chi'^2 = 2\int_0^{\chi} d\chi \int_0^{\chi} k^2 d\chi_0 \quad . \tag{5}$$

The surface potential $\chi(x,b) = \chi_b$ and its derivative $\chi'(x,b) = \chi_b'$ satisfy the field-effect boundary condition

$$U \equiv U_G - U_F - \phi = \chi_b + \chi_b'/h \quad , \tag{6}$$

where U_F is the flat-band voltage, $h = C_o/\varepsilon$, and C_o is the insulator capacitance. (5) and (6) establish a function $U(\chi_b)$. The reference film conductivity for vanishing space charge is

$$\sigma_o = \sigma_{mo} + \sigma_{po} = e(\mu_n n_o + \mu_p p_o) \quad , \tag{7}$$

where n_o and p_o denote the concentrations of mobile electrons and holes at E_C and E_V with the constant mobilities μ_n and μ_p. Under the influence of χ, the electron and hole concentrations become

$$n = n_o \exp(\chi/V_o) \quad , \quad p = p_o \exp(-\chi/V_o) \quad , \tag{8}$$

where V_o is the temperature voltage. (8) changes (7) into

$$\sigma(x,y) = \sigma(\chi) = \sigma_m + \sigma_p = e(\mu_n n + \mu_p p) \quad . \tag{9}$$

The conductance ratio $r = \sigma_{po}/\sigma_{mo}$ defines the transition potential $\chi_T = V_o \log(r)$, yielding

$$\sigma_m(\chi_T) = \sigma_{po} \, , \quad \sigma_p(\chi_T) = \sigma_{mo} \, , \quad \sigma(\chi_T) = \sigma_o \quad . \tag{10}$$

The potential $\chi = \chi_T/2$ leads to $\sigma_m = \sigma_p = \sqrt{\sigma_{mo}\sigma_{po}}$ and the minimum of $\mathcal{E}(\mathcal{X})$. The (normalized) local conductance is given by

$$\sigma(x) = \sigma(\mathcal{X}_b) = \frac{1}{b} \int_\sigma \sigma(x,y)\, d\gamma = \sigma_0 + \frac{1}{b} \int_\sigma \frac{\sigma(\mathcal{X}) - \sigma_0}{\mathcal{X}'}\, d\mathcal{X} \quad . \tag{11}$$

The combination of $\sigma(\mathcal{X}_b)$ and $U(\mathcal{X}_b)$ gives $\sigma(U) = \sigma(\phi)$. Related to $\mathcal{X}_b = \mathcal{X}_T$ are $U = U_T$ and $\phi = \phi_T$ to be reached at $x = x_T$. The definition $U_{GT0} = U_F + U_T$ is consistent with $\phi_T = U_G - U_{GT0}$ and $U - U_T = \phi_T - \phi$. $\mathcal{X}_T$ and U_T or ϕ_T indicate, according to (10) and (11),

$$d\sigma(x)/dU = 0 \quad , \tag{12}$$

i.e. the minimum of $\sigma(x)$. In the case of $\phi_T > 0$, the interval $0 < x < x_T$ spans an electron domain ($\mathcal{X}_b > \mathcal{X}_T$, $\phi < \phi_T$) and $x_T < x < a$ a hole domain ($\mathcal{X}_b < \mathcal{X}_T$, $\phi > \phi_T$). In the case of $\phi_T < 0$, the domain sequence is reversed. Besides two-domain channels containing a coordinate x_T, one-domain channels are naturally also possible. The current density

$$j = \sigma(x)\, d\phi/dx = \text{const.} \tag{13}$$

defines the (normalized) channel conductance σ via

$$\sigma\, U_D x/a = \int_\sigma^\phi \sigma(\varphi)\, d\varphi \tag{14}$$

conditional upon $U_G = \text{const.}$. (14) first provides $\sigma = \sigma(U_G, U_D)$, putting $\phi = U_D$, and then the function $x(\phi)$. The drain current follows

$$I_D = bc\sigma U_D/a \quad , \tag{15}$$

where c denotes the film width. This completes our model. The n-channel portion was described earlier [3].

DENSITY OF GAP STATES

We measured the drain current I_D of our TFT structure (Fig.1) with U_G and U_D both varied between -20V and 20V in 1V steps. Fig.2 presents the input conductance $\sigma(U_G)$ for three different U_D values (solid lines). The minima occur at the transition gate voltages $U_{GT}(U_D)$ separating electron branches (right) from hole branches (left). The I_D measurements were made at 100°C under a vacuum of 10^{-4} mbar. The U_G values were applied with alternating signs. $U_G = 0$ was sustained for 2 min before setting up any new U_G level. Then, after waiting another 2 min,

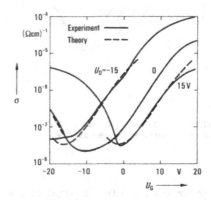

Fig.2.

Input conductance σ, gate voltage U_G, drain voltage U_D

364

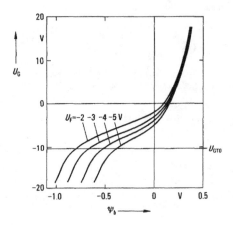

Fig.3.

Gate voltage U_G versus surface potential ψ_b. Flat-band voltage U_F. At $U_G = U_{GT0}$ the function $\sigma(U_G)$ for $U_D = 0$ (Fig.2) passes through its minimum

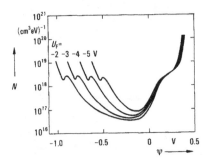

Fig.4.

Effective density of gap states N versus (surface) potential ψ

a U_D cycle was started during which U_G was held const.. By these means we tried to approach steady-state characteristics.

In the case $U_D = 0$, $\chi = \psi$ and $\chi_b = \psi_b$, and $\sigma(U)$ can be readily transformed back to $U_G(\psi_b)$ and N(E) with $E = E_F + e\psi$ [4,5]. As may be seen from Figs. 3 and 4 the result depends on the chosen flat-band voltage U_F. The conductance minimum occurs at $U_{GT}(0) = U_{GT0}$. The property $U_{GT0} < U_F$ indicates an n-doped film. The potential $\psi_b = \psi_T$ depends on U_{GT0} and U_F. In Fig.3 we observe electron accumulation $(U_G > U_F, \psi_b > 0)$, electron depletion $(U_G < U_F, \psi_b < 0)$, and hole inversion $(U_G < U_{GT0}, \psi_b < \psi_T)$. The procedure covers a ψ_b range that depends on U_F (Figs. 3 and 4). This range plus the activation energies of σ for accumulation and inversion should add up to the gap $E_C - E_V \approx 1.75eV$. Thus we estimate $U_F = -4V$ and ratio $r = 9 \cdot 10^{-9}$. Together with the respective function $N(\psi)$ (Fig.4) it is now possible to recover $\sigma(U_G)$ (see dashed curve for $U_D = 0$ in Fig.2). Without further fitting our TFT model delivers $\sigma(U_G, U_D)$ within a certain area of the U_G, U_D plane inside our external window -20V to 20V. Examples are the dashed curves for $U_D = \pm 15V$ (Fig.2). These do not cover the full interval of the measured curves (solid lines) because the function $N(\psi)$ (Fig.4) is too limited. Poor agreement between theory and experiment is indicated for hole inversion (to the left of the minima), but the shift of these minima is reproduced rather well.

CONDUCTANCE AND CHARACTERISTICS

The contour plot of the measured channel conductance $\bar{\sigma}(U_G, U_D)$ given with Fig.5 provides a general survey of the characteristics. The partition of the U_G, U_D plane into 8 sectors outlines the $\bar{\sigma}$ symmetry. The integral (14) follows a vertical path (U_G = const.). Thus any point U_G, U_D indicates, besides the associated $\bar{\sigma}$ value, a $\phi(x)$ interval, and in particular whether a domain boundary $\phi(x_T) = U_G - U_{GT0}$ appears in the channel. The diagonal $U_D = U_G - U_{GT0}$ (Fig.5), in fact, separates one-domain from two-domain channels. Any point in sectors 8,1 (4,5) represents a channel consisting of a single electron (hole) domain. The sectors 2,3,6,7 contain two-domain channels. Thus the $\bar{\sigma}$ surface shows electron and hole hills opposite one another (sectors 8,1 and 4,5) with a two-domain groove in between. The bottom line of the groove (sector boundaries 2/3 and 6/7) marks the transition between the externally apparant electron predominance and the hole predominance. The (omitted) diagonal $U_D = U_G - U_F$ represents the boundary between electron accumulation and depletion. The imperfect point symmetry of the contours reflects the near-equivalence of electrons and holes. The asymmetry of the density of states (Fig.4) and the n-doping of our film ($r < 1$, $U_F > U_{GT0}$) both contribute to the contour asymmetry. As a direct consequence of the ambipolar FET concept (ohmic contacts, thermal equilibrium), the transition groove appears somewhere within the sectors bounded by the straight lines $U_D = U_G - U_{GT0}$ and $U_G = U_{GT0}$. The FET model simulates $\bar{\sigma}$ surfaces having this property.

A section through the surface along U_D = const. (U_G = const.) reveals an input (output) conductance curve. For illustration we demonstrate the transition from electron to hole predominance by selected characteristics. Fig.6 shows a set of measured output characteristics $I_D(U_D)$ for some U_G values. The curves are superlinear for $U_D < 0$, proceed sublinearly with $U_D > 0$, and rise sharply when certain U_D thresholds are exceeded. This latter event is caused by holes entering the channel through the drain. Figs. 7 and 8 show the theoretical $\gamma_c(x)$ and $\phi(x)$ functions along the channel while proceeding along the output characteristic for $U_G = 0$.

We recall that $\gamma_c(x)$ determines the local conductance $\sigma(x)$. Fig.7 shows accumulation

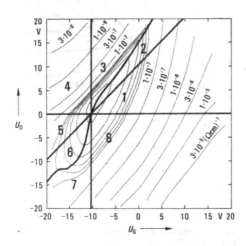

Fig.5.

Contours of channel conductance $\bar{\sigma}$ in U_G, U_D plane. Inner loops $\bar{\sigma} = $ ... to 4.75 in 0.5 steps and units 10^{-5} (Ωcm)$^{-1}$; outer lines as indicated

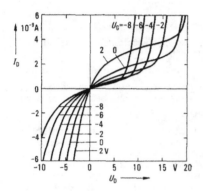

Fig.6.
Output currents $I_D(U_D)$,
gate voltages U_G

$(\chi_b > 0)$, depletion $(\chi_b < 0)$, and inversion $(\chi_b < \chi_T)$. The appearance of $\chi_b = \chi_T$ indicates the $\sigma(x)$ minimum. The $\phi(x)$ curves inflect at $\phi(x_T) = \phi_T$. The transition coordinate x_T decreases with increasing U_D. At a certain x_T position the respective characteristic ascends abruptly.

Our TFT structure does not represent a device intended for an application, but rather serves to demonstrate the ambipolar FET effect. Besides amorphous silicon other semiconductors too should allow its realization.

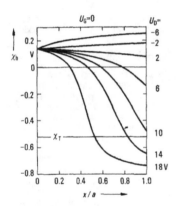

Fig.7.
Surface potentials χ_b,
normalized channel coordinate x/a,
transition potential χ_T

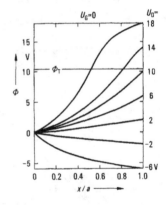

Fig.8.
Fermi potentials ϕ,
channel coordinate x/a,
transition potential ϕ_T

REFERENCES

1. G.W.Neudeck and A.K.Malhotra, Solid-State Electronics 19, 721 (1976)
2. M.Matsumura and Y.Nara, J.Appl.Phys. 51, 6443 (1980)
3. T.Suzuki, M.Hirose and Y.Osaka, Jap.J.Appl.Phys. 21, L315 (1982)
4. T.Stoica, J.Physique 42, C4-407 (1981)
5. M.Grünewald, K.Weber, W.Fuhs and P.Thomas, J.Physique 42, C4-523 (1981)

CONTINUED DEVELOPMENT OF R.F. SPUTTERED a-Si:·H
THIN-FILM TRANSISTORS TOWARDS AN ALL-SPUTTERED DEVICE

J. ALLISON, D.P. TURNER AND D.C. COUSINS
University of Sheffield, Department of Electronic and Electrical
Engineering, Mappin Street, Sheffield, S1 3JD. England.

ABSTRACT

Prototype thin film transistors have previously been fabricated by
r.f. magnetron sputtering of a-Si: H on to CVD SiO_2 using a crystalline
silicon gate. These devices exhibited an on/off current ratio of four
orders of magnitude for gate voltages as low as 10 volts. This
demonstrated the suitability of the sputtered layer for liquid crystal
display applications.
The use of a crystalline substrate negates the advantages of using
a thin film, such as large area capability and low cost, so we have
turned our attention to the provision of a sputtered dielectric.
Several candidate materials have been considered, including SiO_2, Si_3N_4,
Ta_2O_5, AlN and TiO_2. We present the characteristics of our first
all-sputtered transistor utilising an SiO_2 gate, and assess the dielectric
properties and potential of other materials, with emphasis on the
silicon oxynitride system.

INTRODUCTION

Since the first devices were produced by Snell et al(1), thin
film a-Si:H field effect transistors based on the glow-discharge process
have undergone rapid development in many laboratories. Glow discharge
deposited Si_3N_4 is the most commonly used gate dielectric although
advantages have been claimed for a silicon oxynitride layer(2) produced
by glow discharge from a mixture of SiH_4, CO_2 and N_2. Medium scale
transistor arrays have now been fabricated and their capability for
addressing liquid crystal displays demonstrated(3).
The ability to make thin film TFTs by the alternative method of
r.f. magnetron sputtering was demonstrated(4) using as a gate a $0.2\mu m$
layer of SiO_2 grown by CVD onto a single crystal silicon slice of
resistivity 0.001Ω cm. Devices satisfied the requirements to drive
liquid crystal pixels, having on-off current ratio in excess of 10^4 with
operating gate voltages less than 30 volts and a turn on voltage of about
6 volts. The effects of the a-Si:H deposition parameters were subsequently
investigated(6). An optimum hydrogen partial pressure of 0.4mTorr was
found to yield devices having a threshold voltage of 3.3 volts. The
field effect density of states of this material was below $10^{17}cm^{-3}eV^{-1}$.
The transfer characteristics of this device are reproduced as curve 'a' in
figure 1 and are comparable to devices derived from the glow discharge
method. Further details of the device stability, with reference to photo-
induced changes have been submitted for publication.
The characteristics of TFTs have been found to be strongly dependent on
the choice of gate insulator . Ast(3) has shown that, for a poor quality
(low deposition temperature) G.D. Si_3N_4 dielectric trapping at the semiconductor-insulator interface results in a short term instability (of the
order 10^5 seconds) in the source-drain current. This was reduced by use
of a higher deposition temperature or replacement by a sputtered SiO_2
layer. The device yield was strongly correlated to gate leakage and pinholes were caused by fine "dust" particles formed by nucleation in the gas
phase. We have observed similar effects in sputtered silicon oxynitride
films. These may be overcome by using the range of alternative materials

available by sputtering and further reduction of the turn on voltage may be possible by using films with high dielectric constants.

Sample preparation and assessment techniques

The dielectrics described below have been prepared in a Nordiko NM2000 r.f. sputtering system operating at 13.56MHz. Three target positions are available on a rotating table; these include a 6" diameter magnetron with a field of 0.01 Tesla at the surface, and a recently introduced 8" version with field strength 0.1 Tesla. In order to minimise cross-contamination between silicon and dielectric targets an aluminium gauge enclosed in a pyrex cylinder has been placed over each target, these being cleaned between runs. This confinement of the plasma will influence the plasma potential so the bias voltages quoted are not directly comparable with previous results. Also a reduction in uniformity has been found; the sample thickness falling radially to 70% at the edges of the substrate platen.

For reactive sputtering a mixture of 24% N_2, 12% O_2 and 64% Ar has been used. This is a compromise, to allow a range of dielectrics to be included in our initial survey; subsequent development will require a choice of gas pertinent to each target material. The total gas pressure has been maintained at 5mTorr, the reactive gas added at a flow rate of 10sccm should correspond to a final mixture 92.8% Ar : 4.8% N_2 : 2.4% O_2. The insulators have been deposited using power densities around 0.85W/cm . The substrate temperature was maintained at $230^{\circ}C$ in order that the silicon layer could be deposited at its optimum temperature over the insulator. The deposition conditions of the dielectric films are summarised in table 1. Amorphous silicon layers were deposited from the smaller target at our established optimum conditions of P_{ar} = 5mTorr, P_h = 0.5mTorr, P = 100 watts and using a substrate bias of V_b = -25 volts.

Samples for IR measurements were deposited on to polished single crystal silicon, and for other measurements 7059 glass substrates with an evaporated chromium layer were used. The substrate bias was not applied during the first 10% of the insulator deposition to avoid re-sputtering the chromium layer. The a-Si-H was removed in KOH solution prior to dielectric measurements. Silicon islands were provided for the transistors using negative photoresist and an adhesion promoter and etching in a solution of ammonium flouride and hydrogen peroxide. Finally, aluminium was evaporated using a photolithographic step to define source-drain contacts with w = 900μm and L = 10μm. A 200Å chromium evaporation through a contact mask provided 1mm diameter counter electrodes for capacitance and breakdown strength measurements.

The dielectric films have been assessed in several ways. The dielectric constant and dissipation factor were measured using a capacitance bridge at 1KHz. Dielectric thicknesses in the range 0.1 - 1.3μm for these calculations were determined optically by the CARIS method(6) from reflectance curves taken at fixed angles 26.5° and 4.5° over a wavelength range 185-800nm using a Perkin-Elmer 330 uv-visible spectrophotometer. Values were confirmed by stylus measurements of film edge step heights using a Rank-Taylor-Hobson talystep.

As an assessment of dielectric breakdown strength and pinhole density, a breakdown yield measurement was made on each film. The breakdown criterion was that a capacitor dot failed if the leakage current exceeded $10^{-8}A$ for applied fields of 10^5 and 10^6 V/cm. A percentage yield was thus obtained; in excess of 50 dots per film were tested.

The composition and structure of films in the oxynitride system have been inferred from infra-red transmission measurements taken on a Perkin-Elmer dual beam spectrophotometer. Spectra were obtained in the wavelength range 3000-400cm^{-1}.

Etch rates have been determined by stylus measurements of photolithographically defined grooves using two etchants; P-etch - 300 parts H_2O, 10 parts HNO_3 and 15 parts HF(7). and buffered HF - 9 parts 40% (by weight) NH_4F solution, 1 part 48% HF.

Silicon-oxynitride system

Considering firstly the SiO_2 films grown in pure argon, we found that the breakdown yield was a strong function of thickness and a good insulating layer could not be produced for t < 0.5μm. The film described in table 1 was 1.3μm thick at the centre of the platen. The d.c. transistor characteristics of a device on this film are presented in figures 1b and 2. Although gate voltages up to 150 volts were required, the on-off current ratio was still around four orders of magnitude. The transfer characteristics show a very low saturation current which decayed with time. The long term stability has not yet been established. This poor performance may be due to positive charges in the insulator, as has been observed by Lee et al (8) for r.f. sputtered SiO_2 deposited in a system which incorporated pyrex components. In that case the replacement of pyrex by teflon and the addition of an r.f. bias improved the films with a minimum P-etch rate of 2.1Å/sec comparing with our current value of 3Å/sec. We have thus demonstrated the possibility of an all sputtered transistor but its application must await further development of the dielectric.

The infra-red spectra of our Si_3N_4 film sputtered in pure argon is compared in figure 3 with that of bulk β-Si_3N_4 after Hu (9). The position of the peak absorption at 835cm^{-1} is lower than the bulk value of 935cm^{-1} and close to that of pyrolitic films at 830cm^{-1} which implies a lower density film. We did not observe absorption peaks around 2000 or 2100cm^{-1} which may arise from hydrogen contamination or Si ≡ N bonding. A denser film may be expected from an increased bias voltage; Stephens et al (10) have reported a fall in the HF etch rate through two orders of magnitude for reactively r.f. sputtered Si_3N_4 with an increase in bias from zero to -50 volts and the infra red absorption of these film peaks at 850cm^{-1}.

The silicon oxynitride system has been scanned by Frank and Moberg(11) by adding up to 20% of N_2O to a nitrogen plasma during reactive r.f. sputtering. The resulting films were nitrogen rich with the infra red peak position moving from 825 to 1040cm^{-1} as the percentage of N_2O was increased. We attempted to obtain a range of SiOxNy compositions by reactively sputtering from different target materials, but figure 4 shows that the composition is determined solely by the gas composition, and at a N/O/Ar flowrate of 10sccm films appear similar to the SiO_2 film sputtered in pure oxygen with a peak absorption around 1075-1062cm^{-1}. This is in agreement with a SiO_2 film reactively sputtered from a silicon target in a 50 : 50 Ar : O_2 atmosphere which gave a peak at 1075cm^{-1}(11). Despite low concentrations, our etch rate and breakdown measurements show an improving trend with increasing nitrogen levels and our best film was produced at a reduced flowrate of 1.0sccm, from a nitride target. Frank(11) found a steadily increasing dielectric constant ranging from 3.9 for SiO_2 to 9.1 for Si_3N_4. The optimum film had ε_r = 8.3 and showed a dramatic increase in yield and a breakdown field of 10^7V/cm. By comparison our highest value for ε_r = 6.2 also corresponds to a higher yield and low dissipation factor. The lower thickness limit for insulating film was also reduced below 0.2μm. We conclude that sputtered silicon-oxynitride shows good promise for TFTs and our future work will concentrate on higher nitrogen fractions and a close examination of bias effects.

TARGET	REACTIVE GAS FLOW RATE	RF POWER	DEPOSITION RATE	DIELECTRIC CONSTANT	TAN δ	10^5V/cm BREAK-DOWN YIELD	10^6V/cm BREAK-DOWN YIELD	IR ABSORPTION MAXIMUM	P-ETCH RATE	BUFFERED - HF
	sccm/min	W	Å/min	ϵ_r		%	%	cm^{-1}	Å/sec	Å/min
SiO_2	0	250	77	4.0	<0.005	97	50	1075	3.0	1000
SiO_2	10	250	30	-	-	0	0	1065	3.6	-
Si_3N_4	0	150	44	12.4	0.24	0	0	835	0	10
Si_3N_4	10	150	16	4.2	0.022	98	52	1062	4.3	900
Si_3N_4	1.0	150	4.4	6.2	0.013	95	64	-	-	-
TiO_2	10	200	2.5	-	-	-	-	-	-	-
Ta_2O_5	10	250	27	22.6	0.005	95	0	-	-	-
Al	10	300	6.2	9.1	0.018	82	0	-	-	-

TABLE 1   SUMMARY OF PRINCIPAL DEPOSITION CONDITIONS AND INSULATOR PROPERTIES

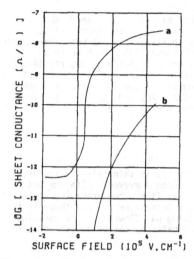

Figure 1. Sheet conductance versus surface field for r.f. magnetron a-Si:H FET's using a) CVD SiO₂ gate and b) magnetron sputtered SiO₂ gate.

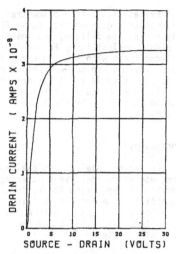

Figure 2. Output characteristic of an a-Si:H with sputtered SiO₂ gate at Vg = 150 volts.

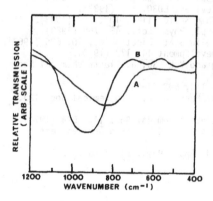

Figure 3. Infrared transmission spectra of Si₃N₄ films. Curve A shows a sample sputtered directly from a Si₃N₄ target in Ar; while curve B shows the spectrum for bulk Si₃N₄.

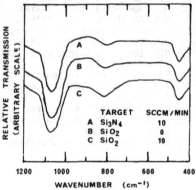

Figure 4. IR spectra of SiO₂ and SiOxNy films. Curve A: sample sputtered from Si₃N₄ target with 10 sccm N₂/O₂ gas. Curves B and C: SiO₂ target without reactive gas and with 10 sccm reactive gas respectively. Curves are offset for clarity.

Alternative dielectrics

Seki et al(12) have demonstrated that magnetron sputtered Ta_2O_5 is suitable as a gate dielectric in TFTs formed by laser re-crystallised polycrystalline silicon, and reactively sputtered AlN has been investigated as a gate dielectric by Fathimulla and Lakhani (13). We have thus included these with TiO_2 in our initial survey. Although the reactive gas concentrations are not entirely appropriate, dielectric films were obtained as noted in table 1. It is encouraging that these films appeared smooth and clean, as compared with the oxynitride films and our present films seem to be limited by their low breakdown fields rather than pinhole density.

Conclusions

The various dielectrics available for use in thin-film transistors using magnetron reactive sputtering have been surveyed. Although not individually optimised fully, these initial films show promising dielectric properties. The best dielectrics in this current series are those in the SiOxNy system, prepared with a low reactive gas flowrate. The usefulness of sputtered dielectrics has been demonstrated by the production of the first all-sputtered TFT.

References

(1) A.J. Snell, K.D. Mackenzie, W.E. Spear and P.G. Le Comber, Appl. Phys. 24, 357 (1981).
(2) K. Ishibashi and M. Matsumura, Appl. Phys. Lett. 41, 454 (1982).
(3) D.G. Ast, IEEE Trans. Electron Devices, ED30, 532 (1983).
(4) M.C. Abdulrida and J. Allison, Thin Solid Films, 102, L43 (1983).
(5) M.C. Abdulrida and J. Allison, Appl. Phys. Lett. 43, 768 (1983).
(6) F. Reizmann and W. Van Gelder, Solid State Electronics, 10, 625 (1967).
(7) J.S. Logan, IBM J.Research and Development 14, 172 (1970).
(8) M.K. Lee, C.Y. Chang, J.S. Tzeng and Y.K. Su, J. Electrochemical Soc. 130, 658 (1983).
(9) S.M. Hu, J. Electrochemical Soc. 113, 693 (1966).
(10) A.W. Stephens, J.L. Vossen and W. Kerr, J. Electrochemical Soc. 123, 304 (1976).
(11) R.I. Frank and W.L. Moberg, J. Electrochemical Soc. 117, 524 (1970).
(12) S. Seki, T. Unagami and B. Tsujiyama, IEEE Electron Devices Lett., EDL5, 197 (1984).
(13) A. Fathimulla and A.A. Lakhani, J. Appl. Phys. 54, 4586 (1983).

EXPERIMENTAL AND THEORETICAL ANALYSIS OF THE ABOVE THRESHOLD
CHARACTERISTICS OF AMORPHOUS SILICON ALLOY FIELD EFFECT TRANSISTORS

M. HACK, M. SHUR*, and C. HYUN* Z. YANIV**, V. CANNELLA**, M. YANG**,
Energy Conversion Devices, Inc., 1675 West Maple Road, Troy, Michigan 48084
*Department of Electrical Engineering, University of Minnesota,
Minneapolis, Minnesota 55455
**Ovonic Display Systems, Inc., 1896 Barrett Street, Troy, Michigan 48084

ABSTRACT

Experimental studies on the above threshold characteristics of
amorphous silicon alloy thin films transistors show that the field-effect
mobility has a weak temperature dependence and that the current-voltage
characteristics are described by a power law dependence on gate voltage,
with an exponent greater than two. These results are in good agreement
with our theoretical model, which demonstrates that in the above threshold
regime, the Fermi level at the semiconductor-insulator interface lies in
the tail states which control device performance. Analysis of the above
threshold characteristics shows the characteristic energy variation of the
exponential tail state distribution to be ~22meV and that the on current
changes by a factor of only two over the temperature range 10-90°C.

INTRODUCTION

Substantial improvements in the development of amorphous silicon thin
film transistors (a-Si TFT) as switching elements in flat panel displays
have been achieved. On/off current ratios of 10^6 with gate voltage swings
of about 10 volts are easily obtained. Ring oscillator delays as short as
210 nS per stage have been reported [1]. The use of a-Si TFTs in logic
circuits is still limited by their relatively slow switching speeds. A
better understanding of the device physics is needed not only to obtain
further improvements in fabrication but also to obtain accurate models
suitable for use as device and circuit design tools. Recently, we have
proposed a new physical mechanism that describes the operation of a-Si
TFTs [2] based on a detailed consideration of the Fermi level movement
through the localized states in the energy gap under the control of the
gate voltage. We showed that the transition from the deep to tail states
leads to two basic modes of device operation. In this work, we shall only
consider n-channel operation. At low gate voltages, the Fermi level at
the insulator-amorphous silicon interface is in the deep localized states
well below the conduction band edge. This is the subthreshold regime
which is characterized by a rapid increase in the drain-to-source current
with increasing gate voltage. Analysis of the sub-threshold
characteristics using our theoretical model has enabled us to determine
both the flat-band voltage and the characteristic energy variation of the
deep localized states [3]. At large gate voltages there is a dramatic
change in the device current-voltage characteristics corresponding to the
interface Fermi level entering the tail states, and at room temperature or
above, most of the charge in the localized states actually lies above the
Fermi level. In this above threshold regime, the behavior of a-Si TFT's
is similar to that of single crystal MOSFET's, although the field-effect
mobility is much lower. As we showed previously [4] the use of a "square
law" model does not accurately describe the above threshold I-V
characteristics of a-Si TFTs, as with increasing gate voltage, the Fermi
level continues to move closer to the band edge.

In this paper we present experimental data on the temperature

Mat. Res. Soc. Symp. Proc. Vol. 49. ' 1985 Materials Research Society dependence of the dc characteristics of a-Si TFT's and derive analytical expressions describing the above threshold characteristics. The temperature dependence gives a good indication as to the validity of any a-Si TFT model because many device parameters depend on the ratio of the characteristic temperature of the localized states distribution to ambient temperature. Our results show that the field effect mobility is not strictly activated in the above threshold regime as the Fermi level position is not independent of temperature. This causes the field-effect mobility to have an approximately linear temperature dependence.

CURRENT-VOLTAGE CHARACTERISTICS

The drain to source current-voltage characteristics may be found using the gradual channel approximation as follows [2]:

$$I_{ds} = q\mu_o(W/L) \int_0^{V_d} n_s(V_{ch})dV_{ch} \tag{1}$$

where q is the electronic charge, μ_o is the band mobility, W and L the gate width and length, n_s the free surface charge density, V_d the potential at the drain end of the channel, and V_{ch}, the longitudinal voltage drop across the channel. In a-Si TFTs, the total charge induced by the gate voltage must include both contributions from the localized states and the free carriers in the conduction band. Hence, the current-voltage characteristics of a-Si TFTs are very dependent on the localized state distribution. A good measure of device performance is the effective field-effect mobility defined as

$$\mu_{fet} = \mu_o(n_s/N_{ind}) \tag{2}$$

where qN_{ind} is the total induced charge given by

$$qN_{ind} = Q_{deep} + Q_{tail} + qN_{ss} + q n_s \tag{3}$$

The three terms represent the total charge induced in the deep, tail and surface states, respectively.
In the above threshold regime, assuming no surface states, the total induced charge can be approximated by

$$qN_{ind} = [2\varepsilon qG_o \exp[A(\alpha)+B(\alpha)X_o]]^{1/2} \tag{4}$$

where ε is the a-Si dielectric constant, $\alpha=T_1/T$ is the ratio of the characteristic energy of the tail states distribution to the thermal energy and X_o is the normalized band bending at the interface. The values of G_o, $A(\alpha)$ and $B(\alpha)$ are derived from the integration of the tail state distribution with the appropriate occupation function (see Ref. [2]) and are given by

$$G_o = \frac{1}{2} g_{ct} \frac{kT_1}{q} \, kT \tag{5}$$

$$A(\alpha) = 1.92\alpha - 0.595 \tag{6}$$

$$B(\alpha) = 0.225\alpha - 1.03 \tag{7}$$

The total surface free carrier density can be expressed as

$$n_s = \frac{N_c kT\epsilon}{qN_{ind}} \exp(-X_o) \tag{8}$$

where N_c is the conduction band effective density of states. The induced charge can be related to the effective gate voltage $V_g = V_{gs} - V_T$ across the insulator by

$$qN_{ind} = C_{ox}V_g \tag{9}$$

where $C_{ox} = \epsilon_{ox}/t_{ox}$ is the oxide capacitance, and V_{gs} is the gate to source voltage. The threshold voltage V_T is given by [2]

$$V_T = \frac{2\pi\epsilon kT_2 kTg_{fo}}{C_{ox}[q \sin(\pi(T/T_2))]^{1/2}} \exp\left(\frac{E_{ct}}{2kT_2}\right) + V_{FB} + \Delta V_{FB} \tag{10}$$

where kT_2 is the characteristic energy of the deep localized states distribution, E_{ct} is the energy from the conduction band edge of the transition from deep to tail states, and V_{FB} is the flat-band voltage. For an energy independent distribution of surface states D_{ss}, the shift in the flat-band voltage is given by

$$\Delta V_{FB} = qD_{ss}(E_{fo} - E_n)/C_{ox} \tag{11}$$

where E_{fo} is the bulk Fermi level and E_n is the neutral level.

Substituting Eqns. (4,8,9) into Eqn. (1) and integrating from 0 to V_{ds}, the above threshold drain-to-source current can be expressed as

$$I_{ds} = q\mu_o \frac{WK}{L\gamma} C_{ox}^{\gamma-1} [V_g^\gamma - (V_g - V_{ds})^\gamma] \tag{12}$$

where $\gamma = -2/B(\alpha)$ and

$$K = \frac{N_c kT\epsilon}{q [2\epsilon qG_o \exp (A(\alpha))]^{\gamma/2}} \tag{13}$$

When V_{ds} exceeds the saturation voltage $V_{dsat} = V_g$, then Eqn. (12) reduces to

$$(I_{dsat})_{ab} = q\mu_o \frac{WK}{L\gamma} C_{ox}^{\gamma-1} V_g^\gamma \tag{14}$$

The subthreshold characteristics have been described in detail in Ref. [2]. The subthreshold drain-to-source saturation current has the following dependence on V_{gs}

$$(I_{dsat})_{sb} = q\mu_o \frac{Wf}{L\beta} C_{ox}^{\beta-1} (V_{gs} - V_{FB})^\beta \tag{15}$$

where $\beta = 2T_2/T$ and f is a parameter dependent on T_2 and T [2]. Both Eqns. (14) and (15) do not account for the bulk conductivity which determines the "off" current and is approximately given by

$$I_{off} = \sigma_D t(W/L)V_{ds} \tag{16}$$

where σ_D is the dark bulk conductivity and t the thickness of the a-Si layer. The total device current is therefore $I_{ds}+I_{off}$.

EXPERIMENTAL RESULTS

Fig. 1 shows the measured output characteristics of an a-Si TFT with gate lengths of 10μm and widths of 250μm at different temperatures together with the corresponding theoretical curves. The device parameters used in all calculations are listed in Table I. The devices were fabricated on glass substrates as follows. The gate electrode pattern was formed on the substrate followed by the deposition of 2000Å of plasma deposited insulator from SiH_4 and N_2O and finally, 2000Å of an amorphous silicon alloy. Source and drain electrodes were defined over n^+ a-Si using conventional photolithographic techniques to form the ohmic contacts. The active areas were covered with a SiO_2 protective layer. These devices exhibited a typical on-current of 200μA/mm of gate width, an on/off current ratio of 10^6 with a gate voltage swing of 10V and an effective field effect mobility of about 1cm²/V-s. Further details can be found, for example, in Ref. 5.

Fig. 2 illustrates the temperature dependence of the drain current at a gate bias of 25V and V_{ds}=25V. The experimental drain current is roughly a linear function of temperature in good agreement with our model, see Eqn. 12. The calculated I_{ds} vs. T curves for different characteristic temperatures T_1 show that a value of T_1=260 K best fits the experimental data. This value of T_1 also agrees with the results of photoconductivity measurements [6], and drift mobility studies [7].

The effective field effective mobility which is derived from the slope of the square root of I_{ds} vs. V_{gs} is approximately a linear function of temperature in the above threshold regime as shown in Fig. 3. The mobility changes from a value of about 0.5cm²/V-s at 10°C to 0.9 cm²/V-s at 90°C and closely follows the theoretically predicted behavior.

The value of the exponent γ in the power law dependence of the drain saturation current on gate voltage (see Eqn. (12)) can be determined from the experimental I-V characteristics using a best least square fitting procedure [5]. The results of such an analysis are shown in Figs. 4. The exponent γ is given by

$$\gamma = \frac{2T}{1.03T - 0.225T_1} \tag{17}$$

As can be seen in Fig. 4, γ is a very weak function of temperature, in agreement with the experimentally derived values, whereas in the subthreshold regime this exponent is a strong function of temperature. Eqn. (10) predicts a slight increase of the threshold voltage with increasing temperature but as shown in Fig. 5, the measured threshold voltages exhibit a somewhat larger change with temperature. This can be attributed to changes in the occupation of surface states with temperature which were not included in our calculations.

To assess whether the above threshold field effect mobility can also be considered to be activated with temperature, $\ln(I_{ds})$ is plotted as a function of 1/T for two different gate voltages in Fig. 6. As can be seen the activation energy is small as expected, and a value of ~60 meV clearly indicates that the Fermi level at the semiconductor-insulator interface lies in the tail states near the band edge. Our results also demonstrate that the apparent activation energy of the field-effect mobility is not equal to the width of the tail states [8], as in the above threshold regime, the Fermi level moves through the tail states with increasing gate voltage.

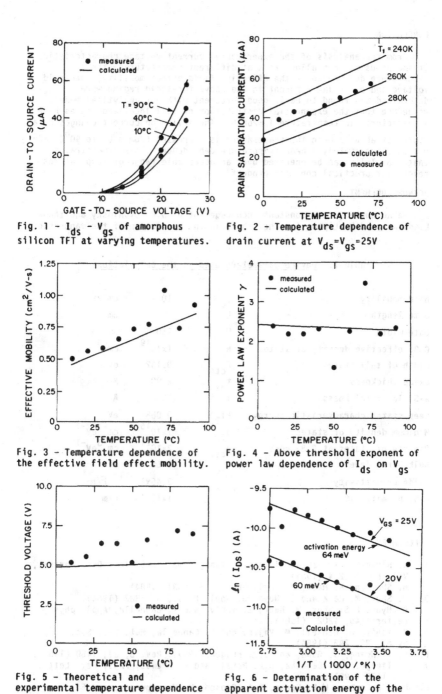

Fig. 1 – I_{ds} – V_{gs} of amorphous silicon TFT at varying temperatures.

Fig. 2 – Temperature dependence of drain current at $V_{ds} = V_{gs} = 25V$

Fig. 3 – Temperature dependence of the effective field effect mobility.

Fig. 4 – Above threshold exponent of power law dependence of I_{ds} on V_{gs}

Fig. 5 – Theoretical and experimental temperature dependence of the threshold voltage.

Fig. 6 – Determination of the apparent activation energy of the field effect mobility.

CONCLUSIONS

From the analysis of the experimental current-voltage characteristics of amorphous silicon alloy TFT's at different temperatures, the temperature dependence of the effective field effect mobility, threshold voltage and power law exponent in the above threshold regime were determined and found to be in good agreement with our analytical model. We deduce that the exponential distribution of tail localized states have a characteristic temperature T_1=260 K. The device on current changes by a factor of about 2 when the temperature is increased from 0°C to 90°C and has an approximately linear temperature dependence. This demonstrates that a-Si TFT's can be operated over an acceptable range of temperatures needed for practical consumer products.

ACKNOWLEDGEMENT

We acknowledge the constant encouragement of S.R. Ovshinsky and thank E. Norman and M. Lipton for their help in the production of the manuscript.

Table I. <u>Device parameters used in the calculations</u>.

Band mobility	μ_o	10	$cm^2/V\text{-}s$
Gate length	L	10	μm
Gate width	W	250	μm
C.B. effective density of states	N_c	1×10^{19}	cm^{-3}
Width of tail states	E_{ct}	0.152	eV
Oxide thickness	t_{ox}	2000	Å
a-Si layer thickness	t	2000	Å
Deep states characteristic energy	kT_2	0.086	eV
Minimum density of states	g_{fo}	7×10^{15}	$cm^{-3}eV^{-1}$
Density of states at C.B. edge	g_{ct}	2×10^{21}	$cm^{-3}eV^{-1}$
Bulk Fermi level	E_{Fo}	0.65	eV
Oxide permittivity	ε_{ox}	3.45×10^{-11}	F/m
a-Si permittivity	ε	1×10^{-10}	F/m

REFERENCES

1. K. Hiranaka, T. Yamaguchi, and S. Yanagisawa, IEEE Elec. Dev. Letters, Vol. EDL-5, 224 (1984).
2. M. Shur and M. Hack, J. Appl. Phys., 55, 3831 (1984).
3. M. Shur, M. Hack and C. Huyn, J. Appl. Phys., 56, 382 (1984).
4. C. Hyun, M.S. Shur, M. Hack, Z. Yaniv, and V. Cannella, Appl. phys. Letters, 45, 1202, (1984).
5. Z. Yaniv, G. Hansell, M. Vijan, and V. Cannella, Mat. Res. Soc., Symp., 33, 293 (1984).
6. C.-Y. Huang, S. Guha, and S.J. Hudgens, Phys. Rev. B, 27, 7460 (1983).
7. T. Tiedje, J.M. Cebulka, D.L. Morel, and B. Abeles, Phys. Rev. Lett., 46, 1425 (1981).
8. P.G. LeComber, W.E. Spear, R.A. Gibson, H. Mannsperger, and F. Djamdji, J. of Non-Crystalline Solids, 59 & 60, 505 (1983).

A HIGH-SPEED AMORPHOUS-SILICON DYNAMIC CIRCUIT

H. Okada, Y. Nara, Y. Uchida, Y. Watanabe[*] and M. Matsumura
Department of Physical Electronics, Faculty of Engneering,
Tokyo Institute of Technology, Meguro-ku, Tokyo 152, JAPAN
(* On-leave from Seikosha Co.)

ABSTRACT

A novel dynamic circuit composed of amorphous-silicon Schottky-barrier diodes and field-effect transistors has been proposed. The circuit response time is as short as the discharging time of a load capacitor through the driver transistor. The circuit having 1μm-long, self-aligned transistors has been predicted theoretically to be able to be operated at multi-MHz rates. Preliminary experimental results are also presented.

INTRODUCTION

All amorphous-silicon (a-Si) electronic devices [1-4] reported to date, are driven by silicon integrated circuits. This is the simplest driving technique but generates serious problems. (1) the external peripheral circuits reduce the total device performance due to wiring capacitance, (2) they also increase final production costs and (3) they decrease reliability due to wire-bonding. The only way to eliminate these problems is to also fabricate the peripheral circuits using a-Si field-effect transistors (FETs) and to integrate them with the a-Si active-matrix. Because of the low electron mobility of a-Si, however, conventional a-Si FET static circuits [5-8] are difficult to operate at 1MHz; the speed necessary for the peripheral circuits.

In this paper, we have proposed a new dynamic circuit with a-Si Schottky-barrier diodes [9] (SBDs) and FETs. The circuit can be operated at multi-MHz rates.

SPEED LIMITATION IN CONVENTIONAL CIRCUITS

Typical FET static circuits are based on an inverter having two FETs, namely, a driver Q_D and a load Q_L. In order to produce a good static transfer function, the mutual conductance g_{mD} of Q_D should be much larger than the g_{mL} of Q_L. During their transient, a load capacitor C_L is charged up by Q_L or discharged by Q_D. The times τ_C for charging-up C_L and τ_D for discharging C_L are given by

$$\tau_C = mC_L/g_{mL},$$

and

$$\tau_D = 2.3C_L/g_{mD},$$

respectively, where the constant m depends on the detailed circuit configuration. Since τ_C is much longer than τ_D, the operation speed is limited predominantly by τ_C. This speed can be only slightly improved by replacing Q_L by a load resistor [7], because its value should be equal to $1/g_{mL}$.

Normally the speed of the MOS FET circuit can be improved by applying a complementary configuration. The complementary a-Si FET circuit, however, does not seem promising because of the extremely low hole mobility in a-Si. The second method of improving the MOS FET circuit is to use a enhancement-depletion configuration. However, the depletion

Mat. Res. Soc. Symp. Proc. Vol. 49. ʿ 1985 Materials Research Society type a-Si FET seems to have a poor electrical performance [10]. The situation can be slightly improved by the use of a quasi enhancement-depletion type circuit [6] as this is also limited by the low hole mobility. The fourth method is to apply a dynamic clocking scheme. In a four-phase dynamic circuit, since there is no dc current path, Q_L can be made larger than Q_D in static circuits. However, because its enlargement increases stray capacitance and thus deteriorates total transfer characteristics Q_L can not be enlarged infinitely but should be made as large as channel width of the driver FETs.

   Thus we conclude that there is no room for a dramatic improvement in the operating speed of a-Si FET static circuits. However the dynamic circuit can be operated at sufficiently high speeds if the current driving capability of the pull-up device is increased even if this causes a slight increase in capacitance. This condition is satisfied by applying a-Si SBDs as discussed below.

DIODE-FET DYNAMIC CIRCUIT

   A proposed dynamic circuit is shown in Fig.1. The inverter has an a-Si SBD D, a capacitor C_L and two a-Si FET drivers Q_1 and Q_2. Non-overlapping four-phase clock pulses ϕ_1, ϕ_2, ϕ_3 and ϕ_4, shown in Fig.2 are applied to their terminals. During the charging period

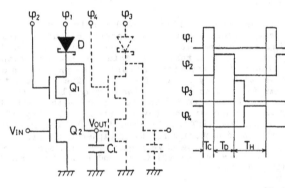

Fig.1 Equivalent Circuit of FET-SBD Circuit.

Fig.2 Clock Pulse Waveforms.

T_C, when ϕ_1 is high and ϕ_2 is low, C_L is precharged through D. When ϕ_1 falls to zero and ϕ_2 turns to a high level, the circuit enters into the discharge period T_D. If the input voltage V_{IN} is high, C_L discharges through Q_1 and Q_2, resulting in a low output voltage , V_{OUT}, because both FETs are driven into an on-condition. While, if V_{IN} is low, charge remains at C_L resulting in a high V_{OUT}, because Q_2 is not turned on. In the holding period T_H when both ϕ_1 and ϕ_2 are low, D and Q_1 are in an off-condition and thus charge remains at C_L. As a result, the half-stage executes a logical inversion. During T_H, the next half-stage is driven by ϕ_3 and ϕ_4. Since its output voltage is kept at a logically correct value while both ϕ_3 and ϕ_4 are low, the following half-stage can be driven again by ϕ_1 and ϕ_2.

   A 0.45μm thick a-Si SBD has a reverse breakdown voltage of more than 15V, can drive current of more than 100A/cm^2 at 5V forward bias voltage, and can rectify an ac signal of more than 10MHz [9]. Thus the 1000μm^2 a-Si SBD is estimated to have a small capacitance of about 0.2pF. However it is possible to charge up the load capacitor of the inverter circuit (may be as large as 6.2pF when the fan out is 4.) within 100ns and τ_C in the new circuit can be reduced to less than τ_D. Thus the minimum response time of the circuit is determined by τ_D and not by τ_C.

ANALYTICAL RESULTS

Device Models

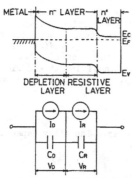
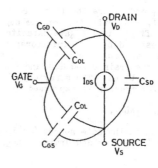

Fig.3   SBD Model.   Fig.4   FET Model.

Similar to conventional circuit simulations, the carrier transit time across the active region of devices is assumed to be much shorter than the circuit response time. Then, the electron density distribution in the device can be approximated by the static one, and the device can be represented by a combination of the static device model and the electrostatic capacitances caused by various charge-storage effects.

An a-Si SBD model is shown in Fig.3. The active n^-a-Si layer is divided into a "depletion" layer and a "resistive" layer. In the depletion layer, the electron density is much less than that in the bulk under thermal equilibrium conditions. While, in the resistive layer, the electron density is higher than, or approximately equal to, the thermal equilibrium value. Thus the depletion layer characteristics can be expressed by the ideal diode current I_D with the saturation current density J_{DO}, and the depletion layer capacitance C_D which can be expressed as the conventional abrupt junction capacitance with an effective donor density of N_R [11]. Since hole injection is neglected in SBDs with low barrier-height metals, such as Mo and Cr, the current in the resistive layer is carried by electrons. Thus, the resistive layer can be represented by the space-charge-limited current I_R [12] and the space-charge capacitance C_R. Both of them are related to the ambient temperature T and the characteristic temperature T_G ($>T$) of the localized state density distribution in a-Si when the localized state density is assumed to fall exponentially with energy from the conduction band edge as,

$$I_R = S_D J_{RO} V_R^{1+T_G/T} / d_R^{1+2T_G/T},$$

and

$$C_R = \varepsilon_S \varepsilon_0 S_D / [(1+T/T_G) d_R],$$

where S_D, d_R, J_{RO} and V_R are the diode area, resistive layer width, proportional constant and voltage applied to the resistive layer, respectively.

An a-Si FET model is shown in Fig.4. We have previously reported on the static FET model [13] where the drain current I_{DS} can not be expressed simply by the conventional MOS FET equation. However, our present concern is in the dynamic switching characteristics of the circuit; especially, in the charging characteristics of C_L through the SBD and in the discharging characteristics of C_L through FETs biased under high gate voltage conditions. Then I_{DS} can be approximated by the conventional MOS FET equation.

The intrinsic gate capacitance per unit area is given by

$$C_I = \varepsilon_I \varepsilon_0 / T_I,$$

where T_I and ε_I are the thickness of the gate insulator and its dielectric constant, respectively. For the purpose of simplifying the analysis, non-linear effects in the gate capacitor C_G are not taken into account, and C_G is represented by two capacitors; one, namely C_{GS} ($=C_I LW/2$), is connected between the source S and gate G, and the other, namely C_{GD} ($=C_I LW/2$), between the drain D and G. L and W are the channel length and width of the FET, respectively. The stray capacitance C_{OL} caused by the gate and source (or drain) overlap is expressed by

$$C_{OL} = C_I W L_{OL},$$

where L_{OL} is the gate-source (or drain) overlap. The stray capacitance C_{SD} between the source and drain is neglected.

Numerical Results

The dynamic performance of various inverters have been compared by computer simulation. The FET is assumed to be of a self-aligned structure with a 5μm long channel, and the field-effect mobility and threshold voltage are assumed to be 1cm^2/Vs and 2V, respectively. The load capacitance C_L is fixed at 6.2pF: about 4 times the gate capacitance of the input transistor whose channel width W is 1mm. Other parameters used in the computation are summarized in Table I. 200kHz operation of the static inverter, the conventional dynamic inverter and the new dynamic inverter are shown in Fig.5, where applied voltages are fixed at 15V. The channel width of the load FET in the static inverter is assumed, as an exception,

Transistor Parameters		
L	5	μm
W	1000	μm
μ	1	cm^2/Vs
V_T	2	V
T_I	0.10	μm
L_{OL}	0.10	μm
ε_I	4.24	

Diode Parameters		
d_{a-Si}	0.45	μm
S_D	900	μm^2
J_{RO}	1.97x10^{-21}	Acm^{-3}/V^{-3}
J_{DO}	1.00x10^{-8}	A/cm^2
T_G	600	K
N_B	1.00x10^{16}	cm^{-3}
ε_S	11.8	
V_{bi}	0.25	eV

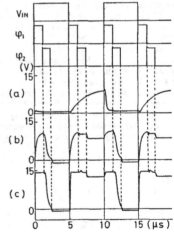

Table I Device and Circuit Parameters Used in the Analysis.

Fig.5 Output Voltage Waveforms of
(a) Static Inverter,
(b) Dynamic Inverter,
(c) FET-SBD Inverter.

to be 100μm wide, in order to obtain a good static inverting function, and S_D in the new circuit is assumed to be 900μm². For the static inverter, the output voltage V_{OUT} under a low input voltage, V_{IN}, condition is a midway to high voltage as shown in (a), and thus it is difficult to operate at 200kHz. For the conventional dynamic inverter V_{OUT} under a low V_{IN} condition approaches a high voltage but is not saturated yet as shown in (b), and thus 200kHz seems to be near the maximum possible operating frequency. While, for the new dynamic circuit, V_{OUT} is saturated, and thus the maximum frequency will be much higher than 200kHz. It should be mentioned that the discharging characteristics in the three inverters are approximately the same and that the difference in the operating speed thus arises from the charging behaviour. In the new dynamic circuit, charging is steeper than discharging, while in the conventional circuits the former is more gradual than the latter. The rapid charging and gradual discharging are the unique feature of the new circuit.

To demonstrate that the new inverter has satisfactory transfer characteristics for driving the following stage, a 4-bit ring-counter operation has been simulated. The initial condition is that just before the clock generator starts only the slave inverter of the first bit is in a high level and that the others are in a low level. Waveforms of 500kHz operation are shown in Fig.6, where V_1, V_2, V_3 and V_4 are the output voltages of each stage. Nearly-ideal voltage-forms are traveling around the inverter-chain directed from the first bit to the final bit without appreciable waveform distortion.

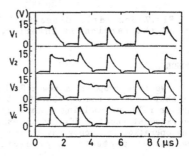

Fig.6 Output Voltage Waveforms of 4-bit Ring Counter.

It should be noted that the circuit delay time is reduced in proportion to the reciprocal drain current flowing through the FETs because the charging time is very short. Thus, if the FET having the highest mobility [14] reported to date (i.e., 1.9cm²/Vs) is integrated, the circuit can operate at about 1MHz. Also if the 1μm long vertical FET [15] having the same mobility is integrated, the circuit can operate at multi-MHz rates. Computer simulation predicted that the circuit under this condition can be operated at more than 10MHz. In this high speed operation, however, the charging time of the capacitor through the diode can not be neglected and an improvement of diode characteristics also becomes important.

EXPERIMENTAL RESULTS

A proto-type dynamic inverter has been fabricated. A schematic cross section is shown in Fig.7. The FET is of a self-aligned type formed by hot-ion implantation and lift-off techniques. The detailed fabrication process will be described elsewhere [16]. L and W were 5μm and about 1mm, respectively and the measured effective mobility was as low as 0.018cm²/Vs. The SBD has Mo as the barrier metal and its active layer thickness and cross sectional area were 0.45μm and 2500μm², respectively.

Input and output waveforms are shown in Fig.8, when the clock frequency is 40kHz. In this inverter, the maximum operating frequency is 80 kHz. The most important feature of the circuit, that is, the rapid charging and gradual discharging is clearly demonstrated. Thus, it can be concluded that by increasing the mobility and by optimizing the circuit parameters, the operation frequency will approach 1 MHz.

384

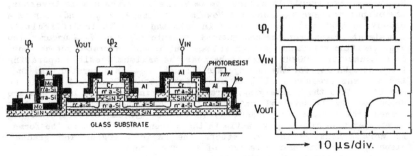

→ 10 µs/div.

Fig.7 Schematic Cross-Section of the Fabricated Inverter.

Fig.8 Clock Pulse (upper trace), Input Voltage (middle trace) and Output Voltage Detected by On-Chip FET Detector(lower trace).

CONCLUSION

We have proposed a new a-Si FET dynamic circuit whose response time is shortened to the discharging time of the load capacitor through the driver FETs. By applying a 5µm long, self-aligned FET, the circuit can be operated at about 1 MHz and by using a short-channel vertical-type FET, the operation frequency may be increased to multi-MHz rates. Experimental results supported this prediction.

ACKNOWLEDGEMENT

A part of this work has been supported by the Ministry of Education, Science and Culture of Japan through the Grants-in-aid for Scientific Project.

REFERENCES

[1] M.Yamano et al., Digest of 3rd Display Research Conf., p.214 (1983).
[2] Y.Okubo et al., Digest of SID '82, p.40 (1982).
[3] T.Hamano et al., Jpn. J. Appl. Phys., suppl.21-1, p.245 (1982).
[4] T.Tsukada et al., Extended Abstructs of 15th Conf. on Solid State Devices and Materials, p.205 (1983).
[5] H.Hayama et al., Proc. IEEE, 68, p.1349 (1980).
[6] Y.Nara et al., IEEE Electron Devices, ED-29, p.1646 (1982).
[7] P.G.LeComber et al., J de Physique, 42, p.c4-423 (1981).
[8] K.Hiranaka et al., IEEE Electron Device Lett., EDL-5, p.224 (1984).
[9] Y.Nara et al., J. Non-Crystal. Sol., 59&60, p.1175 (1983).
[10] O.Sugiura et al., Trans. IECE of Japan, J65-C, p.914 (1982).
[11] T.Suzuki et al., Jpn. J. Appl. Phys., 22, p.785 (1983).
[12] S.Ashok et al., IEEE Electron Device Lett., EDL-1, p.200 (1980).
[13] S.Kishida et al., Jpn. J. Appl. Phys., 22, p.511 (1983).
[14] K.Ishibashi et al., Appl. Phys. Lett., 41, p.454 (1982).
[15] Y.Uchida et al., IEEE Electron Device Lett., EDL-5, p.105 (1984).
[16] H.Okada et al., To be submitted.

AN ACTIVE MATRIX LIQUID CRYSTAL DISPLAY USING
A NOVEL AMORPHOUS SILICON SWITCHING DEVICE

W. DEN BOER, J.S. PAYSON, G. SKEBE*, L. SWARTZ*, and Z. YANIV*
Energy Conversion Devices, Inc., 1675 West Maple Road, Troy, Michigan 48084
*Ovonic Display Systems, Inc., 1896 Barrett Street, Troy, Michigan 48084

ABSTRACT

A 32x32 pixel liquid crystal display using amorphous silicon n-i-n diodes as switching elements has been developed. The display can be operated at duty ratios up to 1:256.

INTRODUCTION

Amorphous silicon alloy devices are currently being investigated for use as switching elements in active matrix liquid crystal displays (LCDs) [1,2]. Most of the research and development effort is concentrated on thin film transistors (TFTs). So far, however, the success of TFT based displays has been limited, mainly due to the difficulty of obtaining a high quality gate dielectric by plasma deposition and due to the presence of crossing conductors on the same substrate, both increasing the probability of defects in the display [3]. Therefore, interest in using sandwich type devices is growing. When a two terminal sandwich device is used, no gate dielectric is required, hence, a higher yield can be expected.
Metal-insulator-metal diodes [4] and a-Si:H back-to-back diodes [1], diode rings [5], n-i-n diodes [6,7] and Schottky diodes [8] have been proposed for incorporation in LCDs. In this paper the fabrication and operation of the first display using a-Si:H n-i-n diodes is described.

CHARACTERISTICS OF a-Si ALLOY NIN DIODES

Steady-state current flow in a-Si:H n^+-i-n^+ sandwich devices is space-charge-limited (SCL) and takes the form [9]:

$$J \sim \frac{V^{T_c/T+1}}{L^{2T_c/T+1}} ,$$

where V is the applied voltage, L the thickness of the intrinsic layer, T the absolute temperature and T_c the characteristic temperature defining the slope of the density of states at the quasi-Fermi level. Values of T_c ranging from 600 K to 1500 K are observed. Figure 1 shows the current-voltage curve of a 30x30 μm n-i-n diode with L=1.3 μm, which was used in the prototype display. The current is nearly symmetrical and $J \sim V^m$ with m=5.5.
Under AC excitation at frequencies between about 10 Hz and about 10 kHz the nonlinearity is enhanced up to m=10 to 20 [10,11]. In Figure 2, the DC curve and AC curves for a triangular sweeping voltage with different amplitudes are schematically shown. The nonlinearity enhancement of the SCL current in the presence of deep traps is a well known effect and was reported 30 years ago for CdS crystals [12,13]. This effect aids the

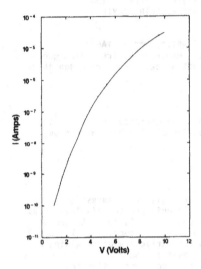

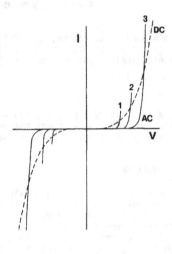

FIGURE 1. DC current-voltage
characteristic of 30x30µm n-i-n
diode with L=1.3µm.

FIGURE 2. Schematical I-V curves
of n-i-n diode at DC bias and
under AC excitation with
increasing amplitude of the
sweeping voltage (1-3).

performance of n-i-n diodes in the present application, since it increases
the charging current to selected pixels and reduces the leakage current
from non-selected pixels.

FABRICATION OF THE DISPLAY

A 32x32 array of 1mm^2 pixels incorporating n-i-n diodes was
fabricated. ITO (300Å), Cr (2500Å), n$^+$ a-Si:H (500Å), intrinsic a-Si:H
(1.3µm), n$^+$ a-Si:H (500Å) and Cr (1500Å) were sequentially deposited on a
glass substrate.
The first mask delineated the 30x30 µm n-i-n diodes (Figure 3a). With
the second mask pixels and lines were patterned (Figure 3b).
Silicondioxide (3000Å) was then deposited and 20x20 µm windows were etched
in the oxide (Figure 3c) and the Cr was removed from the pixels. Finally,
3000Å of metal was sputtered and with the fourth mask the top contacts
were patterned (Figure 3d). Figure 4 shows schematically a top view of
part of the display.
In this configuration, the diode is completely enclosed between
electrodes which have a larger area than the device. Therefore, light
cannot enter the device and leakage due to photocurrents in the intrinsic
layer is eliminated. It may be noted that in most TFT based displays,
special steps are required for the light shield.
On the top glass ITO counter electrodes and lines were patterned. The
completed device was filled with twisted nematic LC material and was
operated in the transmissive mode.

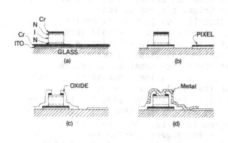

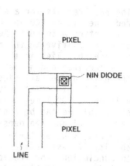

FIGURE 3. Process steps in display fabrication.

Figure 4. Schematical top view of part of the pixel array.

OPERATION OF THE DISPLAY

The simplest possible addressing scheme for this display is conventional multiplexing in a one-line-at-a-time mode. We used a standard frame time of 16.6 msec, a scan voltage of ±5V and a data voltage of 0 and ±5V for off and on pixels, respectively. After each frame time, the polarity of the scan and data voltages is reversed to avoid electrochemical degradation of the liquid crystal material. A contrast ratio of 8:1 was obtained and the display could be operated at duty ratios up to 1:256. Figure 5 shows a photograph of the operating display.

FIGURE 5. Operating 32x32 pixel LCD. In the upper left corner a void in the liquid crystal is seen.

388

DISCUSSION

The requirements for a single two-terminal device to address several
hundred lines in an LCD include: Symmetrical current-voltage
characteristics with sufficient nonlinearity (m>6), a small device
capacitance relative to the pixel capacitance and a threshold voltage
which preferably lies in the range of 2 to 10 V. In addition, a high
yield is required and a small number of process steps is desirable.

The n-i-n diode satisfies most of these requirements. Since the
intrinsic layer is relatively thick (1 to 2 μm), the capacitance of the
diode is about 100 times smaller than the pixel capacitance. The
threshold voltage can be continuously adjusted by modifying the thickness,
the Fermi level and the density of gap states of the intrinsic layer. The
device operates at electric fields below 10^5 V/cm, which reduces the
probability of destructive breakdown at inhomogeneities in the film.

In conclusion, the feasibility of an active matrix LCD with a-Si:H
n-i-n diodes as switching elements has been demonstrated. The n-i-n diode
is a relatively novel device and has to be further optimized for this
application.

ACKNOWLEDGMENT

We gratefully acknowledge the constant encouragement of
S. R. Ovshinsky.

REFERENCES

1. D.G. Ast, Proc. 1982 Int. Display Research Conf., p. 152.
2. S.R. Ovshinsky, Proc. Symposium on Glass Science and Technology,
   Vienna, Austria, July, 1984.
3. A.I. Lakatos, Proc. Flat Information Displays 1984 Conf., October,
   1984, San Jose, California.
4. D.R. Baraff, J.R. Long, B.K. McLaurin, C.J. Miner and R.W. Streater,
   IEEE Trans. El. Devices, ED. 28, 736 (1981).
5. S. Togashi, K. Sekiguchi, H. Tanabe, E. Yamamoto, K. Sorimachi, E.
   Tajima, H. Watanabe and H. Shimizu, Proc. 1984 Int. Display Research
   Conf., p. 141.
6. W. den Boer, M.J. Geerts and M. Ondris, J. Non-Cryst. Solids, 59/60,
   1185 (1983).
7. Z. Yaniv, C.E. Catchpole and W. den Boer, to be published.
8. Y. Nara, Y. Kudou and M. Matsumura, J. Non-Cryst. Solids, 59/60, 1183
   (1983).
9. W. den Boer, J. de Physique 42, C4-451 (1981).
10. M. Silver, E. Snow, M. Aiga, V. Cannella, R. Ross, Z. Yaniv, M. Shaw
    and D. Adler, J. Non-Cryst. Solids, 59/60, 445 (1983).
11. W. den Boer and A.F.P. Pop, Solid State Comm., 45, 881 (1983).
12. R.W. Smith and A. Rose, Phys. Rev., 97, 1531 (1955).
13. A. Rose, Phys. Rev., 97, 1538 (1955).

CONDUCTION AND THRESHOLD SWITCHING IN AMORPHOUS SILICON*

A. SA-NETO
Instituto Venezolano de Investigaciones Científicas, Apartado
1827, Caracas 1010-A, Venezuela.

ABSTRACT

We report on measurements of conduction, threshold switching and high frecuency oscillations (~100 MHZ) on amorphous
silicon produced by RF sputtering in lateral devices with a very
small ($4\mu m^2$ to $40\mu m^2$) metalic contact area. The results show
that the mechanism of threshold switching in this material is
not due to the formation of a metalic filament through diffusion
of the contacts. A small amount of Hydrogen ($PH_2/PA_r=0.025$) is
sufficient to inhibit threshold switching . The "Zero voltage"
resistance is reproducible and seems to be independent of the
switching events.

INTRODUCTION

Threshold and switching phenomena have appeared in a wide
variaty of materials, chalcogenides[1], glasses and oxides[2], and
organic polymer films[3-5]. They have been and continue to be topics of interest because of their possible practical implications as well as for a general understanding of the basic phenomena underlying the observed behaviour. While both threshold
and memory switching have been studied in detail[3-4] and either
Schottky[6], PooleFrenkel[7], or space-limited occurents[8] have
been used to explain the off-state characteristics no self
consistent explanation has been advanced to explain the
switching behaviour, due to among other things the complex
conduction mechanism existent in these materials.

MEASUREMENTS AND FABRICATION TECHNIQUES

In this paper results are presented using a lateral[9] metal
-(a) silicon-metal (LMSM) devices. Our results show that the
off-state characteristics are always ohmic at low applied voltages as threshold is approached, a clear exponential behaviour
sets in until switching occurs. The zero voltage resistances
are found to be consistent with a constant resistivity for the
amorphous silicon of the order 10^7 Ω.cm.

The LMSM devices are fabricated from sputtered chromium
2000 A° thick using standard photolitographic technics; an SEM
photograph of a finished device is shown in Fig. (1). Then silicon is sputtered on the center of the device, see table I for
all sputtering conditions.

The measurements were performed using a HP214 pulse generator and measuring voltage, across the LMSM device and a series

*This work is supported by CONICIT. Pr. SI-1185.

Fig 1. SEM photograph of LMSM device.

limiting resistance, with a tektronix memory oscilloscopy 7633.

EXPERIMENTAL RESULTS -

Initially we attempted to find threshold switching on devices fabricated with hydrogenated amorphous silicon. (table I).

Table I.- Sputtering conditions (Sputtering 13.56 MHZ)

a - Si - H_2 Films	Rate
$A_r + H_2$ (1 H_2 : 40 A_r) Total pressure 410^{-3} Torr, Tsubstrate ~ 200°C RF power - 300 watts	50 A°/min
a - Si	
A_r. Total pressure 410^{-3} Torr, Tsubstrate ~ 200°C RF power - 300 watts	50 A°/min
Cr.	
Ar Total pressure -810^{-3} Torr. Tsubstrate 300°C RF power - 350 watts	200 A°/min

Even though we decreased the hydrogen partial pressure to the minimum that could be accurately measured by our apparatus, ($PH_2/P_{Ar}=0.025$) no switching phenomena was encountered up to fields in the order of 10^6 v/cm, when H_2 was turned off the a - Si - films exhibited just prior to threshold (~2 10^5 v/cm) high frecuency oscillations on the order of 100 MHZ, when the field was slightly increased threshold switching would occur, Fig. 2. It becomes difficult to assert which type or types of mechanism determine the off-state conduction;assuming a Richardson - Schottky type an activation energy of 0.6 ev can be obtained, if we try to obtain a high frequency dielectric constant through this equation we get too high a value for these films. This indicates that the mechanism cannot be described by this simple equation even though a good functional fit can be obtained for about three decades of the I-V curve.

CONCLUSIONS

The I-V characteristics were S shaped similar to those of tellurium-rich chalcogenide glasses[10]. There the phenomenon of threshold switching is attributed to phase separation of multi-component glasses as to create a conducting channel through electronic or thermal means[11]. Here one can guess a mechanism of recrystallization, but we are only prepared to say that in these devices the phenomena of threshold is not due to field aided diffusion of the electrodes[9]; if this was the case one would expect threshold switching even with H_2 added.[12]

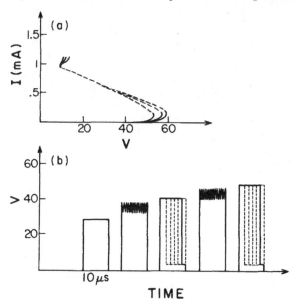

Fig 2.   I-V and pulse characteristics of   LMSM devices.

REFERENCES

1. D. Adler,H.K.Henisch,and N.Mott,Rev.Mod.Phys. 50,209 (1978)
2. G.Dearwaley,A.M.Stonneram, and D.V. Morgan, Rep. Prog.Phys. 33,1129 (1970).
3. H.K.Henisch and W.R.Smith, Appl. Phys.Lett. 24, 589 (1974).
4. H.K.Henisch, J.A.Meyers, R.C.Callarotti, and P.E.Schmidt, Thin Solid Films 51, 265 (1978).
5. P.E.Schmidt, J.G. Mena, and R. Callarotti, Thin Solid Films 55,9 (1978).
6. P.E. Mehendru, N.L.Pathak, Satbir Singh, and Mchendru , Phys. Status Solidi (A) 38, 355 (1976).
7. W.Wollman and H.U.Poll, Thin Solid Films 26,201 (1975).
8. J.Chutia and K.Barua, Thin Solid Films 55, 387-390 (1978).
9. A.Sa-Neto,M.Octavio,R.Callarotti,P.Schmidt and P.Esqueda, I. Appl.Phys. 51(7),July 1980.
10. D.Adler,"Amorphous Semidonductors" (CRC press 1971).
11. D.M.Krell and M.H.Cohen, J.Non-Cryst. Solids 8-10,544 (1972).
12. An other possible mechanism of breakdown was proposed by the referee. He suggested that the observed effects could be indicative of the breackdown of a contact resistance having a resistivity of $10^7 \, \Omega$ cm as compared to the ohmic resistance of unhydrogenated a-Si which is typically on the order of $10^4 \, \Omega$ cm.

PART XII

Photoreceptors

PROBLEMS IN a-Si PHOTOELECTRIC DEVICES;
PHOTORECEPTOR AND VIDICON

ISAMU SHIMIZU
The graduate school at Nagatsuta, Tokyo Inst. Tech., 4259 Nagatsuta, Midori, Yokohama, 227 Japan

ABSTRACT

The history in developing a-Si image devices, photoreceptor and vidicon is briefly reviewed. Some problems to be solved are discussed for the device structures and preparation techniques of Si-base alloys and Si itself from technical point of view.

INTRODUCTION

Recently copying machines installing amorphous silicon (a-Si) drum have been sent to the market as a consequence of efforts for several years since we proposed firstly of its possibility. Image devices such as photo-receptor of electrophotography and vidicon including solid state image sensors have been said ,as well as solar cells, being one of the most promising application of a-Si and related materials,making use of their inherent advantages;
1. easiness to prepare the films with structureless smooth surface,
2.strong optical absorption cover whole visible region,
3.non-toxic and
4.excellent photoconductivity.
Moreover high electric resistivity by the aid of its small mobility make it a promising material for these imaging devices, requiring both high photo-conductivity gain and large resolving power simultaneously. Despite of these distinctive advantages, enthsiastic efforts have been devoted to developing the devices and will be neccessary further. And therefore we will review briefly the progresses in the a-Si image devices and discuss some problems lying further from technological point of view.

HISTORY

In Table 1 marked events performed in developing image devices, i.e. photoreceptor and vidicon are summed up. Within a short duration from the first paper triggering the studies on a-Si presented by Spear and LeComber[1], the idea of a-Si photoreceptor had been proposed in patents, insisting the possibility and their advantages of the a-Si photoreceptor. Couple of years later, we firstly pointed out the device structure to give

Table 1
Marked events in developing a-Si photoreceptor and vidicon

year	Photoreceptor	Vidicon
1977	Patents(K.Kempter; Siemens and Y.Hirai et al, Canon)	a-Si vidicon(Y.Imamura et al (Hitachi)
1979	a-Si photoreceptor with blocking (Tokyo Inst. Tech.)	
1980	a-Si drum (N.Yamamoto et al, Univ. Osaka Pref.) High resistive films; a-Si:H:O:B a-Si:H:N:B and a-Si(excess H)	
1981	High speed deposition (>65 A/s) with higher silanes (K.Ogawa et al, Tokyo Inst. Tech.)	The role of blocking in a-Si vidicon (S.Oda et al Tokyo Inst. Tech.)
1982	a-Si laser printer (Y.Nakayama et al, Univ. Osaka Pref.)	
1983	a-Si drum (reactive sputtering) (K.Tamahashi et al, Hitachi) Compensated a-Si:H:P:B(F.Jansen et al Xerox)	a-Si vidicon scanned by high velocity e-beam(C.Kusano et al, Hitachi)
1985	Laser Copier(R) with a-Si drum(Canon)	a-Si/IT-CCD image sensor (N.Harada et al ,Toshiba)

sufficient charge retentivity on photoconductive a-Si prepared in the optimum condition, which was consisting of triple layers, i.e. P (or B)-doped a-Si for prohibiting reprenishment of carriers from the conductive substrate, photoconductive a-Si and top dielectric layer for passivation.[2] We had verified experimentally the fact that surface charges given by corona on the a-Si were sustained for time longer than that of the dielectric relaxation time anticipated from its electric conductivity as far as inhibiting carriers injection from the electrodes. The first image reconstructed from the a-Si photoreceptor by conventional electrophoto-graphic processing is illusterated in Fig.1. The a-Si drum furnished in a copying machine is also shown in the figure in addition to its schematic diagram.

Successively a practical size of a-Si drum was fabricated by Univ. Osaka Pref. 's group[3] as a single layer photoreceptor with high resistive a-Si achieved by doping foreign elements such as oxygen-boron (a-Si:H:O:B)

Several attempts had been made in fabricating high resistive a-Si by doping as the similar manner i.e., a-Si:H:N:B[4] or a-Si:H:P:B.[5] The films including excess hydrogen(>20 atm%) prepared by reactive sputtering showed high resistivity. The first a-Si vidicon was prepared by Y.Imamura et al (Hitachi) using this film.[6] In the early stage, very slow growth rate in preparing photoconductive a-Si had often used as a strong evidence to support pessimistic view of a-Si photoreceptor. The experimental results establishing the fast growth of photoconductive a-Si of 20 μm thick by

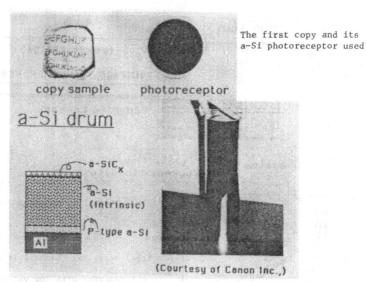

The first copy and its
a-Si photoreceptor used

(Courtesy of Canon Inc.)

Fig.1 a-Si photoreceptor

glow discharge of higher silanes encouraged us greatly.[7] Since then, the industries had extensively devoted their efforts to commercialize the a-Si drum. Consequently we can see a sophisticated copier installing a-Si photoreceptor. According to their catalogues, fascinating performances have been realized as follows;

1. high image quality (16 pel / mm)
2. high durability (> 500,000 sheets guarantee and more than one million sheets achievable without any deterioration),
3. high photoconductivity gain and capability to use a diode laser as a light source.

Turning now to the a-Si vidicon. As analogous to the photoreceptor, utilization of the primary photocurrent by adopting blocking electrodes is expected to offer advantages to the devices. As shown in Fig.2, a hetero-phase blocking using very thin dielectric films (SiNx or SiOx) are adopted at the contact with the illuminating electrode. By the aid of this ,the charge depletion condition is spreaded at whole photoconductive layer without any loss of the light at the electrodes. Consequently the current level in the dark is maintained at low and the maximum gain of unity is attained under low electric field (< 5×10^5 V/cm).[8] High resolution of >800TV lines is reconstructed with a-Si target (2/3 " dia). Despite of these advantages, a-Si vidicon scanned with slow e-beam have not been commercialized yet because of its long decay lag attributed to the high impedance of the e-beam and slow response of the holes in a-Si.

C.Kusano et al (Hitachi) had successfully solved this problem using high velocity scanning beam, where electrons played the majour role.[9] In Fig.3, the lag characteristics of both cases are compared respectively.

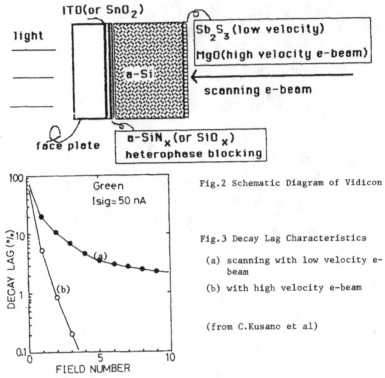

Fig.2 Schematic Diagram of Vidicon

Fig.3 Decay Lag Characteristics

(a) scanning with low velocity e-beam

(b) with high velocity e-beam

(from C.Kusano et al)

Surprisingly fast progress has been performed in the solid state image devices as a strong competitor of the image pick-up tube. In this devices, a-Si is a promising material either. N.Harada et al(Toshiba) had recently reported high quality image sensor consisting of CCD for the signal transfer and a-Si photoconductive layer deposited on it.[10]   To avoid the defect in the image, it was essential to be smoothed the surface of CCD with polyimide before depositing a-Si layer. In the glow discharge method, some morphological inhomogeniety should be made when we prepared the films on a substrate with geometrical structure.   The photoconductive layer adopted in this device was made of multi-layer, i.e. a-SiCx at the bottom to avoid smearing images, photoconductive a-Si and p-type a-SiCx for the blocking at the concact with the transparent electrodes, which is essentially similar to that of vidicon.

PROBLEMS AND SOLUTIONS

A. PHOTORECEPTOR

The multi-layers device structure shown in Fig.2 has been generally adopted, so as to make the best use of distinctive characteristics of a-Si. Problems , if there are, appears at the top layer for the passivation, giving stabvility in the charge acceptance and also in the image quality. Because either the electrostatic latent image by exposing to corona and light illumination or corresponding tonor image is formed on this layer, almost opposite requirements, namely high sensitivity and high stability should be simutaneously satisfied in this top materials. The stability is achievable by means of "insulative" thin a-SiCx layer deposited on the top as a trade-off of the possibility causing deterioration in the gain and image quality resulting from the collection of the space charge at the contact. The most simple solution for this difficulty is given by utilizing a photoconductive alloy, a-SiCx or a-SiNx, with wider gap of about 2.0 eV as the top layer. The stability of both the charge acceptance and the electrostatic latent image was surprisingly enhanced by the aid of its high resistivity without any collection of the space charge because of its excellent photoelectric property. Very rcently we have succeeded in making a-SiCx exhibiting excellent photoconductivity except its very slow growth rate.

One of the most promising application of the a-Si drum is for the laser printer furnished diode laser made of AlGaAs compounds, emitting light at near ir region (700--800 nm in wavelengths). For these purpose, the a-Si drum had met serious trouble, namely, poor sensitivity and gohst image caused by reflected light at the substrate, which are all attributed to its poor optical absorption of a-Si at this region.

a-SiGex with its gap of around 1.4--1.5 eV must be a candidate as the photoconductive layer, absorbing effectively light of long wavelengths within one micron meter thick. Recently, great progresses have been brought in making the Si-base alloys. It is anticipated to be important to control "the chemistry" for avoiding the preferencial attachment of terminator between these páir of elements; Si and C in a-SiCx and Si and Ge in the a-SiGex. We have chosen fluorides instead of hydrides as the sources and succeeded in preparing alloys, a-SiCx, a-SiGex and a-SiNx exhibiting high photoconductivity under controlling hydrogen radicals as "the activator". Typical characteristics are summed up in Table 2.

Now attempts have been made of fabricating image devices with these photoconductive alloys, so as to know the nature of the heterophase contacts between a-Si/a-SiGex and a-Si/a-SiCx, respectively.

Apart from the device structure, the establishment of the mass-production system is considered to be the most important subject for the a-Si photoreceptor. Despite of the distinguished progress undertaken in the fabrication of a-Si drum with high growth rate, prolonged efforts must be devoted to some new techniques being adequate to the production, safely and ecomonically.

B. VIDICON

Marked improvement should be done in its lag-characteristics of a-Si vidicon scanned by slow e-beam. Distinctive reduction is expected in the lag remained at low level for the long time if the hole-transport in a-Si

Table 2   Characteristics of a-Alloys prepared from Fuluorides

	a-SiGe$_x$	a-SiC$_x$	a-SiN$_x$	
gases	SiF_4-GeF_4	SiF_4-C_2F_6	Si_2F_6-NF_3	SiH_4-NF_3
conductivity (S/cm)	10^{-9}	10^{-11}-10^{-12}	10^{-12}	$< 10^{-14}$
range (cm^2/V)	10^{-5}--10^{-6}	5×10^{-7}	10^{-7}-10^{-8}	
free spins (spins/cc)	$< 7 \times 10^{15}$	$< 3 \times 10^{16}$		
E$_o$ (eV)	1.4	1.9--2.1	2.0,	> 6.0

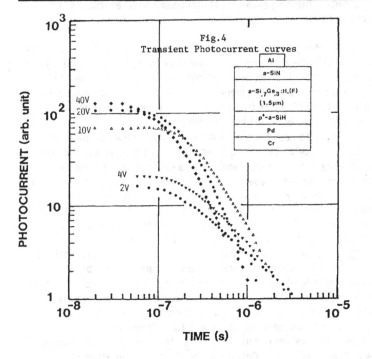

Fig.4
Transient Photocurrent curves

can be improved by the reduction of the localized state located at the valence band. Of particular interest to us is the non-dispersive hole transport attained in the a-SiGex films prepared from the fluorides. Typical current curves as a function of time are shown in Fig.4. These results encouraged us to challenge in making higher quality materials.

For the purpose of the photoconductive layer on the solid state image devices, the quality of a-Si and related materials is expected to be improved further.

AKNOWLEDGMENT

The auther wishes to express his gratitude to Mr.E.Kondo and T.Fukuda (Canon) for the a-Si photoreceptor , to Mr.T.Hirai (Hitachi) for the a-Si vidicon and to Dr.N.Harada (Toshiba) for the solid state device for their kind discussions. The auther also wishes to thank Dr. S.Oda and Dr.J.Hanna for their helpful discussion.

References

1. W.E.Spear and P.G.LeComber; Solid State Commn. 17,1193(1975)

2. I.Shimizu,T.Komatsu,K.Saito,and E.Inoue; J. Non-cryst. Solids, 35&36,773 (1980)

3. N.Yamamoto,Y.Nakayama,K.Wakita,M.Nakano, and T.Kawamura; Jpn. J. Appl. Phys., 20,suppl 20-1,305(1981)

4. H.Watanabe, K.Katoh, M.Yasui and Y.Shibata; J. Non-cryst Solids, 59&60, 605(1983)

5. F.Jansen,J.Mort,S.Grammatica and M.Morgan; J. Appl. Phys.,55, 4128(1984)

6. Y.Imamura,S.Ataka,Y.Takasaki,C.Kusano,T.Hirai and E.Maruyama; J. Appl. Phys., 35, 349(1979)

7. K.Ogawa, I.Shimizu, and E.Inoue, Jpn. J. Appl. Phys.,20, L639(1981)

8. S.Oda,K.Saito,H.Tomita,I.Shimizu and E.Inoue; J. Appl. Phys., 52, 7275 (1981)

9. C.Kusano,S.Ishioka,Y.Imamura,Y.Takasaki,Y.Shimamura,T,Shirai and E.Maruyama; Proc.IEEE Int. Electron Device Meeting (1983) pp 509

10. N.Harada,S.Uya,Y.Hayashimoto,Y.Endo,T.Adachi and O.Yoshida; 1985 IEEE Int. Solid-State Circuits Conf.,pp270

HIGH DEPOSITION RATE AMORPHOUS SILICON ALLOY
XEROGRAPHIC PHOTORECEPTOR

S.J. HUDGENS and A.G. JOHNCOCK
Energy Conversion Devices, Inc., 1675 West Maple Road, Troy, Michigan 48084

ABSTRACT

A new multilayer amorphous silicon alloy photoreceptor has been deposited at rates exceeding 36 µm/hr. using 2.45 GHz microwave glow discharge. The device whose structure is Al/a-Si:H:F (B-300)/a-Si:H:F (B-10)/a-Si:H:F:C is deposited in a powderless plasma deposition process which exhibits gas utilization efficiency approaching 100%. The xerographic performance of a 28µm device is: $V_{sat} \sim 1100$ V for a +7 KV corona; dark half decay time ≈ 5 sec; and photosensitivity ~ 0.3 µJ/cm^2 at λ = 650 nm. Stable, high quality xerographic images are obtained with these photoreceptors.

INTRODUCTION

The usefulness of a-Si alloys for electrophotographic applications [1,2] has been apparent for several years. Despite the well established advantages of these materials over conventional organic and chalcogenide based photoreceptors, commercial manufacturing has been hindered by the poor feedstock gas utilization efficiency (10%) and the low deposition rate (~3-5Å/sec) typically encountered in the a-Si alloy R.F. glow discharge deposition process. Higher deposition rates have been reported using disilane [3] but cost considerations restrict this approach for manufacturing at the present time. Deposition of amorphous silicon alloy films on a research scale has been reported [4] at rates as high as 50Å/sec using high R.F. power density coupled with techniques which control the substrate self-bias but, so far, attempts to significantly increase deposition rate by simply increasing R.F. power density have given unacceptable degradation in material quality [5] and upon further increase in power density increasing gas phase nucleation and powder production.

The plasma chemistry [6,7,8] and film growth mechanisms [9,10] occuring in SiH$_4$ glow discharges have been recently investigated. Ross and Jaklik [11] argue that even under typical RF glow discharge high deposition rate conditions which do not yet result in powder generation, the formation of polymer species in the plasma and subsequent incorporation in the growing film leads to electrically active defect sites.

In this paper we report the development of a low pressure powderless microwave glow discharge deposition process which effectively solves the plasma polymerization problem resulting in the production of a-Si:F:H photoreceptor drums with acceptable xerographic performance at deposition rates exceeding 36µm/hr and with essentially 100% gas utilization efficiency. The development of this technique follows from the work of Ovshinsky [13] regarding the fundamental role of free radical precursors in the growth of amorphous silicon alloy films and has resulted in the successful scale up of a proprietary [14] prototype manufacturing process.

EXPERIMENTAL

The thin films of a –Si:F:H used in this study were prepared in the research scale, microwave glow discharge deposition system illustrated in Figure 1. The sample holder is designed to hold either 1 in^2 glass

slide substrates or "drum test pieces" which are made sections of a 130mm
diameter Al cylinder. The microwave antenna is fabricated from a slotted
section of 1/4" diameter rigid coaxial cable. Microwave power is coupled
into the antenna through a three stub matching network from a coaxial
cable connected to a directional coupler containing appropriate detectors
to measure forward and reflected power. A low power magnetron (125W)
operating at 2.45 GHZ provides the source of microwave power. Depositions
occur at a pressure of 3×10^{-2} Torr with applied microwave power of 115W,
resulting in deposition rates of 100Å/sec on the substrates which are held
at a temperature of 500K. Measurements of the quantity of a-Si:F:H
deposited on the tube walls, substrate holder, and sample substrates
indicates that essentially all of the silicon containing feedstock gas is
deposited as a-Si:F:H with no indication of powder production. Even
higher powderless deposition rates (up to 400Å/sec) have been observed in
a copier drum production prototype deposition machine [14] using low
pressure microwave glow discharge.

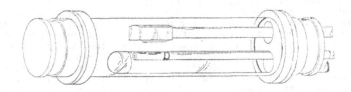

Figure 1.   Research scale microwave deposition chamber

MICROWAVE GLOW DISCHARGE DEPOSITION

High deposition rates require high densities of precursor free
radicals in the plasma. However, as previously mentioned, to obtain low
defect density amorphous silicon alloys it is necessary to suppress
formation of polymeric species. This is accomplished in our system both
by operating at low pressure which reduces polymer forming collisions
between excited and feedstock species and by utilizing fluorinated
feedstock gases which ab initio suppress polymer formation because of the
strong silicon fluorine bond. Microwave excited plasmas can generate high
free radical densities at low gas pressure using much lower power than
plasmas created by lower frequency sources. The lower microwave power
allows one to avoid the large self bias and consequent high energy ion
bombardment which would result from the use of high power RF excitation to
achieve the same free radical density.

MATERIAL PROPERTIES

The fabrication of electrophotographic devices requires simultaneuous
optimization of several materials properties. In particular, since the
device must be $\approx 25\mu m$ thick, deposition parameters which produce low
stress material must be selected. This must be done as well as maintaining
electronic properties adequate to insure proper photocopier performance.

Undoped films typical of high deposition rate material which have been
optimized for photocopier application were deposited on 7059 glass and

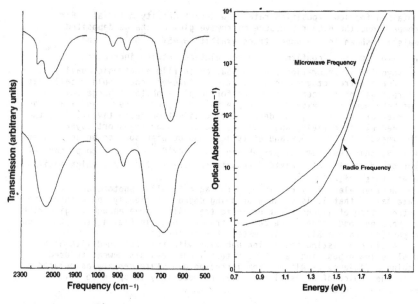

Fig. 2
I.R. Absorption Spectra
(lower trace p=50μm; upper
trace p=400μm)

Fig. 3
Sub Bandgap
Optical Absorbtion

single crystal Si substrates and characterized electrically and
optically. These ≈ 3μm thick undoped films were deposited from a gas
mixture of 30 SCCM SiH_4, 7 SCCM SiF_4, and 5 SCCM H_2.

The principle features seen in the I.R. absorption spectrum in Fig. 2
are the Si-H absorption modes. One can see that the incorporated hydrogen
exists in both the monohydride, and the dihydride or polyhydride
configuration. For comparison, the I.R. spectrum of films deposited at a
higher gas pressure is also shown. The expected trend towards greater
incorporation of dihydride and polyhydride resulting from plasma
polymerization is clearly seen.

Fig. 3 shows the sub-bandgap absorption characteristics of both high
deposition rate a-Si:F:H and typical low defect density low deposition
rate material. One sees that the Urbach tails resulting from localized
tail state to band edge transitions in the two films are essentially
identical. The optical absorption data for $\alpha < 10 cm^{-1}$, due to transitions
from midgap localized states to the band edge shows differences in the two
films. Here we see enhanced absorption in the high deposition rate film
indicating a somewhat larger mid-gap defect density. Experience with this
spectroscopic technique using simultaneous electrical measurements [15]
have shown us that this enhanced absorption in the high deposition rate
material results from a density of states at the Fermi level, $g(E_f) \simeq 10^{17}$
$eV^{-1} cm^{-3}$. This value is higher than the value $g(E_f) \simeq 10^{16} eV^{-1} cm^{-3}$
typical of good quality R.F. glow discharge a-Si alloy. This increased
deep trap density is also borne out by the co-planar photoconductivity
$\sigma_p = 4 \times 10^{-6} \Omega^{-1} cm^{-1}$ and dark conductivity, $\sigma_D = 1 \times 10^{-10} \Omega^{-1} cm^{-1}$ measured for the
high deposition rate film under AM1 illumination and at room temperature.
This value of σ_p is approximately an order of magnitude lower than that

obtained for low deposition rate, low defect density material. For comparison, the higher pressure microwave glow discharge deposited material shown in the upper trace exhibits $\sigma_p = 5 \times 10^{-8} \Omega^{-1} \text{cm}^{-1}$ and $\sigma_D = 9 \times 10^{-11} \Omega^{-1} \text{cm}^{-1}$ confirming the correlation between increasing polyhydride incorporation and degradation in film electronic quality.

For electrophotographic applications, however, one should recall that a practical photoreceptor requires for complete photodischarge only that the Schubweg, $\mu\tau\mathcal{E}$ exceed the sample thickness, L. Here, μ and τ are the carrier drift mobility and deep trapping life time respectively, $\mathcal{E}$ is the applied electric field, and L is the photoreceptor transport layer thickness. For $L \simeq 20\mu m$ and at typical photocopier surface potentials of $\simeq 10^3 V$, this requirement, $\mu\tau\mathcal{E} > L$ implies $\mu\tau > 4 \times 10^{-9} \text{cm}^2/V$. A co-planar AM1 photoconductivity of $\sigma_p = 4 \times 10^{-6}$ results from $\mu\tau = 2 \times 10^{-8} \text{cm}^2/V$, which easily meets this criterion.

Another electrical requirement for an a-Si alloy photoconductor material is that it is capable of being doped efficiently and without the introduction of a density of unwanted trap sites large enough to give rise to hopping conduction. One can see from the data in Fig. 4, that the high deposition rate a-Si:F:H also meets this criterion.

Again contrasting the doping behavior with that obtained with high quality low deposition rate R.F. material one sees a somewhat reduced doping efficiency resulting from increased gap state density associated with the choice of deposition parameters necessary to optimize film mechanical properties.

ELECTROPHOTOGRAPHIC PERFORMANCE

An electrophotographic device structure of the type Al/a-Si:H:F (B-300)/a-Si:H:F (B-10)/a-Si:H:F:C was fabricated using the procedures described in Section 2. Here the gas phase ratio of boron to silicon is

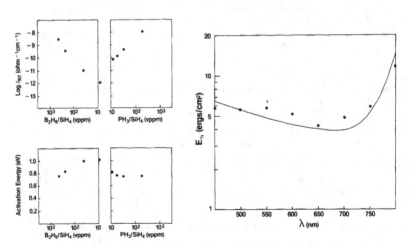

Fig. 4. Room temperature conductivity and conductivity activation energy as a function of gas phase dopant concentration.

Fig. 5. Electrophotographic spectral sensitivity for device described in text.

indicated in parenthesis in units of ppm. The a-Si:C:F:H top
protective layer was prepared from the gas mixture: 10 SCCM SiH_4; 4 SCCM
SiF_4; 22 SCCM CH_4. This top protective layer, deposited using the same
net microwave power, gas pressure, and substrate temperature as for the
other layers, has a Si/C ratio of (as determined by Auger analysis) 6.7;
an optical bandgap of E_{opt} = 2.15eV; and room temperature dark
conductivity of σ_D = 4×10^{-13} $\Omega^{-1} cm^{-1}$ and is thermally activated, with an
activation energy of $\Delta E=0.94eV$. This layer is used to eliminate the
humidity dependent instability in charge acceptance which has been
reported [1] when the naturally occuring oxide blocking layer forms at the
free surface of an unprotected device.

Electrophotographic performance of this device is typical of other
high quality a-Si alloy devices [1] reported in the literature. Using 7KV
positive corona charging we obtain for a 28μm thick device a saturation
surface potential, V_{sat} = 1100V with a dark half decay time ≈ 5 sec.
Under negative corona the saturation surface potential varies from device
to device from 10V to 75V. This unipolar charging characteristic is
typical of devices made with boron doped a-Si alloy back blocking layers.
The wavelength dependance of photosensitivity is shown in Fig. 5. For
this data, the initial surface potential was established at V_0=500V and
consequently, the photosensitivity obtained is very near that predicted
for a unity gain xerographic device. The fall off at long wavelength
follows the optical absorption co-efficient of the material. Loss of
sensitivity at short wavelength results due to absorption in the 2.15 eV
optical gap top protective layer, and can be controlled by minimizing the
thickness of the layer. Good photosensitivity for strongly absorbed light
demonstrates that, for the surface potentials used in this measurement,
$\mu\tau\mathcal{E} > 28μm$. A direct measure of $\mu\tau$ was obtained by measuring the initial
photodischarge current as a function of surface potential. For conditions
of weak light exposure this data can be interpreted using the "Hecht
Formula" approximation:

$$J = neF (\mu\tau\mathcal{E}/L) [1-exp(-L/\mu\tau\mathcal{E})] \qquad (1)$$

Here η is the photogeneration quantum yield and F is the photon flux.
Fitting our data to this expression gives $\mu\tau$ = 5 x 10^{-9} cm^2/V.
We see that for V_{sat} = 1100V and d = 28μm this gives $\mu\tau\mathcal{E}$ = 39μm which
exceeds the sample thickness as we had previously inferred from co-planer
photoconductivity data.

CONCLUSION

Amorphous silicon alloy
electrophotographic devices have
been deposited at rates exceeding
36μm/hr. High quality copies
made with these devices, using the
conventional dry toner process
are shown in Fig. 6. They exhibit
resolution in excess of 6 line
pairs/mm with no indication of
image flow as reported by other
authors [16].

The low pressure microwave
glow discharge deposition process
used to fabricate these devices
has been succesfully scaled up to
fabricate these full size copier

Fig. 6. - Photocopy of resolution
chart obtained with high
deposition rate device.

drums and consequently overcomes a major barrier in the commercial manufacturing of amorphous silicon alloy photoreceptors.

ACKNOWLEDGMENTS

The authors wish to thank J. S. Payson for photothermal deflection spectroscopy measurements and to gratefully acknowledge helpful discussions with S. R. Ovshinsky, S. Guha, S. S. Chao and W. den Boer. They also wish to thank C. J. Petrides and L. R. Peedin for help in sample preparation and to thank M. L. Reid and C. A. Smolinski for assistance in sample characterization.

REFERENCES

1. I. Shimizu, T. Komatsu, K. Saito and E. Inoue, J. Non-Crystalline Solids 35 and 36, (1980) 773.
2. S. R. Ovshinsky, M. Izu, U.S. Patent number 4,217,374.
3 K. Ogawa, I. Shimizu and E. Inoue, Jpn. J. Appl. Phys. 20 (1980) L639.
4. T. Hamasaki, M. Ueda, A. Chayahara, M. Hirose and Y. Osaka, Appl. Phys. Lett. 44 (1984) 600.
5. R.A. Street, J.C. Knights and D.K. Biegelson, Phys. Rev. B18 (1978) 1880.
6. G. Turban, Y. Catherine and B. Grolleau, Plasma Chemistry and Plasma Processing 2 (1982) 61.
7. P.E. Vanier, F.J. Kampas, R.R. Corderman and Rajeswaran, J. Appl. Phys. 56 (1984) 1812.
8. Ivan Haller, J. Vac. Sci. Technol. A1 (1983) 1376.
9. Bruce A. Scott, Jeffrey Reimer and Paul A. Longway, J. Appl. Phys. 54 (1983) 6853.
10. J.P.M. Schmitt J. Non-Crystalline Solids 59 and 60 (1983) 649
11. R.C. Ross and J. Jaklik Jr., J. Appl. Phys. 55 (1984) 3785.
12. This deposition technology and materials are protected an issued U.S. Patent, Ovshinsky et al. number 4,504,518.
13. These concepts are most recently discussed in Stanford R. Ovshinsky, Proc. Int'l Ion Engineering Congress, Kyoto 1983, 817.
14. G. Fournier and E. Bjornard, unpublished report.
15. S. Payson and S. Guha, private communication.
16. F. Jansen, J. Mort, S. Grammatica, M. Morgan, and I. Chen, J. Non-Crystalline Solids 66 (1984) 357.

CHARACTERISTICS OF THE BLOCKING LAYERS IN THE A-Si:H PHOTORECEPTOR

H.KAKINUMA, S.NISHIKAWA, T.WATANABE AND K.NIHEI
Research Laboratory, Oki Electric Industry Co., Ltd., Higashiasakawa-cho 550-5, Hachioji-shi, Tokyo 193, Japan

ABSTRACT

The effects of characteristics of the blocking layers in multilayered a-Si:H photoreceptors on electrophotographic properties are reported. The structure of the photoreceptor is $Al/a-Si:H(p)/a-Si:H(i)/a-Si_{1-x}C_x$. The thickness and boron doping ratio of the bottom p-type layer and the thickness and compositional fraction x of the surface $a-Si_{1-x}C_x$ layer are varied. It is shown that the dark decay time is particularly sensitive to the thickness and doping ratio of the p-type layer. These dependences are discussed assuming a space charge layer in the p-type layer. It is also shown that the thickness and x of the $a-Si_{1-x}C_x$ layer greatly influence the residual voltage and effective quantum efficiency of photocarrier generation.

INTRODUCTION

Hydrogenated amorphous silicon (a-Si:H) and its alloys are suitable materials for electrophotographic photoreceptors because of their higher photosensitivity and hardness compared with conventional materials such as chalcogenides and ZnO. However, the resistivity of undoped a-Si:H is not high enough to maintain a sufficient amount of charges on the surface. There are two main methods to overcome this. One is to produce high-resistivity ($>10^{13}\Omega^{-1}cm^{-1}$) films by adequate doping of oxygen, nitrogen and boron [1]. The other is to lay blocking layers at the substrate/a-Si:H interface and/or on the surface [2]. The latter method has superior merit in that the electrophotographic properties are less sensitive to variation in electrical properties such as conductivity of the intermediate a-Si:H layer which occupies most of the thickness of photoreceptors. Instead, it is expected that the characteristics of the blocking layers have a large influence on the electrophotographic properties.

We have previously reported on the fabrication of practical a-Si:H drums by the glow-discharge method [3], employing a multilayered structure of $Al/a-Si:H(p)/a-Si:H(i)/a-Si_{1-x}C_x$. The drums showed excellent electrophotographic properties and provided clear images. In this paper, we describe the thickness, doping and composition effects of the two blocking layers on the electrophotographic properties.

EXPERIMENTAL

The structure of photoreceptors in this study was as same as that mentioned above and is schematically shown in Fig.1. All the layers were deposited on aluminum plates by the RF decomposition of adequate mixtures of SiH_4, B_2H_6 and C_2H_4. Detailed preparation conditions are described elsewhere [3]. The thickness and gaseous doping ratio $R = [B_2H_6]/[SiH_4]$ for the p-type layer and the thickness and compositional fraction x for the $a-Si_{1-x}C_x$ layer were varied. The thickness of the intermediate a-Si:H layer was 3 μm.

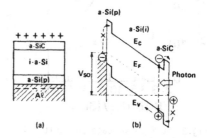

Fig.1 (a) Structure of the multilayered photoreceptor. (b) Schematic energy diagram of the photoreceptor.

The initial surface voltage Vso, dark decay time t_d, light exposure $E_{1/2}$ and quantum efficiency η were measured for these photoreceptors with various blocking layers. The measurements were carried out using a system with a turn table, a corona charger, an electrometer and light sources. The photoreceptor mounted on the turn table was initially charged positive by the +7 kV corona charger and Vso was successively measured in the dark by the electrometer. The values t_d, $E_{1/2}$ and η were derived from decay curves in the dark for t_d and under constant light exposure for $E_{1/2}$ and η, respectively.

We defined t_d as the time at which the surface voltage decayed to Vso/2, and $E_{1/2}$ as the light energy required to make the surface voltage decrease to Vso/2 under constant intensity (10^{12}photons/cm^{2}s) of exposure. Therefore, $E_{1/2}$ is inversely proportional to the photosensitivity. The effective η value can be calculated from the equation, $\eta=K(dV/dt)/(d \cdot F)$ [4], where dV/dt is the initial photo-decay rate, d is the total photoreceptor thickness, F is the photon flux and K is a constant.

RESULTS AND DISCUSSION

P-type layers

Figure 2 shows the R dependence of Vso/d, t_d and $E_{1/2}$ for a constant d_p (=0.4 µm). Among the three quantities t_d is most sensitive to R. It decreases rapidly with increasing R. In contrast, the other two are much less sensitive to R.

Figure 3 shows the d_p dependence of Vso/d, t_d and $E_{1/2}$. The t_d value again is the most sensitive quantity and takes the maximum value at d_p=0.6 µm. The Vso also shows a similar dependence but its change is much smaller. It has been reported that a phosphorus doped n-type layer for negative charging shows a similar thickness dependence of t_d [2]. Thus this thickness dependence appears to be a general nature of blocking layers in the photoreceptors. Therefore we next discuss a possible interpretation of the thickness and doping dependence of t_d.

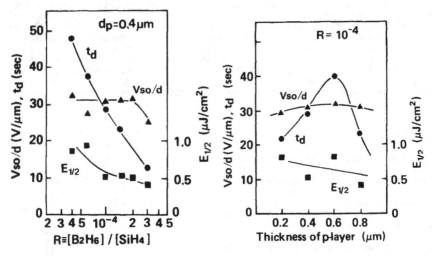

Fig.2 Surface voltage per unit thickness Vso/d, dark decay time t_d and exposure $E_{1/2}$ as a function of gaseous boron doping ratio R for p-type layer.

Fig.3 Surface voltage per unit thickness Vso/d, dark decay time t_d and exposure $E_{1/2}$ as a function of p-type layer thickness d .

As shown above, the properties of p-layer are responsible for the magnitude of t_d, indicating that dark decays originate from electron injection from the substrate. However, it is clear that the electron is not injected over the barrier E_C-E_F (E_C: the energy level at the conduction edge, E_F: the Fermi level) at the interface because t_d decreases with increasing doping level (Fig.2). This is inconsistent with the fact that E_C-E_F increases with doping. Alternative processes should be considered.

When high reverse electric fields are applied to the photoreceptors, the holes in the p-type layer will be swept out into the substrate leaving ionized acceptors. The negative charges in the substrate are neutralized by these released holes. The width of the hole depletion layer depends on the doping level of p-layer. The maximum depletion width w_p can be estimated to be 0.4 μm for R=10^{-4} from the relation $w_p N_A^- = Q_s/e$ (N_A^- is the density of ionized acceptors per unit volume and Q_s is the surface charge per unit area.) using $Q_s/e = 2 \times 10^{12}$ cm^{-2} and a typical doping efficiency of 10^{-3} [5]. However, it should be noted that the doping efficiency is dependent on film deposition conditions, and consequently w_p varies with them. The w_p value of our photoreceptors for R=10^{-4} is considered to be about 1 μm according to the following discussion.

When d_p is small (less than 0.6 μm in our samples), a large fraction of negative charges still remain in the substrate because of the lack of released holes ((a) in Fig.4), and would tunnel through the thin p-type layer by the strong electric field. This tunneling process would be enhanced with the assistance of gap states [6]. Increasing boron doping will increase the number of defects [5] and will increase the tunneling current, which is in agreement with the result shown in Fig.2.

As d_p increases (but less than w_p) the tunneling current will decrease, but, in place of this, a generation current from the depletion layer [7] will increase and dominate. Thus, the leakage current takes a minimum value, consequently t_d takes a maximum value at a certain d_p. When d_p exceeds w_p , the generation current will saturate ((b) in Fig.4). The d_p values larger than 1 μm might correspond to this saturation region in Fig.3.

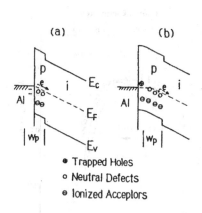

(a)   (b)

• Trapped Holes
○ Neutral Defects
⊖ Ionized Acceptors

Fig.4   Schematic presentation of p-type blocking layer for (a) a small d_p where tunneling current is prevailing, and for (b) a large d_p where generation current is prevailing. w_p denotes the maximum hole depletion width.

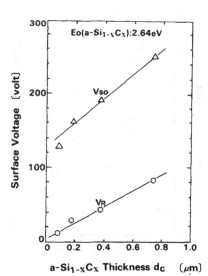

a-Si$_{1-x}$C$_x$ Thickness d_C (μm)

Fig.5   Initial surface voltage Vso and residual voltage V_R as a function of a-Si$_{1-x}$C$_x$ thickness d_C .

A-Si$_{1-x}$C$_x$ layers

The top a-Si$_{1-x}$C$_x$ layer is a very efficient passivating layer from adsorption or oxidation which bring about a gradual degradation of Vso [8] as well as a hole blocking layer. The variations of this layer have little influence on t_d different from the p-type layer.

Figure 5 shows the d_c dependence of Vso and V_R for material with a optical gap Eo of 2.64 eV. The V_R and Vso increase linearly with d_c. The increment of Vso is approximately equal to that of V_R.

Figure 6 shows the Eo dependence of V_R for an approximately constant thickness of 1 µm. V_R increases exponentially with Eo.

These results indicate that most of V_R is due to the charges at both sides of the a-Si$_{1-x}$C$_x$ layer. This is also supported by the fact that the reciprocal dielectric constant $1/\epsilon_r$ is proportional to V_R (Fig.6), and so the simple capacitor equation $V_R = qd_c/\epsilon_0 \epsilon_r$ holds. The q value deduced from this equation and Fig.6 is 2.4×10^{12}cm^{-2}, and is very close to Qs. This is due to a much lower conductivity of the a-Si$_{1-x}$C$_x$ layer. The conductivity ($>10^{12}$ Ω^{-1} cm^{-1}) of the layer with the wide Eo of 2.64 eV is at least two orders of magnitude larger than that of the a-Si:H layer. The dielectric relaxation time in the order of 1 s for the layer is one order longer than the photodecay time for F=10^{12}photons/cm·s . As Eo decreases, the conductivity increases and the relaxation time becomes smaller, resulting in the reduction of V_R (Fig.6).

To reduce V_R, a-Si$_{1-x}$C$_x$ with smaller Eo is desirable. However, the light absorption in the layer with small photosensitivity may occur for small optical gaps. Figure 7 shows the initial discharge rates and the deduced effective quantum efficiencies η for various a-Si$_{1-x}$C$_x$ layers with different Eo as a function of the incident photon energy. Decreases in η are clearly observed for incident light with larger energy than Eo. As Eo increases the critical energy (a, b, and c in Fig.7) above which η decreases also increases. However for red (around 1.9 eV) light sources, η is rather independent of Eo.

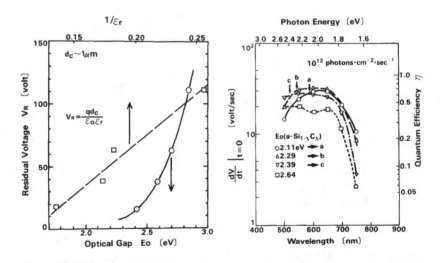

Fig.6 Residual voltage V_R vs. optical gap Eo of a-Si$_{1-x}$C$_x$ layer. The reciprocal dielectric constant $1/\epsilon_r$ of corresponding a-Si$_{1-x}$C$_x$ is also plotted.

Fig.7 Spectral initial photodecay dV/dt$|_{t=0}$ and deduced quantum efficiencies η. The marks a, b and c indicate optical gaps of a-Si$_{1-x}$C$_x$ layer.

SUMMARY

We have shown that the electrophotographic properties greatly ·depend on the characteristics of the blocking layers in multilayered a-Si:H photorecep-tors. The dark properties such as the surface voltage and dark decay time, particularly the latter, proved to be very sensitive to the thickness and boron doping ratio of the bottom p-type layer. In contrast, photo-discharging proper-ties, i.e., the residual voltage and effective quantum efficiency were shown to be greatly dependent on the thickness and compositional fraction x, i.e., the optical gap of the top $a-Si_{1-x}C_x$ layer. Optimization of these blocking layers is impotant to realize high-performance a-Si:H photoreceptors.

REFERENCES

[1] K.Wakita, Y.Nakayama and T.Kawamura: Photogr. Sci. and Eng. 26 (1982) 183.
[2] I.Shimizu, T.Komatsu and E.Inoue: Photogr. Sci. and Eng. 24 (1980) 251.
[3] H.Kakinuma, S.Nishikawa, T.Watanabe and K.Nihei: Jpn. J. Appl. Phys. 22 (1983) L801.
[4] R.M.Schaffert: "Electrophotography" (Focal Press, 1966) p258.
[5] R.A.Street: Phys. Rev. Lett. 49 (1982) 1187.
[6] A.J.Harris, R.S.Walker and R.Sneddon: J. Appl. Phys. 51 (1980) 4287.
[7] S.M.Sze: "Physics of semiconductor devices"(Weiley and Sons, 1969) p102.
[8] S.Nishikawa, H.Kakinuma, T.Watanabe and K.Kaminishi: J. Non-Cryst. Solids 59/60 (1983) 1235.

PART XIII

Image Sensors

AMORPHOUS SILICON LINEAR IMAGE SENSOR

T.Ozawa, M.Takenouchi, S.Tomiyama
Electronic Imaging & Devices Research Lab.,Fuji Xerox Co., Ltd. 2274 Hongo, Ebina-shi, Kanagawa 243-04, Japan

ABSTRACT

Document width linear image sensor arouse strong and wide interest as a compact image reading component for facsimile and other document readers. Among those, a-Si:H is excellent in stability and light sensitivity. The photosensing element of those sensors have a sandwich-like structure which consist of Cr/i-a-Si:H/ITO. In this structure, i-a-Si:H/ITO shows good shottky contact resulting high photo to dark current ratio and fast photoresponse. We have been developing linear image sensor of A4 and B4 size with resolutions of 8dot/mm and 16dot/mm and had applied successfully to facsimile and image input terminals.

INTRODUCTION

Applications of a-Si:H to solar cells and electrophotography are well known and they are already on mass production.

a-Si:H can be formed into very large area and combined with its superb photoelectric characteristics, it is suitable for a photosensitive material of document width linear image sensor. Document width image sensor can be placed very close to the document and read a document by using a rod lens array and LED array as an illuminator. It is not necessary to have a space for long optical length of conventional CCD image sensor, so that a very compact document reader can be realized.

We have been developed A4 and B4 size a-Si:H linear image sensors with the resolution of 8dot/mm (1),(2), which are now being mass produced.

This paper reviews structure, characteristics and application of amorphous silicon linear image sensor.

STRUCTURE OF PHOTOSENSING ELEMENT

The structure of photosensing elements is shown in Fig.1.

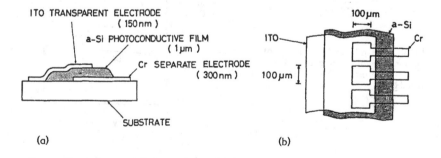

Fig.1 Structure of a-Si:H photosensing element.(a) Schematic cross section, (b) plane view. The sensor consists of chromium separate electrodes, a-Si:H photoconductive layer and ITO transparent electrode.

It consists of chromium separate electrodes and an indium tin oxide (ITO) transparent electrode with a-Si:H film between them. A ceramic plate coated with a glazed glass or a glass plate(CORNING7059) is used for a substrate. Chromium is deposited with 300 nm thickness and separated to 100 micron square by using photolithography. This electrode restricts photosensitive area for each element. The a-Si:H film of 1 micron thickness is deposited by capacitance coupled RF glow discharge with 100 % SiH₄ gass. The deposited area of a-Si:H on the substrate is restricted to a few mm width by using a metal mask. ITO of 150 nm thickness is also deposited by DC sputtering. After the ITO is deposited, photosensing elements are annealed for 30 minutes at 200 °C. This process has an effect of reducing dark current by one order of magnitude.

An energy band diagram of the sensor is shown in Fig. 2. In this case an ITO electrode is biased negative to chromium electrodes. Optical band gap of a-Si:H is approximately 1.7 eV. Both ITO and chromium electrodes have shottky contacts to a-Si:H, and ITO/a-Si:H contact act as an electron barrier and Cr/a-Si:H as a hole barrier.

PHOTOELECTRIC CONVERSION CHARACTERISTICS

Fig.3 shows I-V characteristics of the photosensing element. Dark current is suppressed to less than 10^{-12} A by both ITO and chromium barriers. Especially barrier of ITO/a-Si:H contact mainly suppressed dark current. Photocurrent is saturated at relatively low voltages and it doesn't depend on the applied voltages. It is because that carriers generated in the a-Si:H film are totally collected on both ITO and chromium electrodes. This means that photocurrent doesn't depend on thickness variations and the sensor can be operated at low voltages such as 5V, which is desirable for productivity and design of power supply. Photo response of the sensor is less than 50 micro second for both rise and fall time. It is fast enough to read a regular size document in a few second.

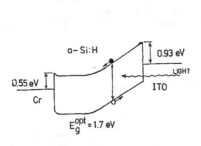

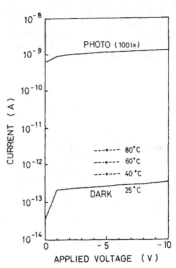

Fig.2 Energy band diagram of a-Si:H sensor. Optical band gap is 1.7 eV. Both ITO and chromium has shottky contact to a-Si:H.

Fig.3 I-V characteristics of a-Si:H sensor. Dark current is less than 10^{-12} A.

Fig.4 shows light transfer characteristics of the sensor. The photo current is proportional to the light intensity and it can discreminate gray scale of images.

Fig.5 shows spectral response of the sensor. Maximum sensitivity was at the wave length of 620 nm and it covers whole visible light.

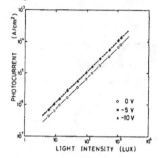

LIGHT INTENSITY (LUX)

WAVELENGTH (μm)

Fig.4 Light transfer characteristics of a-Si:H sensor. Photo current is proportional to the light intensity.

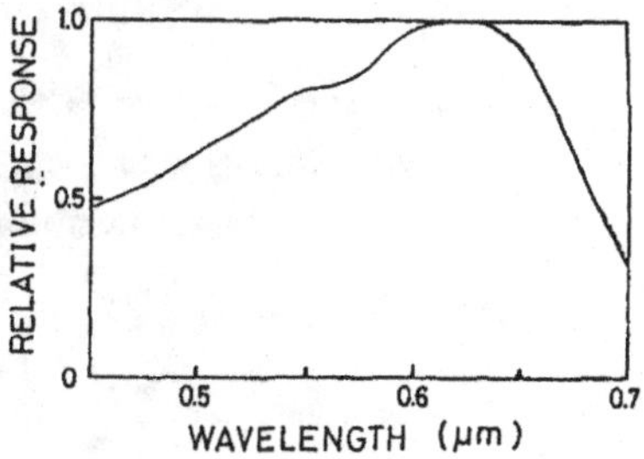

Fig.5 Spectral response of a-Si:H sensor.

SIGNAL DETECTION

The sensitivity of a photosensing element with the area of 100 micron squre is about 10 pA/lx as from Fig.3. So the photocurrent is I nA at the 100 lx illumination which is obtained by optical systems such as a rod lens array and LED array. This value is too small to detect in real time mode and so charge storage mode which integrates the photocurrent during line scanning is used and enables signals larger.

Fig.6 shows a fundamental circuit for detecting signals of a-Si:H image sensor. A-Si:H photosensing element corresponds to a current source. Capacitor consists of capacitor of photosensing element, wiring capacitor and input capacitance of an amplifier.

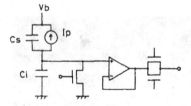

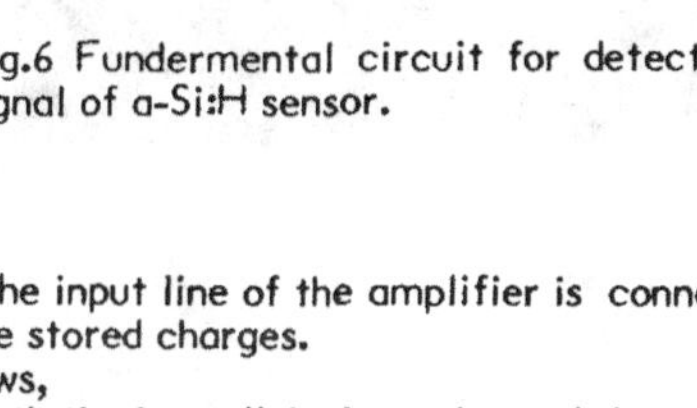

Charges generated in the a-Si:H layer are stored in this capacitor. According to the amount of stored charges in the capacitors, voltage at the input line of the amplifier varies.

A voltage follower amplifier of high input impedance detects this voltage and its output is switched one by one by analog switches leading to make a sirial

Fig.6 Fundermental circuit for detecting signal of a-Si:H sensor.

video signal. After the signal is detected the input line of the amplifier is connected to the ground by a MOSFET to discharge the stored charges.

Characteristics of this operation is as follows,

1. Output voltage is proportional to both the input light intensity and the storage time interval

2. Output voltage is high enough so that it doesn't require external amplifier.

3. Sensitivity and SN ratio are high.

A custom LSI has been designed based on this operation. It had 128 channel input terminals, voltage follower amplifiers, analog switches and shift registers which are made of CMOS.

APPLICATION TO A LINEAR IMAGE SENSOR

Fig.7 shows B4 size, 16 dots/mm resolution a-Si:H linear image sensor.

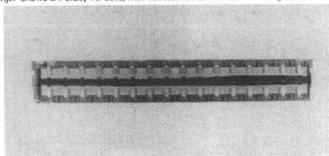

Fig 7. Photograph of B4 size 16 dot/mm resolution a-Si:H linear image sensor. Substrate size is 270 x 40 mm.

Substrate is CORNING 7059 with the size of 270 x 40 mm. 4096 dots photosensing elements are located at the center of the substrate and 32 chips LSIs are mounted both sides of sensors. Sensing area of an element is 50 micron square. Both the photosensing elements and the LSIs are connected by chromium and gold thin film circuit.

Fig.8 shows output signal wave forms at the condition of 100 lx illumination and 5 msec line scan time. Nonuniformity of the output are mainly caused by the nonuniformity of charge storage capacitor which depends on the length of sensing elements to ICs connection.

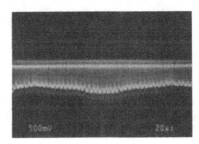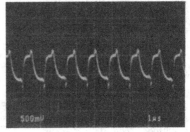

Fig.8 Output signal wave form of B4, 16 dot/mm a-Si:H linear image sensor.

Table.1 shows the electrical characteristics of B4, 16 dots/mm a-Si:H image sensor. Supply voltage is plus and minus 5V and direct drive by CMOS digital circuit is possible. Maximum operating clock frequency is 5 MHz. It can attain the scanning speed of less than 1 msec/line.

Table.I Electrical characteristics of B4,16 dot/mm
a-Si:H image sensor

supply voltage	+5	V
	-5	V
maximum clock frequency	5	MHz
output saturation voltage	-3.5	V
saturation exposure	2.5	lx·s
sensitivity	1.0	V/lx·s
output nonuniformity	± 20	%
element number	4096	dot
element size	50	micron square

RELIABILITY OF THE IMAGE SENSOR

Standards of reliability test for a-Si image sensor has not yet been decided, however typical test items for ICs and hybrid circuits have been applied to it tentatively.

Test result shows that a-Si:H image sensor has good reliability against humidity, high temperature and continuous operation.

CONCLUSION

a-Si:H linear image sensor has been developed. It can be applied to compact facsimile and image input terminals.

A4 and B4 size, 8 dots/mm linear image sensor is now mass produced and 16 dots/mm sensor have been developed. Cost reduction study is now under way. One of the promising ways for cost reduction is use of a-Si:H TFT as scanning circuit. To compensate the slow speed of TFTs, new methods such as multiplexing have to be considered.

ACKNOWLEDGMENT

The authers would like to thank Dr.Y.Itoh for his advice and support, members in EIDRL for the experiments.

REFERENCES

(1) T.Hamano et al:An Amorphous Si High Speed Linear Image Sensor,Proceedings of the 13th Conference on Solid State Devices, Tokyo(1981); Jpn.J.Apply.Phys., Supplement 21-1, p245 (1982)
(2) T.Ozawa et al: Design and Evaluation of A4 Amorphous Si Hybrid Image Sensor, IMC 1982 Proceedings, Tokyo, p132, (1982)

AMORPHOUS Si:H HETEROJUNCTION PHOTODIODE AND ITS APPLICATION TO A COMPACT SCANNER

S. KANEKO, Y. KAJIWARA, F. OKUMURA AND T. OHKUBO
Microelectronics Research Labs., NEC Corporation
4-1-1 Miyazaki Miyamae-ku Kawasaki, 213 Japan

ABSTRACT

This paper reports heterojunction photodiode properties and its application to a compact scanner. The photodiode has ITO / p-a-SiC:H / a-Si:H / metal structure. This diode has high photo to dark current ratio and small photocurrent saturation voltage, because of the excellent blocking characteristics for a heterojunction with large built in potential. Moreover, a-Si:H / metal contact has been investigated. A contact linear image sensor has been fabricated using the heterojunction photodiode array and compact optical system. Performance tests showed excellent results. Good reproduced images have been obtained.

INTRODUCTION

A contact linear image sensor has been developed to produce a compact facsimile or OCR [1][2][3]. Conventional equipment uses a small size IC sensor and high magnification lens system, which requires about 200 mm optical path length to scan a document. On the other hand, a contact linear image sensor, whose width is the same as the document, does not require this high magnification lens system. Therefore, compact equipment can be built.

Hydrogenated amorphous silicon (a-Si:H) is suitable material for application to this long linear image sensor, because of its large area producibility, high photosensitivity and material stability. In this sensor, blocking structure photodiodes, such as Schottky barrier diode [3][4], p-i-n diode [5] and MIS diode [6] have been employed in order to obtain small dark current and quick photoresponse. Among these blocking diodes, the Schottky barrier diode has the simplest structure to fabricate a photodiode array. However, Schottky barrier formation is sometimes unstable, due to surface plasma damage during indium tin oxide (ITO) sputtering deposition [7].

In order to obtain a stable and simple structure blocking diode, heterojunction photodiode, in which p type amorphous SiCx:H is inserted between undoped a-Si:H and ITO electrode, has been developed. This paper reports the heterojunction photodiode characteristics and its application to compact scanners.

DIODE STRUCTURE AND PREPARATION

Figure 1 shows elementary structure of the Schottky barrier diode and the heterojunction photodiode. In the Schottky barrier diode, undoped a-Si:H was produced by plasma deposition of SiH on a metal coated glass substrate. Then ITO was magnetron-sputtered. The undoped a-Si:H was 1-1.5 µm thick. In the heterojunction diode, 300-400 A thick p-a-SiCx:H deposition followed the 1-1.5 um thick undoped a-Si:H deposition. The p-a-SiCx:H was produced by plasma deposition of SiH_4, CH_4 and B_2H_6 mixed gas.

Figure 2 shows optical gap energy and conductivity for undoped and boron doped a-SiCx:H. Increase in carbon concentration causes an increase in optical gap energy and a decrease in the film conductivity. The conductivity increases by boron doping. In the heterojunction diode, amorphous SiCx:H was produced by the CH_4 / SiH_4 + CH_4 = 0.5 gas mixture. This amorphous SiCx:H

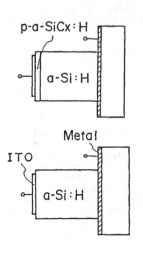

Fig. 1 Heterojunction diode and Schottky barrier diode structure

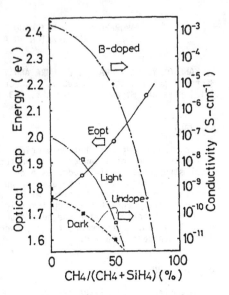

Fig. 2 Optical gap energy and conductivity for undoped and boron doped a-SiCx:H

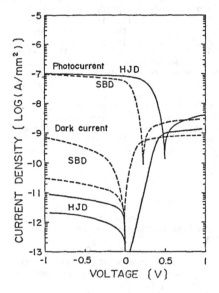

Fig. 3 I-V characteristics in dark and lighted conditions for heterojunction diode and Schottky barrier diode.

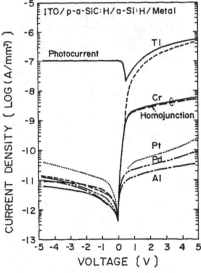

Fig. 4 I-V characteristics for heterojunction diode using various metals as lower electrode.

has the 1.97 eV optical gap and 10^{-5} S-cm^{-1}dark conductivity. Activation energy, measured from temperature dependence of dark conductivity, was 0.4 eV.

DIODE CHARACTERISTICS

Figure 3 shows I-V characteristics in dark and lighted conditions for a Schottky barrier diode and a heterojunction diode. In both diodes, Cr metal was used for the lower electrode. The photocurrent shows 10^{-7} A/mm under 150lx Y-G LED light (570 nm peak intensity wavelength) illumination at -1 V reverse voltage. The photocurrent saturation, which is caused by unity collection efficiency for incident light, is observed at small voltage. In the heterojunction diode, the photocurrent saturates without any applied voltage. This small saturation voltage is due to high internal field at the heterojunction and gives small operation voltage for image sensors.

The darkcurrent for the Schottky barrier diode shows a relatively large value (10^{-11}-10^{-9} A/mm) at -1V reverse voltage. On the other hand, the heterojunction diode exhibits very small dark reverse current (10^{-12} - 10^{-11} A /mm^2) and a photo to dark current ratio as high as 10^4 has been obtained. This dark current keeps constant value up to 280°C thermal annealing in air, while dark current of the Scottky barrier diode is changed by this thermal annealing.

Figure 4 shows I-V characteristics for the heterojunction diode using various metals as lower electrodes. The reverse dark current shows almost the same value for the lower electrode material. It is different from the previously reported result, where the dark reverse current depends on the lower electrode material (3). The forward current depends on the metal electrode material. Large forward current and good rectification have been obtained using low work function material except for Al.

Figure 5 shows an energy band diagram for the heterojunction diode without applied voltage. High built-in potential fully depletes the undoped a-Si:H layer. Photogenerated carrier drifts towards each electrode due to the high internal field in the undoped a-Si:H layer. Thus, photocurrent saturates without applied voltage. The small reverse current and stable characteristics for the heterojunction indicate that heterojunction does not suffer significant plasma damage during ITO sputtering and that the heterojunction is a good electon barrier. The dependence on the lower electrode material indicates that the reverse current is not hole injection current from the lower electrode, while forward current is mainly electron current flow through a-Si:H/metal contact. The temperature dependence of the I-V characteristics for diodes using Cr and Pd as lower electrode was measured. Plots of ln (J_F / T^2) vs 1/T are shown in fig 6, where J_F is a forward current. These closely match the thermionic emission theory with barrier heights ϕ_B of 0.9 eV for a Pd electrode and 0.72 eV for a Cr electrode. In Fig 6, plots of ln (I_R) vs 1/T are also shown, where I_R is a reverse current. These give almost the same activation energy, 0.85 eV, and indicate that reverse current is mainly generation current.

Figure 7 shows spectral responses for the heterojunction diode and a homojunction diode using p-a-Si:H. The heterojunction diode has high sensitivity in the shorter wavelenghth due to light window effect of p-a-SiCx:H. Both diodes have high sensitivity for the LEDs used in the contact linear image sensor.

APPLICATION TO A COMPACT SCANNER

Figure 8 shows a contact linear image sensor exploded view. This sensor consists of LED arrays , a rod lens array and a long linear image sensor. LED arrays illuminates a document and reflected signal light is introduced

426

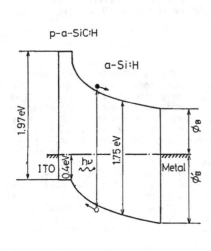

p-a-SiC:H

a-Si:H

1.97 eV

0.4eV

ITO

$h\nu$

1.75 eV

Metal

ϕ_B

ϕ'_B

Fig. 5 Energy band diagram

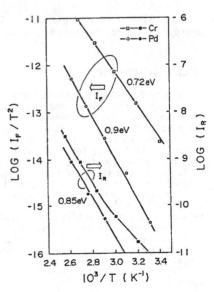

---□--■-- Cr
---○--●-- Pd

0.72 eV

I_F

0.9 eV

I_R

0.85 eV

$LOG \ (I_F / T^2)$

$LOG \ (I_R)$

$10^3 / T \ (K^{-1})$

Fig. 6 Temperature dependence of forward current (I_F) and reverse current (I_R).

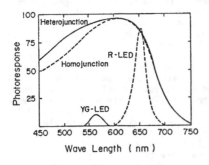

Heterojunction

R-LED

Homojunction

YG-LED

Photoresponse

Wave Length (nm)

Fig. 7 Spectral response for heterojunction diode and homojunction diode. Light emission spctra for YG-LED and red LED are also shown.

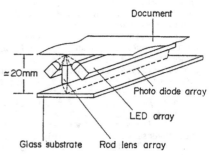

Document

≈ 20mm

Photo diode array

LED array

Glass substrate   Rod lens array

Fig. 8 Contact linear image sensor exploded view

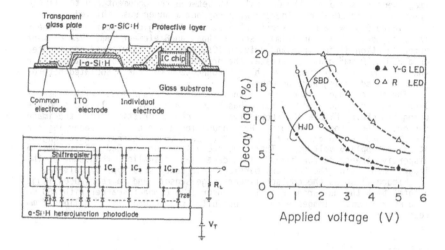

Fig. 9 Cross sectional view and
driving circuit

Fig. 10 Decay lag charac-
teristics at first field

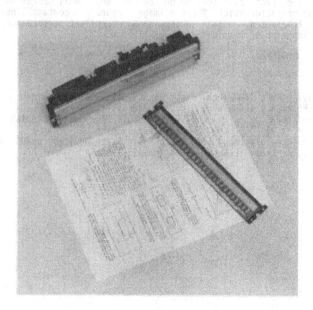

Fig. 11 Contact linear image
sensor and reproduced image

to the linear image sensor through the rod lens array. Active length of the image sensor is 216 mm. The Optical path between the document and the linear image sensor is 20-40 mm. Therefore, a compact scanner was achieved.

Figure 9 shows a cross sectional view and driving circuit for the linear image sensor. The linear image sensor has an 8 element /mm, 1728 element heterojunction diode array and driving ICs. Both were fabricated and mounted on a 230 x 30 mm² glass substrate. The driving IC consists of shift register and 64 MOS FET switches, which are connected to the associated heterojunction diode using wire bonding.

The image sensor is operated in the charge storage mode with 1 msec / line scanning speed. 30dB SN ratio has been obtained at 0.15 lx·sec exposure using Y-G LED. Figure 10 shows image lag characteristics for the image sensor. The image sensor using heterojunction diode has 3.5% decay lag for Y-G LED and 6 % decay lag for Red LED (650 nm peak intensity wave length) at -5 V applied voltage. The large decay lag for the Red LED is due to dispersive transport for hole. This decay lag is smaller than that of the image sensor using the Schottky barrier diode under this applied voltage.

Figure 11 shows the contact linear image sensor and a reproduced image printed by a thermal printer. This sensor is a compact scanner with 260 mm x 50 mm x 30 mm. Images were successfully reproduced with 8 lines / mm resolution.

CONCLUSION

A heterojunction photodiode, using p type a-SiCx:H has been investigated and applied to a compact scanner. The heterojunction is stable and has good electron barrier with high built in potential. It was also clarified that there is no hole injection at a-Si:H / metal contact. These result in high photo to dark current ratio for the photodiode and small operation voltage and good lag characteristics for the image sensor. A contact linear image sensor was fabricated. This sensor shows excellent performance. Good reproduced images have been obtained.

REFERENCES

(1) K. Komiya et al. IEDM Tech. Dig., pp 309-312, 1981
(2) S. Kaneko et al. IEDM Tech. Dig., pp 328-331, 1982
(3) T. Hamano et al. Jpn. J. Appl. Phys. Suppl. 21-1 pp 245-249 1982
(4) K. Suzuki et al. IEEE CHMT-7 pp 423-428 1984
(5) H. Yamamoto et al. Proc. 15th Conf. Solid State Devices, pp205-208 1984
(6) M. Sakamoto et al. IEEE CHMT-7 pp 429-433 1984

INFLUENCE OF TRANSPARENT ELECTRODES ON IMAGE SENSOR PERFORMANCE

K. KEMPTER, H. WIECZOREK AND M. HOHEISEL
Siemens AG, Research Laboratories, D-8000 München 83, Fed. Rep. of Germany

ABSTRACT

The short response times required for image sensors demand blocking contacts at the sensor cell. It was found that the junctions between transparent electrodes (ITO or a thin palladium film) and the metallic back electrode with a-Si:H form blocking contacts yielding photocurrent decay times of the order of some microseconds. The two different time regimes observed for the decay are interpreted as being limited by the drift and the release of holes respectively.

I. INTRODUCTION

Image sensors having the size of an A-4 document depend on the availability of large-area photoconductors. These materials have been common for more than 20 years as photoreceptors in the field of electrophotography. However, image sensors require, as an additional feature, a response time shorter than 100 µs, in contrast to the 100 ms required in electrophotography.

This speed requirement cannot be met by a simple photoconductor; rather, it demands a photodiode with blocking contacts. Furthermore the photoactive material must show a high carrier mobility and a minimum of dispersive transport. Amorphous hydrogenated silicon (a-Si:H) is considered a highly suitable material for such image sensors. To obtain the diode behaviour with carrier depletion necessary for high speed response some of the sensors discussed in the literature employ a p-i-n structure /1/ or additional blocking layers (mostly Si_3N_4)/2/. Others relay on the blocking effect of the contacts /3, 4/.

This paper presents some results which were obtained in an attempt to use sandwich cells consisting merely of a metallic back electrode, undoped a-Si:H and a transparent top electrode. The blocking effect of the contacts stems from the junctions between a-Si:H and the electrodes. These junctions act as a Schottky barrier or as a heterojunction. The results were selected in accordance with their relevance to image sensor performance with special emphasis on the time response of the sensor cell. Measurements of the photocurrent decay relevant to image sensor performance differ in three aspects from the usual decay investigations reported in the literature /5/. First: the starting point of the decay is the steady-state photocurrent, because the time period between illumination changes of an image sensor is in the range of milliseconds. Therefore, the current decay figures the transport processes during the transition from one equilibrium distribution of the carriers to another. Second: considering the large dynamic range of a sensor, the current decay must be traced down to within a few percent of the initial photocurrent value. This means that the slow tail of the decay is of essential importance. Third: the sensor has a sandwich geometry rather than the gap configuration of the electrodes usual for carrier life-time measurements.

II. EXPERIMENTAL METHODS

The samples used were sandwiched between a metallic back electrode and a transparent top electrode; see FIG. 1.

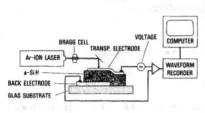

FIG. 1 Experimental system and electrode arrangement in the sample.

The back electrode consisted of a thin film of a low work-function metal, mostly titanium. The 1 μm thick undoped a-Si:H film was deposited by RF decomposition of pure silane in a capacitively coupled plasma reactor. The deposition parameters were optimized by minimizing the subgap optical absorption, measured by the Constant Photocurrent Method (CPM) /6/. The absorption value $\propto$ of a typical a-Si:H samples (T_s = 220° C) at hv = 1 eV was in the range of 0.1 cm^{-1}. A sample particularly rich in defects was deposited at T_s = 150° C, yielding an $\propto$ (1 eV) of 1 cm^{-1}. The transparent top electrode was either a semitransparent layer of Pd (10 nm thick) or a 150 nm thick ITO film. All electrodes were evaporated by an electron gun. The patterning of the top electrode was performed either by photolithography in a lift-off process (electrode area 0.2 cm^2) or by a shadow mask during evaporation (electrode area 1.1 cm^2). Patterning by evaporation through a mask avoids contact between the a-Si:H surface and the developer solution containing sodium ions. Sodium penetrating into the a-Si:H may act as a donator. In order to investigate the influence of the oxide layer, the surface of the a-Si:H film was etched with hydrofluoric acid to remove the oxide, or the oxide thickness was increased by an oxygen plasma.

The photocurrent decay was measured after 5 s illumination period with the light of an argon-ion laser (λ = 514 nm) of usually 7.10^{14} photons/cm^{2}s. The laser beam was switched off by a Bragg cell whose fall time was 50 ns. The light-to-dark ratio was 1000:1. The photocurrent was fed to a low-impedance operational preamplifier, which was connected to an HP 5180 waveform recorder. The dynamic range of the measuring circuit was 200:1 down to the dc level. The subsequent analysis was performed with an HP 9836 computer. The shortest decay times measured were a few microseconds. The varying capacitance of the sample under varying bias caused a small influence only to the fastest part of the measured decay because of the low impedance of the amplifier. The dark current was measured before and after every experiment and subtracted to yield the photocurrent.

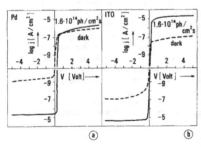

FIG.2 Dark and photocurrent as a function of the voltage applied to the top electrode . The value given is the photon flux incident onto the top electrode.
  a) Semitransparent palladium as top electrode
  b) ITO as top electrode.

III. EXPERIMENTAL RESULTS

The current - voltage characteristics yield general information on the nature of the contacts. Typical results both for dark and photocurrent are shown in FIG. 2. The voltage polarity given applies in each case to the (illuminated) top electrode. The observed characteristics represent the behaviour of a back-to-back diode with current saturation for both polarities. In the case of the negative-biased top electrode, carrier blocking seems to be more effective, because the (reversed) dark- and photocurrent displays a better saturation. This saturation is more pronounced with a palladium than with an ITO top electrode. The open-circuit voltage of the illuminated sandwich was 0.2 V and 0.1 V with Pd and ITO electrodes, respectively. The smaller photocurrent of the Pd-covered sample is due to the stronger light absorption

of the palladium in comparison with ITO.
Two typical examples of measured photocurrent decay curves are displayed in FIG. 3.

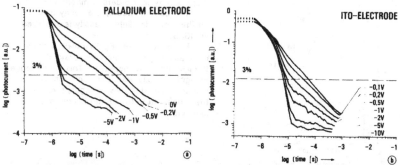

FIG. 3 Photocurrent decay vs. time for various negative voltages. The incident photon flux was $7 \cdot 10^{14}$ photons/cm^2 s.
a) Semitransparent palladium as top electrode
b) ITO as top electrode.

Most of the measured current decays show two parts: a steep drop in the beginning, followed by a slow tail. The steep drop becomes more pronounced as the bias voltage increases. For the following, the decay time will be defined as the time period from the illumination cutoff to the point where the current has dropped to 3 percent of the initial steady-state value.

The results for the 3-percent decay time obtained for different samples are summarized in FIG. 4 as a function of the bias voltage. Most of the samples show a tremendous variation of the decay time extending up to six orders of magnitude. On both sides of the narrow peak near zero bias two broad minima extend in both voltage directions. The shortest decay times were found at moderate negative bias. The absolute value of these decay time minima in the microsecond region may be affected by the measuring circuit. Thus the faster response of the Pd contact (2 μs) seems to result from the smaller resistivity of this electrode in comparison to the ITO electrode.

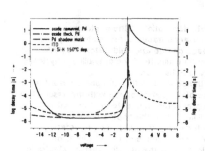

FIG. 4 Decay time for a current drop to 3 percent of the initial value as a function of the applied bias voltage measured on samples with different top electrodes and deposition temperatures of the a–Si:H.

The increase of the decay time at higher negative bias occurred only with those samples which showed a steep increase of the current at high bias (breakdown of blocking).

At positive bias, the decay times are generally much longer than at negative bias. Furthermore a great difference between ITO and Pd electrodes was found in this voltage regime.

The influence of the interface can also be seen from the voltage dependence of the sample with the thick oxide interlayer (produced by plasma treatment). A larger negative voltage is necessary to reach the minimum decay time with this sample. The different surface treatments of the interface affected predominantly the reproducibility of the blocking effect of the contacts. Evaporation of the top electrode through a shadow mask renders possible higher negative voltages than photolithographic patterning.

432

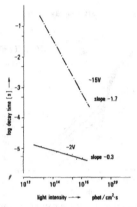

The sensor consisting of a–Si:H deposited at 150° C shows the strong influence of the bulk material. The minimum decay time is increased by four orders of magnitude.

The dependence of the decay time on the preceding light intensity is displayed in FIG. 5.

For a moderate negative bias, a small decrease of decay time was obtained as the light intensity increased. This dependence on light intensity was found to be much higher at elevated bias voltages, i. e. in the range where the contacts are no longer blocking.

FIG. 5   Decay time for a current drop to 3 percent of initial value as a function of the light intensity, measured with a Pd top electrode,
  a) at a voltage of –2V,
  b) at a voltage of –15 V (breakdown of blocking).

IV. DISCUSSION

There is a great deal of complexity in a complete description of the photocurrent because many different effects may coincide, for instance trapping, recombination, nonuniform carrier distribution, carrier injection, space charge and field distortions /7/. The following discussion attempts to give a very rough estimate of the prevailing physical processes which may be responsible for the observed behaviour.

The I–V characteristics shown in FIG. 2 permit the conclusion that the photocurrent is of the primary type at least with moderate negative bias, in accordance with the assumption of others /8/. In this case electrons and holes are prevented from entering the photo-conductor at the electrodes. We assume that the limitation of the photocurrent decay is primarily effected by the holes, because their drift mobility is smaller by two orders of magnitude than the mobility of the electrons /9/. The influence of the electrons on the decay grows as the penetration depth of the light below the negative–biased electrode decreases.

During buildup of the steady–state photocurrent, most of the excess photogenerated holes will be distributed to localized trap states and a few to extended states. After the illumination is terminated, the photocurrent decay reflects the drift of the excess holes to the negative–biased electrode. This current may be limited by the field–dependent drift time of the holes or by the release time of the holes from the traps. Right at the beginning of the decay when the shallow traps (nearest to the valence band) are emptied, the release time is very short and the current is limited by the drift. The drift time of the holes is estimated at one microsecond, assuming a hole mobility of 10^{-3} cm^2/V.s /9/, an internal field of 10^4 V/cm, and a drift length equivalent to the light penetration depth of 0.14 μm. /10/.

During decay, the emptying of the traps proceeds to ever lower energetic states (more distant from the valence band) so that a point will be reached where the release time exceeds the drift time of the holes; from this point onward the decay will be release-limited. The constant slope of the slow tail is ascribed to the release of holes from a continuous distribution of trap states.

The transition from the initial steep current drop (drift-limited) to the slow tail occurs after about 5 μs. This yields a value of 0.45 eV for the energetic depth of the traps at which release limitation begins (escape frequency was assumed to be 10^{13} s^{-1}).

The pronounced increase of the decay time near zero bias is explained as a consequence of the field drop which increases the drift time of the holes.

The increase of the decay time observed with some samples at high negative bias must be correlated with the injection of electrons from the transparent electrode. Thus, with an additional secondary photocurrent flowing, the long recombination lifetime of the electrons determines the decay time.

For positive bias, the generally longer decay times are explained by the greater drift length of the holes from the illuminated electrode to the back electrode. In this case, the pronounced field dependence supports the model of the transit time limitation. The rather long time scale of this transit can be explained by the dispersive nature of the hole transport /11/. No consistent explanation can be offered so far for the different decay times associated with palladium and ITO electrodes.

To explain the observed decrease of the decay time with light intensity, we shall first consider the release-limited case at moderate negative bias. Increasing light intensity shifts the quasi-Fermi level of the holes closer to the valence band edge. Therefore an increasing part of the trapped holes profit by a shorter release time. Hence the photocurrent drops in a shorter time. The observed slope of the intensity dependence should be correlated with the slope of the relevant trap state density vs. energy.

The much steeper intensity dependence of the decay at high reverse bias (secondary photocurrent) must be due to a variation of the recombination lifetime of the electrons. A decrease of electron lifetime with increasing density of excess carriers has already been found by Snell, Spear and LeComber /12/.

Except for this last case, carrier recombination was not considered, the reason being the generally primary nature of the photocurrent with both polarities of the bias as a consequence of the blocking contacts. The strongly absorbed light is an additional factor making for a marked nonuniformity of the electron and hole distribution across the a-Si:H /7/. As a consequence, recombination is suppressed in most parts of the sample. This is supported by the almost linear dependence between photocurrent and light intensity observed in these samples.

V. SUMMARY

It could be shown that image sensor cells with a very fast response (2 μs) may be realized using suitable electrodes as blocking contacts in conjunction with optimized a-Si:H bulk material.

Interface and surface treatments are of importance to the blocking effect of the electrode (and its reproducibility), which is the basic requirement for the short response time.

The pronounced dependence of the observed photocurrent decay on the bias voltage can be qualitatively explained by a model which is based on drift- or release-limited hole motion.

ACKNOWLEDGEMENT

The authors would like to thank H. Doneyer, W. Müller and R. Primig for preparing the samples, and H. Harms for providing the CPM results.

434

REFERENCES

1. H. Yamamoto, T. Baji, H. Matsumaru, Y. Tanaka, K. Seki, T. T. Tanaka, A. Sasano and T. Tsukada, Ext. Abstr. of 15th Conference on Solid State Devices and Materials, Tokyo 1983, pp. 205 - 208

2. S. Kaneko, F. Okumura, H. Uchida, M. Kanamori, M. Sakamoto, T. Itano, U. Kajiwara and T. Saito, J. Non-Cryst. Sol. 59 & 60 (1983) 1227

3. T. Hamano, H. Ito, T. Nakamura, T. Ozawa, M. Fuse and M. Takenouchi, Jap. J. Appl. Phys. 21 (1982) Suppl. 21-1, pp. 245

4. K. Suzuki, Y. Suda, S. Takayama, T. Nakai, K. Mori, O. Takikawa and T. Saito, IMC Tokyo 1984 Proceedings

5. N. Matsumoto et al., Jap. J. Appl. Phys. 20 (1981) Suppl. 20-2 pp. 179. J. M. Hvam and M. H. Brodsky, Phys. Rev. Lett. 46 (1981) 371. T. Kagawa et al., Phys. Rev. B 26 (1982) 4714. A. A. Andreev et al. Solid State Commun. 52 (1984) 589. W. B. Jackson et al. Solid State Commun. 47 (1983) 435.

6. M. Vanecek, J. Kocka, J. Stuchlik and A. Triska, Solid State Commun. 39 (1981) 1199.

7. R. S. Crandall in Semicond. and Semimet. (Edit. R. K. Willardson, A. C. Boer) Vol. 21 B (Edit. J. I. Pankove) Academic Press 1984 pp. 245

8. W. B. Jackson, R. J. Nemanich and N. M. Amer, Phys. Rev. B 27 (1983) 4861.

9. R. A. Street, Appl. Phys. Lett. 41 (1982) 1060.

10. P. J. Zanzucchi, C. R. Wronski and D. E. Carlson, J. Appl. Phys. 48 (1977) 5227

11. T. Tiedje, J. M. Cebulka, D. L. Morel, B. Abeles, Phys. Rev. Lett. 46 (1981) 1425

12. A. J. Snell, W. E. Spear, and P. G. LeComber, Phil. Mag. B 43 (1981) 407.

PART XIV

Heterojunction and Polycrystalline Devices

THE USE OF AMORPHOUS SILICON EMITTERS IN BIPOLAR TRANSISTORS

M. GHANNAM, J. NIJS, R. DE KEERSMAECKER AND R. MERTENS
IMEC, Kapeldreef 75, 3030 Heverlee - Belgium

ABSTRACT

For the first time an operating heterojunction bipolar silicon transistor has been realized with phosphorous doped amorphous silicon (a-Si) emitter. The deposition of a-Si is a relatively simple technique. The current gain (β) of 14 at a base Gummel Number (G.N.) of $1.35 \ 10^{13} \ s/cm^4$ is higher than that obtained with normal diffused emitter bipolar transistors with the same G.N. for the base. This adds a degree of freedom to the design of bipolar structures according to the compromise between base resistance and current gain. Crucial points that have to be looked at further are : interface recombination at the a-Si/c-Si transition and emitter resistance.

INTRODUCTION

The injection of minority carriers into the emitter is the most important base current component in modern VLSI-bipolar structures, because with the trend to horizontal and vertical miniaturization the base region has become very thin.

Therefore use of improved emitter structures in bipolar devices, in which the flow of current carriers from the base to the emitter is suppressed relatively to the flow from emitter to the base, can offer significant advantages both for high speed and high voltage transistors. Indeed, it gives an additional degree of freedom in design of base width and doping for optimisation of the compromise between base resistance and current gain. Also a lower base sheet resistance (for the same current gain !) yields a smaller β-spreading over the wafer. An additional advantage in the use of improved emitter structures arises when trench isolation is used. Then the base must be doped slightly higher, because otherwise population inversion or strong depletion can occur at the trench walls, creating leakage between emitter and collector or strong recombination at the base/oxide interface. With this more heavily doped base it is extremely important to have low injection into the emitter in order to obtain reasonable current gains, because the collector current is lower and base recombination current is higher.

Different groups have published their work on polysilicon emitters [1,2,3,4,5,6], pseudo-heterojunction MIS-emitter [7] and SIPOS-heterojunction [8]. These groups have obtained interesting results of current-gain improvement, but some problems are still unsolved. The polysilicon emitters are not really heterojunction transistors and still have charge storage in the emitter. Also the presence and role of an interfacial oxide is not completely understood and this interfacial oxide degrades the device characteristics at high current and high frequency operation. The pseudo-heterojunction MIS emitter is an MIS tunnel transistor with a thin oxide on top of a shallow implanted base region. One must not underestimate the technological problems associated with this device such as control of the oxide thickness, the critical sintering step etc. Also, despite of the high current gain, the emitter resistance and emitter capacitance are very high, resulting in poor high current and high frequency performance. Also the doped SIPOS-emitter has some drawbacks. A long high temperature annealing step is necessary to transform the semi-insulating SIPOS-layer to a semi-conducting layer with resistivity of 100 Ωcm. Because of that, most likely an n^+ single crystal emitter part is formed by drive-in, eliminating the advantage of reduced emitter delay, associated with real heterojunction transistors. Also the high resistivity of the emitter is a problem.

Real heterojunction structures are known in the III-V compound technology. When using a real heterojunction bipolar transistor, injection into
the emitter is suppressed, hence the emitter can be lowly doped and therefore the capacitance is reduced. Indeed, when the emitter material has a
wide bandgap the intrinsic carrier concentration is reduced in this region.
In this paper a phosphorous doped amorphous hydrogenated silicon emitter is
used on top of a crystalline base. The deposition of amorphous silicon by
plasma glow discharge is a simple, well known low temperature (250-300°C)
process. Doping can be obtained by introducing doping gases in the gas atmosphere and unlike poly and SIPOS, no high temperature treatment has to be
done. If enough phosphine ($\geqslant$ 0.6 %) is introduced together with SiH_4 a conductivity of $10^{-2} \, \Omega^{-1} \, cm^{-1}$ can be reached. The band structure of the n^+-a-
Si:H emitter and the p-c-Si base before and after contact are shown in Fig.
1. One can see that a large hole barrier is formed preventing injection
into the emitter. The barrier for electrons is nearly unchanged.

EXPERIMENTAL

Windows are made in the thermally grown SiO_2 at the surface of a boron
doped base with thickness of 2.4 μm and sheet resistance of 160 $\Omega/\square$. The
Gummel number of the base

$$GN = \int_0^{W_B} \frac{N(x)}{D_n(x)} \frac{n_i^2}{n_{ie}^2(x)} \, dx$$

equals $1.35 \; 10^{13}$ s/cm^4. In this formula W_B is the base thickness, $N(x)$ the
position dependent doping, $D_n(x)$ the position dependent electron diffusion
coefficient, n_i the intrinsic carrier concentration and n_{ie} the effective
intrinsic carrier concentration which can be different from n_i due to bandgap narrowing effects.

A thin layer of n^+ a-Si:H is deposited by the rf-glow discharge technique on the wafer and simultaneously also on a glass plate to determine
thickness, phosphorous content and activation energy $E_A = E_C - E_F$, with E_C
the conduction band edge and E_F the fermi level. The gas mixture consists
of 99.4 % SiH_4 and 0.6 % PH_3. The rf power was maintained at 10W, the operating pressure between 100 and 500 mTorr and the substrate temperature
around 250°C. The obtained film thickness was estimated at 0.4-0.5 μm.
The phosphorous content was measured with SIMS to be about 10^{20} cm^{-3}, from
which only 1 % is electrically active. Dark conductivity measurements in
function of temperature were used to determine the activation energy E_A =
0.18 eV and the room temperature conductivity amounts $10^{-2} \, \Omega^{-1} \, cm^{-1}$. Using
more phosphine in the reactor did not result in a higher conductivity any-

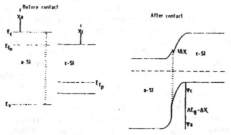

*Fig. 1 : Energy band diagram of a-Si emitter and c-Si base before and after
contact.*

more. n^+-a-Si:H islands are formed in the windows cut in the oxide, by dry etching in a SF_6 plasma. Since processing steps at temperatures higher than 300°C have to be avoided in order to prevent hydrogen evolution, the standard sintering step at 450°C cannot be used. A double metal layer (500 Å Ti/1 μm Al) is now evaporated to assure good ohmic contact to the n^+ a-Si:H and to the base (this was checked on a diffused base resistance : a linear I-V relationship was obtained). The finished device is shown schematically in Fig.2. Finally a low temperature (290°C) annealing step in forming gas (N_2/H_2) for 25 min. has been found necessary to obtain the highest possible current gain, possibly because this step improves the base contact.

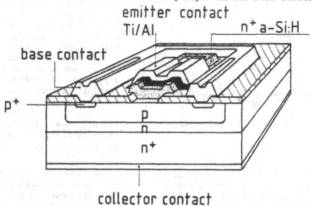

Fig. 2 : *Cross section of the heterojunction bipolar transistor device.*

RESULTS AND DISCUSSION

1. A maximum current gain β_{max} of 14 is observed at a collector current I_c of 3 mA. The area of the emitter metal contact being $\simeq 2 \ 10^{-5}$ cm^2, this corresponds to 150 A/cm^2 collector current density. The dependence of current gain versus collector current is shown in Fig.3 and the I_c-V_{CE} characteristics at low collector current (around 100 μA) and higher collector current (1-10 mA) are shown respectively in Fig.4 and 5. The "snap back" effect in Fig.5 for I_b = 0 results from the dependence of BV_{CEO} with the current gain β. Indeed when collector current rises, β increases, decreasing BV_{CEO}.

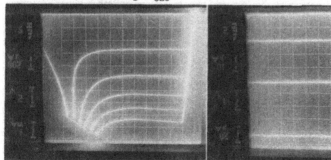

Fig. 4 : *I_c-V_{ce} characteristics at collector currents around 100 μA.*

Fig. 5 : *I_c-V_{ce} characteristics at collector currents around 1-10 mA*

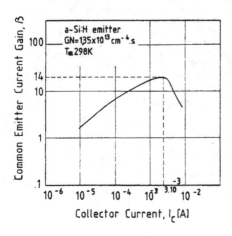

Fig. 3 : *Dependence of the current gain β versus collector current for the heterojunction bipolar transistor. The emitter area is 2 10⁻⁶ cm².*

2. As can be observed from Fig. 5 the I_c-V_{ce} characteristics at relatively high collector currents show a slightly negative slope. At these current levels thermal effects may be important and therefore an explanation could be a negative temperature coefficient of the current gain. If this is the case, it would indicate that the injection into the emitter is not dominating the base current, because then, as in conventional bipolar transistors, the current gain would show a positive temperature coefficient. Indeed, the activation energy of current gain in conventional bipolar transistors equals ΔE_{Ge}-ΔE_{Gb} respectively being the bandgap narrowing in emitter and base.

3. Fig.6 shows the maximum current gain versus GN of the base for conventional bipolar transistors (HOMOJUNCTION),

for the SIPOS-emitters and for the result obtained here (▲). It can be seen that a 5-6 fold improvement is obtained in comparison with the conventional transistors.

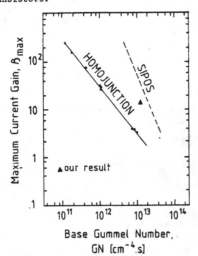

Fig 6 : *Maximum current gain versus base gummel number for conventional bipolar transistors (homojunction), sipos and our result (▲).*

Some problems still have to be looked at and therefore suggestions can be given for further work :
- the emitter resistance is relatively high, decreasing the current gain very rapidly beyond I_c of 3 mA. An outcome could be to transform the amorphous silicon to microcrystalline silicon at moderate temperatures or further to polycrystalline at higher temperatures. From the sign of the slope of the I_c-V_{CE} curves (pos. or neg.) one could eventually have a check of the fact whether or not a real heterojunction is maintained. A negative slope would most likely indicate the presence of a real heterojunction. On the other hand, going to higher temperature, the amorphous silicon can be used as diffusion source for the formation of shallow polysilicon emitters.

- interface recombination at the a-Si/c-Si interface is a main concern, because this can rule out completely the advantage of lower injection. Most likely the important decrease in the current gain as current level decreases (Fig.3) is caused by significant interface recombination. Therefore different passivation schemes should be tried in order to keep this recombination very low.

CONCLUSIONS

Higher common emitter current gain has been obtained using novel emitter bipolar transistors. The first heterojunction silicon bipolar transistor using phosphorous doped a-Si as wide bandgap emitter material has been realized. The additional process step is very simple. This transistor shows an improvement in maximum current gain of 5-6 times relative to the conventional bipolar transistor at a given base Gummel number of $1.35 \ 10^{13}$ s cm^{-4}. This offers interesting prospects for better design of bipolar transistors. Some problems still have to be investigated further such as emitter resistance and interface recombination.

REFERENCES

[1] M. Takagi, K. Nakayama, Ch. Tevada, and H. Kamioko, J. Jap. Soc. Appl. Phys. 42, 101 (1972).
[2] J. Graul, A. Glasl, and H. Murrmann, IEEE J. Solid-State Circuits, SC-11, 491 (1976).
[3] H.C. De Graaff and J.G. De Groot, IEEE Trans. Electron. Devices, ED-26, 1771 (1979).
[4] E. Wieder, IEEE Trans. Electron. Devices, ED-27, 1402 (1980).
[5] T.H. Ning and R.D. Isaac, IEEE Trans. Electron. Devices, ED-27, 2051 (1980).
[6] A.A. Eltoukhy and D.J. Roulston, IEEE Trans. Electron. Devices, ED-29, 1862 (1982).
[7] M.A. Green and R.D. Godfrey, IEEE Electron Dev. Letters, EDL-4, 225 (1983).
[8] N. Oh-uchi, H. Hayashi, H. Yamoto and T. Matsushita, IEDM Technical Digest, 522 (1979).

HIGH PERFORMANCE POLYCRYSTALLINE SILICON THIN FILM DEVICES

WILLIAM G. HAWKINS
Xerox Corporation, Webster Research Center, Webster, New York 14580

ABSTRACT

The goal of this work was to produce a fabrication process for high performance polycrystalline silicon thin film MOS devices. We have fabricated p-channel devices with mobilities of 35 cm^2/V-sec and n-channel devices with mobilities of 50 cm^2/V-sec by tailoring the process for depositon of the channel layer, by gate oxidation of the channel at high temperature, and by use of plasma hydrogenation. Under optimal conditions deduced from the study, device threshold voltages are close to zero. Leakage currents in the off-state are less than 0.1 pA/μm of channel length. Fabrication of the devices requires four mask levels and employs standard process steps. Therefore, polycrystalline silicon devices are attractive candidates for a variety of electronics applications, including thin film logic over large area.

INTRODUCTION

The use of polycrystalline silicon (poly-silicon) as a MOSFET thin film semiconductor material is attractive because of the simplicity and availability of deposition equipment, because thickness uniformity is achievable in all process steps, and because the fabrication techniques for bulk silicon MOSFET's can be used to manufacture thin film devices. Implementation of poly-silicon MOSFET's as elements in three-dimensional integrated circuits and in large area logic applications has been limited as a consequence of low channel mobility (typically less than 10 cm^2/V-sec)[1], high threshold voltage (10-20 volts is not uncommon)[2], and high junction leakage at the source of the device.[3] Low channel mobility limits frequency response of polysilicon devices, while high threshold voltage precludes device turn-on at the supply voltage of logic circuitry. Polysilicon devices must offer a substantial performance advantage over benchmark amorphous silicon devices which currently achieve channel mobility of 1 cm^2/V-sec and one to two volt threshold voltages. The junction leakage exhibited by polysilicon devices gives rise to high off currents and "anomalous leakage" when gate bias is reduced below the voltage necessary to turn off the device.[3]

Performance improvement of polysilicon devices was recently reported.[1] In that work, two important advances were made. Doping can control channel carrier concentration when it exceeds the background defect level in the film.[4] A fairly high doping level is necessary to gain control of channel carrier concentration. High channel doping level, however, results in a narrow maximum depletion layer width. Consequently, only a thin but highly doped channel can be fully depleted under appropriate gate bias, thereby producing low leakage while simultaneously allowing the threshold to be controlled by the channel doping level. The second advance was to use accumulation mode devices instead of inversion mode. The high density of states in the band gap of polysilicon necessitates large changes in gate voltage to move the Fermi level to the edge of the valence band. By employing accumulation mode devices, it is only necessary to move the Fermi level from mid gap to the valence or conduction band instead of from accumulation to inversion as in the case of inversion mode devices. Lower threshold voltage results from use of accumulation mode devices.

In the present work, devices were fabricated with a top gate to allow for a self-aligned source-drain implant and to enable large area integration, while previous work employed bottom-gated devices.[1] By proper control of the fabrication sequence, we have achieved mobilities of 35 cm^2/V-sec for p-channel and 50 cm^2/V-sec for n-channel devices. In previous work, a mobility of 9 cm^2/V-sec was achieved for p-channel mobility.

EXPERIMENTAL

The substrates for all work reported here were 10Ω-cm boron doped, (100), 75mm silicon wafers. The 660nm SiO_2 layer on which devices were fabricated was formed by a 90-minute pyrogenic wet oxidation at 1050°C. Deposition temperature for the channel was 612°C (growth rate = 10nm/minute) for polycrystalline silicon or amorphous silicon was deposited at 575°C (growth rate = 5nm/minute) immediately following the oxidation. Silicon film thickness was 150nm. The deposited silicon film was implanted with boron to a dose of $3x10^{12}$. The implanted film was then patterned using conventional lithography followed by wet etching in a commercial resist compatible polysilicon etch. A 105nm gate oxide was formed at 1000°C, 1050°C or 1100°C in dry oxygen. The channel was masked with 1.5μm of positive photoresist before ion implantation to form the source and drain of aluminum gate devices. A 600nm polysilicon film was deposited and delineated by wet etching for self-aligned structures. (The polysilicon etch is highly selective with respect to SiO_2.)

The source-drain implant was boron at a dose of $5x10^{15}cm^{-2}$. The standard source-drain implant activation step was 1000°C for 30 minutes in N_2, after 1000°C, 1050°C, and 1100°C were all compared in an initial experiment. Vias were then etched through the gate oxide covering the source and drain. A 1μm Al/1% Si layer was deposited by flash evaporation at a rate of 2nm/second. The metal was patterned by wet etching. Following resist stripping, the devices were sintered in N_2 for 15 mintues at 400°C. No further improvement in device performance was seen with longer sintering. Following sintering, some devices received a plasma hydrogenation at 300°C for various time intervals.[5] The hydrogenation system was operated at a pressure of 400 mTorr in a mixture of 50% H_2 and 50% N_2. An RF power of 350 watts was used to generate atomic hydrogen. The fabrication sequence requires only four lithography steps. All etching is highly selective and can be done by wet chemical techniques. Therefore the process should be easily adaptable to large area fabrication. Figure 1A and B shows an aluminum and polysilicon gated device after completion.

Devices were characterized by measurement on a curve tracer to determine channel mobility and threshold voltage at both low source-drain voltage and in saturation. Leakage characteristics were determined on a picoampmeter.

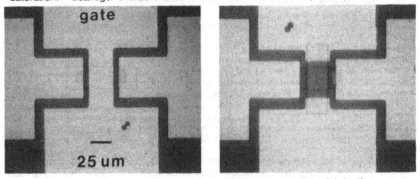

Figure 1. Aluminum (A) and polysilicon (B) gated devices after fabrication.

RESULTS AND DISCUSSION

Several batches of wafers were processed. It was found that within wafer, wafer to wafer and batch to batch uniformity was good. Within a wafer, differences in characteristics from device to device could not be observed on a curve tracer. The characteristics were stable with time with no drift occurring over a several hour period. Different wafers within a batch exhibited a few per cent variation.

The first wafer batch was all p-channel devices. Gate oxidation was carried out at 1000°C in dry oxygen. Both aluminum and polysilicon gated devices were fabricated. The source-drain implant anneal schedule was either 1000°C, 1050°C, or 1100°C for 30 minutes in N_2, following insertion and ramping from 800°C. The channel was deposited as polysilicon. Table I lists mobility and threshold voltage derived from linear characteristics at low source-drain voltage. These results reveal that the

Table I

Ion Implant Anneal Temp	Al Gate		Poly-Si Gate	
	V_{th}(volts)	μ_p(cm^2/V·sec)	V_{th}	μ_p
1000°C	-26.7	3.69	-27.0	2.40
1050°C	-26.5	4.30	-26.9	2.62
1100°C	-26.2	4.67	-26.8	2.71

channel mobiltiy of Al gated devices is enhanced with respect to poly-Si gated devices by 50%, that the threshold voltages are far from zero, and that the threshold voltage is not affected by the identity of the gate material. The improved mobility associated with Al gates is most likely a consequence of hydrogenation of surface states and possibly some defect states in the channel caused by generation of atomic hydrogen due to reaction between Al and surface moisture followed by diffusion through the gate oxide. It will later be shown that further hydrogenation dramatically improves aluminum and polysilicon gated devices. A second point to be drawn from Table I is that high temperature inert annealing does not improve channel mobility substantially. There is only a 12% difference in channel mobility between the 1100°C anneal and 1000°C anneal. Part of the improved mobility could be due to grain growth at the higher temperature, but some is an experimental artifact caused by diffusion of the source and drain implant into the channel. If the entire 12% is due to enhanced diffusion, the channel geometry change would correspond to 1.3μm diffusion of the contacts into the channel at 1100°C. Since the channel layer is only 100nm thick, diffusion of boron 1.3μm would entail many interactions with the silicon-SiO$_2$ interfaces, where segregation favors movement of boron into the oxide. Such segregation would reduce source and drain penetration into the channel. Our channel length is 25μm, so it can be concluded that the accuracy of channel mobility measurements will be high if source-drain activation is carried out at 1000°C for 30 minutes.

Hydrogenation of the poly gate and aluminum gate devices was then carried out for various times. The mobility and threshold voltage was monitored as a function of exposure time using device characteristics in saturation. Figure 2 shows the effect of hydrogenation on the devices listed in Table I. Initially, devices have low mobility and high negative threshold. After 30 minutes exposure to the plasma, the aluminum

446

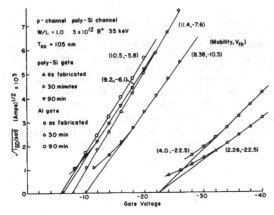

Figure 2. Effect of plasma
hydrogenation on polysilicon
and aluminum gated devices.
Al gated devices saturate
with hydrogen quickly, but
poly gated devices become
equivalent on longer
treatment.

gated device is already saturated with hydrogen, while almost 90 minutes of exposure
is necessary to bring the poly-gated and aluminum gated devices into equivalence. It
was possible to increase the threshold voltage to about -6 volts and increase mobility
to 10 cm^2/V-sec before no further improvement occurred on extended exposure.

A second batch of devices was fabricated to access the effect of channel
deposition parameters and channel oxidation temperature on device performance. In
particular, it is well known that the chemical vapor deposition of silicon from
reduction of silane at low pressure (250 mTorr) can produce either an amorphous or
polycrystalline film depending on whether the deposition is carried out at around
575°C or 612°C, respectively.[6] In addition, the initial experiments showed that
raising the anneal temperature used to activate the source and drain from 1000°C to
1100°C did not affect channel mobility. It might be anticipated that the choice of
oxidation temperature for gate oxidation would also be of little consequence.   To
examine both effects, devices with both polysilicon and amorphous silicon channels
were oxidized in dry oxygen at 1000°C, 1050°C and 1100°C. Figure 3 shows mobility
plots resulting from oxidation of an amorphous deposited channel at three
temperatures and oxidation of polycrystalline deposited channels at 1050°C. These
devices had aluminum gates and were hydrogenated for 30 minutes. Comparison of
Figure 2 with Figure 3, shows that mobility of polycrystalline deposited channels
improved from 11.4cm^2/V-sec to 17.6 cm^2/V-sec as oxidation temperature was
raised from 1000°C to 1050°C. Therefore, high inert anneal temperature does not
improve channel mobility but high oxidation temperature does.   A possible
explanation for the observation is that oxidation generates silicon self-interstitials.[7]
The self-interstitials are silicon atoms which diffuse rapidly in the silicon lattice until
they recombine at a surface, with a vacancy or at a defect. In polysilicon, it is likely
that a large fraction of self-interstitials recombine at grain boundaries and other
defects. It has previously been proposed that self-interstitials generated during high
concentration   phosphorus   diffusion[8]   into   polysilicon   aid   grain   growth.
Consequently, n^+ polysilicon has large grain size in comparison with undoped
polysilicon which has undergone the same thermal processing.[9]

A second exciting result which can be seen from Figure 3 is that amorphous
deposited channel layers yield devices with channel mobility which is a factor of two
higher than otherwise identically processed polycrystalline deposited channels. In
addition, the enhancement of mobility which results from higher oxidation
temperature can be seen for amorphous deposited channels.

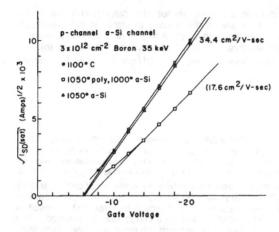

Figure 3. Higher oxidation temperature and deposition of the channel as amorphous silicon each improve device performance by a factor of two!

The work presented so far demonstrates that the combination of deposition of silicon as an amorphous layer, oxidation of the channel at 1050°C and plasma hydrogenation for 60 minutes yields devices with superior channel mobility to those previously reported.[1] An equally important performance criteria for many applications is leakage and subthreshold conduction. Figure 4 shows a log plot of current versus gate voltage for a device which was produced using optimal processing. The results show that the hydrogenation step dramatically improves subthreshold conduction while reducing device leakage currents by about one order of magnitude. It can also be seen from the Figure that anomalous leakage currents under reverse gate bias are low in comparison with previous results.[3]

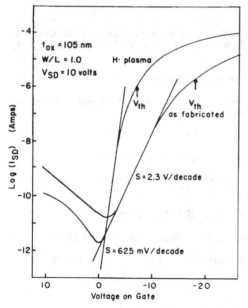

Figure 4. A plot of log I_{SD} versus V_g illustrates that subthreshold conduction is dramatically improved as a consequence of hydrogenation. The improvement in S is a factor of 4.

The subthreshold conduction observed versus the calculated value to be expected allows an estimate to be made of the density of states in the band gap. The extended linear region in Figure 4 would suggest that a fairly uniform density of states exists. The slope of the subthreshold conduction yields a density in the mid $10^{16}cm^{-3}$ range for hydrogenated samples and mid $10^{17}cm^{-3}$ for "as fabricated" devices.

CONCLUSION

We have demonstrated that substantial improvements in polycrystalline silicon device performance can be achieved by deposition of the channel layer as an amorphous film, by oxidation of the channel at higher temperature and by hydrogenation of completed devices. We are currently examining the behavior of n-channel devices and have observed channel mobilities in excess of 50 cm^2/V-sec for electron mobility. These results will be presented shortly. In addition, hydrogenation of devices in the absence of a plasma environment would be expected to result in further improvement in performance since radiation damage can be avoided. The observation that oxidation is effective at improving channel mobility while inert annealing is not suggests that oxidation generated excess silicon self-interstitials greatly accelerate channel grain growth.

ACKNOWLEDGEMENTS

The author is indebted to S. F. Pond and other management of Xerox Webster Research Center for supporting this work. The help of M. Poleshuk in fabrication of the first batch of devices is appreciated.

REFERENCES

1. S. D. S. Malhi, P. K. Chatterjee, R. F. Pinizzotto, H. W. Lam, C. E. C. Chen, H. Shichijo, R. R. Shah and D. W. Bellavance, IEEE Electron Device Letters EDL-4, 369 (1983).

2. T. I. Kamins, Solid State Electronics 15, 789 (1972).

3. S. Onga, Y. Mizutani, K. Taniguchi, M. Kashiwagi, K. Shibata and S. Kohyama, Japanese J. Appl. Phys. 21, 1472 (1982).

4. J. Y. W. Seto, J. Appl. Phys. 46 (1975).

5. T. I. Kamins, IEEE Electron Device Letters EDL-1, 159 (1980).

6. G. Harbeke, L. Krausbauer, E. F. Steigmeier, A. E. Widmer, H. F. Kappert and G. Neugebauer, Appl. Phys. Lett. 42, 249 (1983).

7. D. A. Antoniadis, J. Electrochem. Soc. 129, 1093 (1982).

8. R. M. Harris and D. A. Antoniadis, Appl. Phys Lett. 43, 937 (1983).

9. Y. Wada and S. Nishimatsu, J. Electrochem. Soc. 125, 1499 (1978).

Author Index

Subject Index

Printed in the United States
By Bookmasters